ARTIFICIAL REEF EVALUATION
With Application to Natural Marine Habitats

Edited by
William Seaman, Jr., Ph.D.

Taylor & Francis
Taylor & Francis Group

Boca Raton London New York Singapore

A CRC title, part of the Taylor & Francis imprint, a member of the
Taylor & Francis Group, the academic division of T&F Informa plc.

COVER ARTWORK: Designed by John Potter.

Library of Congress Cataloging-in-Publication Data

Artificial reef evaluation : with application to natural marine habitats / edited by William Seaman.
       p.   cm. — (Marine science series)
    Includes bibliographical references.
    ISBN 0-8493-9061-3
    1. Artificial reefs. I. Seaman, William, 1945–   II. Series.

SH157.85.A7 A755 2000
639.9′2—dc21   99-053387

Visit the CRC Press Web site at www.crcpress.com

Preface

Artificial reefs represent a popular and accessible technology for modifying aquatic ecosystems. Their deployment in coastal waters of all inhabited continents and around oceanic islands of the tropics has increased significantly in the past two decades. In each situation, we can ask, "Is the artificial reef or reef system satisfying the purposes for which it was designed and built?" Indeed, this central question must be addressed if the technology is to be applied responsibly.

Increased interest in artificial habitats comes from a diversity of commercial, recreational, conservation, and management sectors. Their goals can be quite different and include exploitation of fishery stocks for food, restoration of benthic habitats, and conservation of biodiversity. This book addresses the global and growing need for an applications-oriented review, compilation, and demonstration of technical methods to evaluate the performance of artificial reef habitats.

The goal of the contributors to this volume is to increase the scope of research for this field. Our expectation is that reliable data will enhance future reef planning and development and contribute to informed decisions about when or when not to build reefs. Further, an opportunity to develop uniformity in study methods, link databases from different areas, and compare research results may be at hand due to the growth of global information exchange.

Marine Science Series

The CRC Marine Science Series is dedicated to providing state-of-the-art coverage of important topics in marine biology, marine chemistry, marine geology, and physical oceanography. The series includes volumes that focus on the synthesis of recent advances in marine science.

CRC MARINE SCIENCE SERIES

SERIES EDITOR

Michael J. Kennish, Ph.D.

PUBLISHED TITLES

Artificial Reef Evaluation with Application to Natural Marine Habitats, William Seaman, Jr.

The Biology of Sea Turtles, Volume I, Peter L. Lutz and John A. Musick

Chemical Oceanography, Second Edition, Frank J. Millero

Coastal Ecosystem Processes, Daniel M. Alongi

Ecology of Estuaries: Anthropogenic Effects, Michael J. Kennish

Ecology of Marine Bivalves: An Ecosystem Approach, Richard F. Dame

Ecology of Marine Invertebrate Larvae, Larry McEdward

Ecology of Seashores, George A. Knox

Environmental Oceanography, Second Edition, Tom Beer

Estuarine Research, Monitoring, and Resource Protection, Michael J. Kennish

Estuary Restoration and Maintenance: The National Estuary Program, Michael J. Kennish

Eutrophication Processes in Coastal Systems: Origin and Succession of Plankton Blooms and Effects on Secondary Production in Gulf Coast Estuaries, Robert J. Livingston

Handbook of Marine Mineral Deposits, David S. Cronan

Handbook for Restoring Tidal Wetlands, Joy B. Zedler

Intertidal Deposits: River Mouths, Tidal Flats, and Coastal Lagoons, Doeke Eisma

Marine Chemical Ecology, James B. McClintock and Bill J. Baker

Morphodynamics of Inner Continental Shelves, L. Donelson Wright

Ocean Pollution: Effects on Living Resources and Humans, Carl J. Sindermann

Physical Oceanographic Processes of the Great Barrier Reef, Eric Wolanski

The Physiology of Fishes, Second Edition, David H. Evans

Pollution Impacts on Marine Biotic Communities, Michael J. Kennish

Practical Handbook of Estuarine and Marine Pollution, Michael J. Kennish

Practical Handbook of Marine Science, Third Edition, Michael J. Kennish

Seagrasses: Monitoring, Ecology, Physiology, and Management, Stephen A. Bortone

Trophic Organization in Coastal Systems, Robert J. Livingston

Acknowledgments

This book had its genesis in a project sponsored by the Florida Department of Environmental Protection in the U.S., which developed guidelines for evaluating artificial reef performance in that state. Although Florida waters contain more than half of all artificial reefs in the country, documenting attainment of objectives has not received emphasis. Clearly this book required a global search for information, which is why its core group of authors from Florida was expanded. Sixteen authors, from six countries, wrote this book.

Principal thanks go to colleagues who reviewed drafts of this book. The senior authors periodically met to discuss all chapters and offer detailed comments on selected contents. Technical review was provided by M. Barnett, W. Figley, C.R. Gilbert, P.J. Sanjeeva Raj, G. Relini, L.M. Sprague, and R.B. Stone.

Assistance with organization of the overall volume and typing of part of the manuscript was provided by J. Whitehouse and T. Stivender, partly supported by Florida Sea Grant College Program Grant NA76RG0120 from the U.S. National Oceanic and Atmospheric Administration. J. Potter designed the cover and chapter title pages.

Finally, we thank Phillips Petroleum Company for major sponsorship of preparation of the book manuscript. The encouragement and interest of W. Griffin and B. Price of Phillips are gratefully acknowledged.

William Seaman, Jr.
Gainesville, Florida, U.S.A.

The Editor

William Seaman, Jr., Ph.D., is Professor of Fisheries and Aquatic Sciences and Associate Director of the Florida Sea Grant College Program at the University of Florida. Educated as a zoologist, Dr. Seaman has been involved with research, teaching, and extension/outreach concerning organisms and aquatic habitats in coastal environments. He also works extensively with development and management of university marine resource programs in diverse subjects, including biotechnology, aquaculture, and estuarine ecosystems.

Dr. Seaman earned his bachelor's degree from Cornell University, Ithaca, New York, and his master's degree and doctorate from the University of Florida, Gainesville, Florida. He has held short-term research appointments at the Hawaii Institute of Marine Biology and the Musée Océanographique, Monaco.

Working with the science and technology of artificial reefs from the local through the international levels, Dr. Seaman has documented the status and trends of reef development in Florida and led a team that prepared guidelines for assessment of reef performance in that state. He has made keynote presentations at European and international conferences on artificial reefs. He was senior editor for *Artificial Habitats for Marine and Freshwater Fisheries,* a book on research in this field, and has chaired one national symposium and one international conference concerning reefs. A current priority is to assist global audiences in formulating and adopting consistent practices to document performance of artificial reefs for fishery, aquaculture, habitat restoration, and other purposes.

Dr. Seaman is past president of the Florida Chapter of the American Fisheries Society. He also has worked extensively with environmental education, with recent focus on extending science-based information to interests concerned with coastal habitats.

Contributors

Stephen A. Bortone, The Conservancy of Southwest Florida, Naples, Florida 34102, U.S.A.

Paul H. Darius, Laboratory of Statistics and Experimental Design, Faculty of Agricultural and Applied Biological Sciences, Katholieke Universiteit Leuven, Haverlee, Belgium

Gianna Fabi, Istituto di Ricerche Sulla Pesca Marittima, IRPEM-CNR-Molo Mandracchio, 60125 Ancona, Italy

Annalisa Falace, Dipartimento di Biologia, Universita degli Studi di Trieste, 34127 Trieste, Italy

Patrice Francour, Laboratoire Environment Marin Littoral, Faculté des Sciences, Université Nice-Sophia Antipolis, 06108 Nice CEDEX 2, France

Stephen M. Holland, Department of Recreation, Parks and Tourism, University of Florida, Gainesville, Florida 32611, U.S.A.

Antony C. Jensen, School of Ocean and Earth Science, Southampton Oceanography Centre, Southampton University, Southampton SO14 3ZH, United Kingdom

William J. Lindberg, Department of Fisheries and Aquatic Sciences, University of Florida, Gainesville, Florida 32653, U.S.A.

Margaret W. Miller, University of Miami and National Marine Fisheries Service, Miami, Florida 33149, U.S.A.

J. Walter Milon, Department of Food and Resource Economics, University of Florida, Gainesville, Florida 32611, U.S.A.

Kenneth M. Portier, Department of Statistics, University of Florida, Gainesville, Florida 32611, U.S.A.

Giulio Relini, Istituto di Zoologia, Universita di Genova, Genova 16126, Italy

Melita A. Samoilys, Fisheries, Department of Primary Industries, Townsville, Queensland 4810, Australia

William Seaman, Jr., Department of Fisheries and Aquatic Sciences, and Florida Sea Grant College Program, University of Florida, Gainesville, Florida 32611-0400, U.S.A.

Y. Peter Sheng, Department of Coastal and Oceanographic Engineering, University of Florida, Gainesville, Florida 32611, U.S.A.

David J. Whitmarsh, Centre for the Economics and Management of Aquatic Resources, University of Portsmouth, Portsmouth PO4 8JF, United Kingdom

Contents

In Memory of Helene

CHAPTER 1

Purposes and Practices of Artificial Reef Evaluation

William Seaman, Jr. and Antony C. Jensen

CONTENTS

1.1 SUMMARY

This chapter is an introduction to the book. The second section defines artificial reefs and identifies 13 broad uses of the technology. The third section focuses on the need for expansion of evaluation practices to include the documentation of how effectively artificial habitats meet the benefits projected for them. A reader's guide to six chapters on study design, research methods, and evaluation of performance is provided. The final section addresses issues of consistency and comparability in research, and their connection to management of natural resources. Themes of the book include the need for clear definition of reef objectives as the basis for scientific evaluation, the opportunity for linkage between studies of natural and artificial reefs, and the utility of different types of information that can be developed consistent with capability for research.

1.2 INTRODUCTION

To ancient peoples who used their observations of the attraction of fishes to foreign objects as a means of harvesting food, it was a simple matter to evaluate performance of rocks and logs introduced to the aquatic environment. Either fishes appeared and were captured or not. In modern times, placement of structure underwater as habitat is done for several additional reasons, including commercial seafood harvesting, sea angling, recreational diving, aquaculture, environmental restoration, natural resource management, and scientific experimentation. This book responds to the growing global need for a compendium of consistent, reliable, and comparable practices with which to technically evaluate how effectively artificial reefs meet the diverse purposes for which they are built.

Of all technologies available to applied and nonscientific interests concerned with modifications to the aquatic environment, artificial reefs may be the most accessible and widely used. In contrast to other practices whose implementation and evaluation may be exclusively the domain of specially trained professionals — such as culture and release of organisms, construction of wetlands, or impoundment of water bodies — artificial habitats are routinely built and sometimes monitored by lay-level or semitechnical interests worldwide, thereby augmenting the efforts of those formally educated to work in this subject area. The practices illustrated in Figures 1.1 and 1.2 typify the range of users, methods, and materials involved. Typically, there has been a lack of emphasis on evaluation of reef performance as it pertains to the benefits expected to accrue to ecological and human systems.

Meanwhile, the senses used by the earliest seafarers as a matter of survival and exploration now are acutely heightened by sophisticated electronic devices that enable monitoring of ocean environments and analysis of elaborate datasets. With more extensive and diverse exploitation, alteration, or management of coastal habitats has come the need to measure a variety of physical, chemical, geological, biological, economic, and social attributes. Both natural reefs and the manmade benthic structures intended to either mimic them or modify ecological and social processes and productivity are receiving greater scrutiny. This book combines proven assessment methodologies of the aquatic sciences with the need for evaluation of the growing system of reefs globally.

Figure 1.1 The global diversity of artificial reefs includes natural and manufactured materials used by artisanal fishing interests (A, B, India, photographs courtesy of P.J. Sanjeeva Raj); commercial fishing interests (C, Korea, photograph courtesy of Soon Kil Yi); and recreational fishing interests (D, United States, photograph courtesy of J. Halusky).

1.2.1 Purpose and Scope of the Book

The aim of this book is to provide a comprehensive, multidisciplinary guide to the strategies and methods for evaluating performance of artificial reef habitats in coastal and ocean waters worldwide. By performance, the authors particularly refer to the more neglected aspect of documentation of benefits to humans that represent the ultimate reason for reef construction (see Section 1.3.1). This is in contrast to the more traditional subject addressed by numerous published reports of, for example, physical stability of reefs or ecology of species.

A principal theme is that the objectives for which the reef is built must be defined clearly as the basis for designing an appropriate and scientifically valid study. A second theme is that although a large amount of data about artificial reefs exists, conclusions have not been drawn in all cases regarding the reefs' effectiveness relative to the objective(s) for which they were established. We emphasize framing questions about reef performance and guiding users of this book to the kinds of successful analyses necessary to answer them. Readers should gain an appreciation that a definite technical process is necessary for objectively assessing attributes of reef systems and familiarity with the tools for conducting that process. The book addresses physical, engineering, biological, and economic assessment. Such methodological tools (see Chapters 3 to 6) derive from the types

A

B

Figure 1.2 Reef design and construction practices bear on the design of evaluation studies. Increasing use of modular structures is reflected in reefs used in fisheries and research (A, Italy, photographs courtesy of G. Relini; B, United States, photographs courtesy of W. Lindberg and F. Vose).

of research traditionally reported in the primary literature, yet also can generate the data necessary for "objectives-driven" investigation (see Chapter 7) whose audience includes:

- Fishing interests that build and use reefs
- Reef design and construction businesses
- Habitat restoration and biodiversity conservation projects
- National and local government reef research, technology, and management programs
- Academic and consulting scientists
- Recreational skin diving interests
- Sustainable tourism and economic development interests
- Volunteer citizen monitoring and research divers
- Natural resource agency/ministry budget offices
- General public, educators, news media, and related user groups.

The question "Is an artificial reef meeting its objective?" is entirely appropriate and indeed vital to advance this field of science and environmental conservation, develop new technologies, and present evidence to determine its cost-effectiveness and biological productivity. While the methods presented in this book are derived from current and established research practices, it is not our intent to exhaustively review and analyze all research literature as an end in itself. Nor do we identify all the geographic locations and interests involved with reef technology, although a cross section of historical and current activities is described.

Much of the methodology presented here also is relevant to natural aquatic habitats, such as coral reefs and rocky outcrops. Scientists from other fields of study of natural aquatic environments and human systems will recognize techniques that originated prior to being used in artificial reef research (e.g., Stoddart and Johannes 1978; English et al. 1997). For example, fish censusing practices created to evaluate coral reef ecosystems have been adapted to artificial reefs. Similarly, economic assessments can be performed for natural and for artificial reefs. The data originating from proper use of these techniques can apply to understanding, comparison, and prediction of reef performance. Indeed, the ability to manipulate artificial reef systems (see Chapter 2) offers an opportunity to create powerful study designs that may not be feasible for natural reef systems.

1.2.2 Definition and Context for Artificial Reefs

An *artificial reef* is one or more objects of natural or human origin deployed purposefully on the seafloor to influence physical, biological, or socioeconomic processes related to living marine resources. Artificial reefs are defined physically by the design and arrangement of materials used in construction and functionally according to their purpose (Table 1.1). Items used in reef construction add vertical profile to the benthic environment. They may be either assembled expressly as a reef or acquired after being used for another, usually unrelated, purpose.

The definition of artificial reef has been changing in the modern era of reef-building, which is only 50 years old. Accidental shipwrecks have been classified at times as artificial reefs. Recent proposals have suggested the incorporation, secondarily, of various objectives sought with artificial reefs to structures already deployed for other purposes. A notable example is the recognition that harbor breakwaters can be designed to achieve "ecofriendliness" (e.g., Ozasa et al. 1995). This volume stops short of discussing such nonbiological structures as breakwaters or beach protection groins. Thus, Table 1.1 does not list purposes such as coastline protection, harbor stabilization, or recreational surfing, since they are built with other primary objectives. However, the methods in subsequent chapters certainly would apply to their evaluation.

The most prominent use of reefs has been for enhancement of fishery harvests in two ways. First, almost immediately after reef deployment, the attraction of mobile organisms to the structure is anticipated by interests seeking to increase catch or its efficiency. Secondly, there is an expectation that ecologically the artificial reef will resemble (or exceed) local natural environments over the long term as assemblages including sessile organisms associate with its surface, structure, and surrounding water column and eventually increase biomass at the site (Figure 1.3). This latter aspect has led to refinement in the last decade of the historical view of reefs as simply fish attractors. Increased focus is now placed on the informed design of the structure to establish or expand populations of plants and animals by fulfilling species life history requirements. Such a focus allows definition of criteria for measuring success of reefs in meeting objectives. (See Chapter 2 and the

Table 1.1 Uses of Artificial Reefs in Marine Environments

Enhance Artisanal Fishery Production/Harvest
Increase Commercial Fishing Production/Harvest
Aquaculture Production Sites
Enhance Recreational Fishing by Hook-and-Line and Spear
Recreational Skin Diving Sites
Submarine Tourism Sites
Control Fishing Mortality
Manipulate Organism Life History
Habitat Protection
Conservation of Biodiversity
Mitigation (off-site) of Habitat Damage and Loss
Restore or Enhance Water and Habitat Quality (on-site)
Research

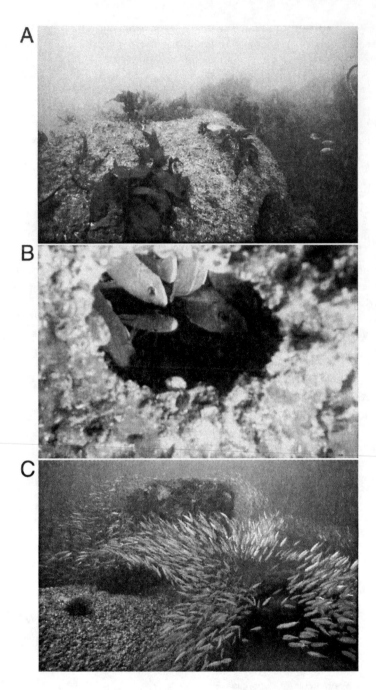

Figure 1.3 Artificial reefs that are physically stable offer sites for A, growth of plants (photograph courtesy of Soon Kil Yi); B, shelter of fishes (photograph courtesy of R. Brantley); and C, gathering of pelagic organisms (photograph courtesy of J. Halusky).

reports of a symposium on the issue of attraction and production by the American Fisheries Society [AFS, 1997] and a conference of the European Artificial Reef Research Network [Jensen 1997a].)

    The concept that an artificial reef emulates natural reef ecology is reflected in its definition as "a submerged structure placed on the substratum (seabed) deliberately, to mimic some characteristic of a natural reef" by the European Artificial Reef Research Network. This reflects contemporary hypotheses that the ecological processes of artificial reefs are (or should be) functionally equivalent

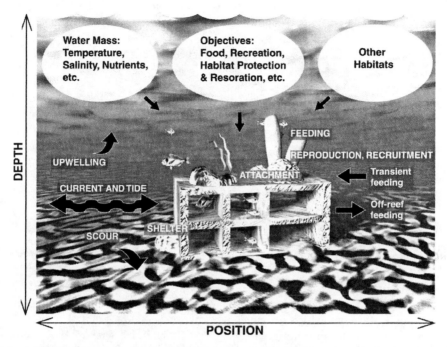

Figure 1.4   Artificial reefs are influenced by, and influence, natural and human forces in the aquatic environment. Evaluation depends on the objectives for the reef as defined by user interests and may address physical, chemical, biological, and economic factors.

to those of natural benthic systems in an area. It excludes emergent structures such as those used in coastal/shoreline defense.

However, mimicking of natural reef characteristics may not be essential for some purposes of artificial reefs, e.g., providing non-consumptive recreation for scuba divers and passengers in submarines who visit ships sunken especially as observation sites (one objective is to alleviate crowding on coral reefs in popular tourist destinations), or creating physical barriers that restrict fishing boat access to designated seabed areas (in order to allocate resources according to ecosystem management objectives). As the popularity and uses of artificial reefs increase, it is important to include socioeconomic and resource allocation functions in their definition.

It is essential to the thorough design of evaluative studies of reef performance that their function in the marine ecosystem be understood. There are behavioral and energetic bases for introducing foreign objects onto the seafloor or into the water column. Some fundamental dynamics are depicted in Figure 1.4. For a survey of technical information in this field the reader is referred to recent general references (e.g., Seaman 1995); compilations of research results, most extensively prepared after international conferences (e.g., *Bulletin of Marine Science* 1994; Sako 1995; Jensen 1997a); and research review and synthesis articles (Seaman and Sprague 1991; Jensen et al. 1999).

Knowledge of natural reefs also can add to the understanding of artificial reefs, for example, concerning ecology (Sale 1991); fishery and applied aspects (Polunin and Roberts 1996); and overall distribution, ecology, and management (McManus and Ablan 1997). Comparison of natural and artificial reef ecology is made by Bohnsack et al. (1991).

1.3 EVALUATION OF REEF ENVIRONMENTS

Continued and growing demand for reefs by a variety of government, public, and private interests prompts legitimate questions about their effectiveness. These questions relate to reef ecology, fishing and environmental impacts, construction practices, and economic aspects. Harvesters of fishes, managers of environmental resources, investors in reef construction, conservation interests, users of related resources, and the public-at-large all have an interest in evaluating the performance of artificial reefs. The types of questions that might be asked include:

- What organisms (will) occur at a reef?
- What species can be harvested?
- Is overharvest a possibility?
- Will the reef interfere with existing uses or resources of the sea?
- What economic benefits and costs are expected?
- Are reefs useful in ecological restoration?
- (When) will the reef sink into the seafloor?
- Can research questions be answered at the reef(s)?

The myriad of other questions that could be listed range from highly applied and economically oriented to those concerning basic physical and biological dynamics of the substrate. Although undocumented, the history of reef-building includes many examples of failures based on anecdotal reports.

This section introduces the case for the type of evaluation of artificial reef performance that focuses on benefits to be expected. Monitoring and evaluation (e.g., Schmitt and Osenberg 1996) are an integral part of overall natural resources and environmental program planning. They can be used to (1) identify ways to improve these activities; (2) determine whether they are worthwhile expenditures of community or corporate resources; or (3) help decide whether to continue or terminate an existing activity. Experiences in characterizing the performance of public works such as sandy beach renourishment and dam construction — where studies of productivity, cost, benefit, nonmonetary values, etc. may be required in some countries — can provide a framework for understanding why it is essential to assess artificial reefs. The motivation for this book is embodied in a report by the National Research Council (1996, pp. 6–7) on the need for understanding ecological systems:

> Indicators that are needed to measure the current status of ecological systems, to gauge the likelihood of meeting society's environmental goals, or to anticipate problems resulting from economic growth are not available. We are spending much money to collect data that are neither complete nor always relevant to the decisions that society needs to make about land use, transportation, industrial activity, agriculture, and other human activities. The system of monitoring the state of the environment needs to be improved to make it more relevant to decision-makers, and there is need for a more-sophisticated and better-informed discussion of what needs to be measured and why.

1.3.1 The Case for Evaluation

Despite the achievements reported at the Fifth International Conference on Aquatic Habitat Enhancement, Grove and Wilson (1994, p. 266) observed that "ecological and/or human benefits are the ultimate objective of habitat enhancement: yet these are frequently not documented." Others support this conclusion; for example, McGlennon and Branden (1994) pointed out the lack of studies to evaluate "enhanced fishing objectives," in contrast to more numerous published comparisons of artificial and natural reefs for fish abundance and diversity. Meanwhile, in Florida, where half of U.S. artificial reefs are located, over 70% of participants at the first statewide Artificial Reef Summit meeting assigned a "high priority" to the need for evaluation, with the consensus that there should be monitoring as part of the state artificial reef program (Andree 1988).

A relatively small proportion of published papers address the actual documentation of benefits (in the sense of Grove and Wilson 1994). For example, of 41 papers from the fifth international reef conference (*Bulletin of Marine Science* 1994) having to do with "function and ecology," "fishing enhancement," "mitigation and restoration," "monitoring and assessment," and "artisanal fishing" — subjects more relevant to documentation of benefits — 15 evaluated at least some aspect of benefits. The rest were more traditional and focused on basic or limited characterization of life history details such as diet and behavior. An example of research with the former orientation is seen in Italy where reefs, explicitly built to enhance fisheries, provide substrate for mussel cultivation, and limit access of trawling vessels, were evaluated for achievement of specific management objectives (e.g., Fabi and Fiorentini 1994; Bombace et al. 1994).

Despite the existence of thousands of artificial reefs worldwide, most published reports of their "performance" are limited in scope as either characterization of one particular aspect of a reef intended to produce benefits or description of monitoring or experimentation at a reef built for scientific purposes. In the first instance, a series of ecological analyses (e.g., fish diet) of a reef built for recreational fishing in the U.S. was made by Lindquist et al. (1994) and others. Here the aim was to produce useful ecological information, but not to document benefits. In the latter instance, the presence and absence of shelter for lobsters and their response to variation in design were examined using experimental structures in Israel (Barshaw and Spanier 1994) and Mexico (Lozano-Alvarez et al. 1994).

In part, a lack of evaluative data concerning benefits is due to the focus of much research on basic ecological questions or a limited number of taxa. Further, this field is young. Relatively low budgets have been afforded for research generally and assessment in particular. Anecdotally, we observe a consensus among scientists and the serious conservation-minded community of reef-builders that, no matter how solidly reefs are constructed, typically their environmental and fishery objectives are general, vague, or weakly defined. Traditionally, reefs have been built to enhance fishing or to improve the environment. But it is difficult or impossible to objectively measure whether these manmade habitats meet their intended purpose or to quantify the benefits they produce, when the stated objectives (if any) are overly broad or vague.

1.3.2 Framework for Reef Evaluation

1.3.2.1 Goals and Assessment Concept

A scientifically rigorous assessment of the performance of an artificial reef in meeting its intended purpose begins with understanding its goal. Ordinarily the goal will be a broad statement of intent to provide or enhance usage of the reef, such as for increasing fishery harvest, recreational opportunities, or protection of habitat. Sometimes the terms "goal" and "overall objective" are synonymous.

As depicted in Figure 1.5, the goals and general objectives for reef usage dictate an Assessment Concept for reef evaluation. The variety of uses listed in Table 1.1 generally are described in terms of either increasing consumptive uses, such as greater production or attraction of fishery species for harvest, or achieving non-consumptive purposes including recreation and restoration of ecosystems. The Assessment Concept for artificial reef evaluation encompasses the (1) specific objectives of the reef and (2) characteristics that must be measured to determine the degree of success in reaching a particular objective. These elements form the basis for subsequent steps in the assessment strategy.

Thus, while a goal (or general objective) for an artificial reef may be stated as "to increase fishery harvest," or "to restore seagrass habitat," or "to protect nursery grounds," specific objectives are needed if the reef is to be evaluated rigorously for performance and if benefits are to be documented. For example, an objective "to increase biomass of species A by weight X" is more amenable to rigorous and quantitative description than an objective "to increase fish populations." Realistically, the specificity of objectives will vary among geopolitical systems so that a continuum exists.

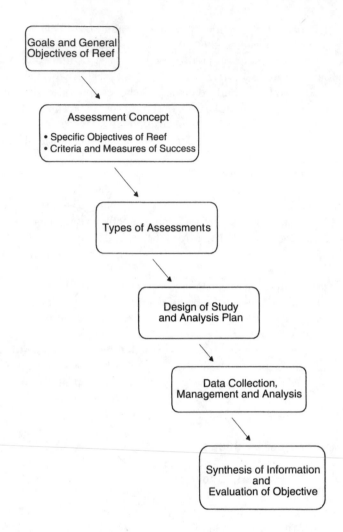

Figure 1.5   Framework for organization of artificial reef performance evaluation.

Based on the ideal of an articulate, focused, and quantitative objective, the features of a reef to be measured can be defined. To determine success in the example just given, one would measure the abundance and weight of species A at the reef site before and after reef construction, in a manner consistent with its life history and incorporating sound statistical practices. How well the criterion of growth of fishes belonging to species A is met would help decide the effectiveness of this reef. The capability of different disciplines to gather sufficient data may vary. At one extreme, physical scientists may be very precise in determining settlement of reef materials, whereas biologists may lack complete knowledge of basic organism and assemblage processes.

Chapter 7 presents detailed examples of reef objectives and approaches to their evaluation. Clearly, this is an evolving field.

1.3.2.2 Basis for Artificial Reef Objectives and Measures of Success

Effective evaluation must be "objectives-driven." For this reason, Chapters 3 through 6 discuss reef objectives from the perspective of particular scientific disciplines as a basis for designing evaluation studies. The purposes for which artificial reefs are constructed have increased in the last 15 years (Figure 1.6). The following review of items summarized in Table 1.1 indicates the scope of reef deployment practices worldwide.

Figure 1.6   One of the newest purposes for artificial reefs is to provide sites for submarine viewing and nature-based tourism, especially in clear tropical waters. (Photograph courtesy of R. Brock.)

1.3.2.2.1 Artisanal and Commercial Production and Harvest of Aquatic Organisms — Artisanal harvest of fishes and invertebrates is the oldest use of artificial reefs and the most widespread use geographically. The broad goal is production of foodstuffs to maintain the diet and sometimes, at least partly, the economy of individuals, families, and communities in local coastal areas. Such practices exist in about 40 countries, including India where indigenous technology to deploy natural materials such as weighted trees has existed for 18 centuries (Sanjeeva Raj 1996). Other centers of activity include tropical coastal waters of the Caribbean Sea, the western Pacific Ocean, and western Africa. Use of the technology continues to spread as in the establishment of a lobster fishery in Mexico in the 1970s (Lozano-Alvarez 1994) and experimentation with juvenile fish habitats in Malaysia in the 1990s (Omar et al. 1994).

Harvest of marine fishes for commercial purposes has drawn the largest investment of funds for construction of artificial reefs, notably due to the commitment of the Japanese government since 1952. Japan's expenditure for reef construction to provide seafood supply, stimulate the economy, and promote social goals includes 600 billion yen from 1995 to 2000 (Simard 1997). By the mid-1980s, 9.3% of Japan's coastal seafloor fishing ground area less than 200 m deep included "improvements" (Yamane 1989).

Aquaculture is an emerging use of artificial reefs. Aquatic organisms are captive or manipulated at some stage of the life cycle as a form of "marine ranching." European interest in this area has been led by Italian researchers working in the Adriatic Sea. The Adriatic is shallow and eutrophic, with a high primary productivity, but its predominantly sedimentary seabed provides limited settlement surfaces for bivalves. In a 25-year program, artificial reefs have become multipurpose complexes that support aquaculture initiatives (e.g., Fabi and Fiorentini 1990), primarily for oysters, *Ostrea edulis* and *Crassostrea gigas,* and mussels, *Mytilus galloprovincalis.* Fisheries applications of reefs in conjunction with fish cages in Russia were proposed by Bougrova and Bugrov (1994).

1.3.2.2.2 Recreational Fishing and Diving and Tourism — Recreational uses of artificial reefs include harvest of fishes by angling or spearing. The largest efforts to promote game-fishing have been in the United States and Australia (Branden et al. 1994; Christian et al. 1998). Since neither country has a formal national government program, data are not available to describe the overall effort, such as for exact area covered or materials used. In the U.S., individual local and state governments and private organizations sponsor this work independently, sometimes with federal (e.g., Wallop-Breaux Fund) assistance, often using volunteer effort and donated materials. As of 1991, over 650 artificial reef sites were in U.S. coastal waters (Berger et al. 1994).

Diving at artificial reef sites is a more recent focus in countries such as Australia and the U.S. For example, obsolete coastal patrol vessels were deliberately sunk off Florida to divert divers from overly popular nearby coral reefs. Even more recently, tropical areas seeking to boost tourism in coastal areas with clear water, as a means of economic development, have proposed or built reefs colonized by species from coral reefs that attract divers. Submarine tour companies have built reefs in the Bahamas and Hawaii. Published reports assessing performance (e.g., diver satisfaction, economic impact) are not yet available.

1.3.2.2.3   Resource Allocation and Protection — Of lesser prominence globally, but important in some local areas, are reefs used as physical barriers to restrict access to certain areas. They serve a dual purpose of protecting habitat and controlling fishing mortality. To counteract trawl damage to seagrass beds, artificial reefs designed as barriers have been deployed in Italy, France, and Spain (e.g., Relini and Moretti 1986). The seagrass acts as a nursery area for many commercially important species that provide income for artisanal fishermen. Seagrass is easily damaged by trawling activity; the value of the habitat is reflected in legislation prohibiting trawling in waters shallower than 50 m in most of the Mediterranean Sea. The static gear of artisanal fishermen is protected from damage by trawls. It poses less of a threat to the seagrass habitat, is generally more targeted, and provides income to coastal communities. In this setting the reefs play a role in fishery resource partitioning.

1.3.2.2.4   Conservation, Species Manipulation, and Habitat Restoration — Artificial reefs are now used as tools in maintaining or restoring biodiversity in marine systems. For example, in coastal waters of Monaco, specially designed concrete boxes, "caves," were deployed to provide substrate for colonization by red coral, *Corallium rubrum* (Allemand et al. 1999). This economically important species has been exploited heavily in the Mediterranean basin; after a ban on fishing, manufactured concrete structures were deployed in the Undersea Reserve of Monaco.

Artificial reefs also are gaining acceptance as a means of repairing or replacing aquatic habitat damaged or destroyed by human actions (Grove and Wilson 1994). Some of the more elaborate projects started in the U.S. in the 1990s include the deployment of (1) natural materials (i.e., boulders) in the Pacific Ocean off California to mitigate destruction of kelp beds by cooling water discharge from an electricity-generating station; and (2) multiple designs using both manufactured and natural materials in the Atlantic Ocean off Florida to restore coral habitat destroyed by a sand dredge. One measure for success is ecological similarity between the resources lost and those intended to replace them, a form of "in-kind" mitigation (Ambrose 1994).

Nutrient removal by artificial reef systems is a technology in development. Nonpoint source, nutrient-rich inputs into the marine environment have the potential to cause eutrophic conditions or increase phytoplankton populations. Workers in Russia, Finland, Poland, and Romania, for example, have been investigating the potential for artificial reefs to act as biofilters. The aim is to remove, via communities of filterfeeders, the phytoplankton that are symptomatic of excess nutrient loading in coastal waters and maintain acceptable water quality (e.g., Gomoiu 1992; Laihonen et al. 1997)

1.3.2.2.5   Research — Research to enhance performance of artificial reefs is a cornerstone of sound planning, beginning perhaps with rigorous understanding of their physical design (e.g., Nakamura 1980). However, a second line of investigation views research as an explicit purpose for artificial reefs, distinct from the preceding four categories. It seeks to develop a fundamental scholarly understanding of them. Study of ecology and, to a lesser degree, physical and socio-economic structure, function, and impact since the mid-1980s has resulted in an extensive literature of more "basic" papers about artificial reefs. Findings are not immediately transferred to reef design. The number of reports dealing with "study reefs" to develop basic knowledge was at least 45% of all presentations at the four international conferences for this field since 1983 and increased to 60% in 1995 (Seaman 1997). The opportunity artificial reefs afford for manipulative experimentation is attracting more scientists to use those habitats in research.

1.3.2.2.6 Integrated Reefs — Part of the evolution of artificial reef technology includes expansion of purposes from the original centuries-old intent to attract fishes for subsistence harvest, to the satisfaction of multiple objectives. This may be seen in experimentation to satisfy life history requirements of both juvenile and adult species of fishes at a particular reef site (e.g., West et al. 1994) and especially to provide different interests with unique benefits simultaneously, such as habitat protection, commercial fish production, and bivalve culture (Bombace et al. 1994).

1.3.3 Approaches to Assessment Information

The breadth of modern reef interests surveyed in the preceding section gives rise to a premise of this book — one must know what needs to be documented about a reef before launching any measurement program, even when enthusiasm and resources to do the work are high. Because the fiscal, personnel, and time resources available for reef evaluation usually are so limited, a reef manager must stress the need to make the right kinds of measurements at the right level of effort. Going to sea — whether kilometers offshore and out of sight of land or in a coastal lagoon — can be quite expensive. Using an overqualified staff can be costly; using someone ill-equipped to gather data can jeopardize an entire study.

The diverse interests concerned with artificial reefs will ask different questions about the performance of their systems. This prompts us to describe below a progression of three increasingly more detailed and extensive approaches to developing knowledge about a reef, discussed in Chapters 3 to 6. The technical requirements of each increase correspondingly. Each level is valuable, for each provides answers to certain questions. The intent is to give reef program leaders and others a perspective of what evaluation practices are necessary or realistic given the level of skill available for field study. Defining the objective(s) of an artificial reef or reef system becomes the cornerstone for the monitoring and assessment process, as discussed above. Based on the reef's objectives, questions are asked, and they determine the kind and amount of information needed. Table 1.2 gives an overview of this scheme. A Type One, short-term study can gain an instantaneous assessment of one or more characteristics (as defined in Chapter 2), using less intense data gathering methods and basic statistical analysis. In a Type Two approach, some comparison of at least selected characteristics would be conducted, and analysis over time and space could be achieved. Finally, in a Type Three study, an explanation of cause-and-effect and prediction would occur based on the most intense investigation and data analysis.

Table 1.2 Approaches for Obtaining Information for Evaluation of Reef Performance

| Characteristic of Investigation | Type | | |
	One: Descriptive	Two: Analytic and Comparative	Three: Interactive and Predictive
Intensity of data gathering	Low–Moderate	Moderate–High	High–Very high
Rigor of training required	High	High	High
Duration of study	Short	Short–Long	Short–Long
Typical nature of information	Condition of reef, initially or at another point in time	Development of reef system; processes; changes	Comparison with other systems; efficiency; prediction
Scope of inferences	Instantaneous snapshot; presence–absence	Pattern; comparing over time and space	Cause and effect; explain pattern
Complexity of data analysis	Simple, basic statistics	High	Very high

These levels are not meant to be exclusive, restrictive, or binding upon reef measurement projects. Rather, they offer a conceptual way of looking at how to deal with increasingly complex information about reefs. This approach is used throughout the book. Each chapter adapts the concept in differing ways to reflect the peculiarities of the discipline and subject area involved.

1.3.3.1 Description

The most fundamental information about an artificial reef is gathered at this level. A representative principal question is, "Has the reef material been deployed as originally proposed and what is the initial overall character of the reef?" To answer this question, an immediate "postdeployment" study would be conducted. All disciplines and fields of study may be employed here: engineering, to determine materials spacing and settlement; biology, to identify early fish, plant, and invertebrate colonizers; and economics, to describe initial use of the reef by people. The time period for such a study is relatively short, and the result is a "snapshot," possibly qualitative.

A Type One study can be conducted other than just immediately after deployment of the reef. For example, Haroun et al. (1994) characterized the site where a reef was to be deployed. Meanwhile, a series of postdeployment studies might be reported as site maps, lists of fish and invertebrate species, or an inventory of reef users.

1.3.3.2 Analysis and Comparison

The typical question at this level is, "What is the structure of the reef, either physically or in terms of the ecological system on and adjacent to it, and what are the user interests that are present?" To answer this question, a study, generally at more than one point in time following reef development, is conducted. This study would provide more than a snapshot and allow comparison of the reef with itself over time. For example, Bombace et al. (1994, Figure 1.2) depict periods spanning 5 years of pre- and postdeployment sampling at seven reef sites. It is likely that scientific techniques of greater intensity and complexity would be used in contrast with a Type One study.

After the reef has been in the water for some time, it tends to a physical and ecological equilibrium (although biologically this may take years), that is, some shifting and settlement of materials may occur, therefore, physical data can help evaluate usefulness and stability of materials. Meanwhile, since an assemblage of plant and animal species will be forming, ecological data can be used to address whether target species are present. Human utilization patterns, such as fishing catch and effort, can now be documented. These data can be organized as a composite description of the structure and function of the reef system, and can be used to discern trends over time. The 18-year database reported by Stephens et al. (1994) represents an extremely long period of research for this field.

1.3.3.3 Interaction and Prediction

Typical questions at this level are, "How does the reef interact with the adjacent environment and how does it compare with other artificial reefs, other habitats, or even alternate fishery technologies and management practices?" In other words, what is its linkage with the ecosystem? How does it compare with other fishery/environmental practices? What is its economic viability? This approach would allow predictions to be attempted. A Type Three study would require the most time, personnel, and resources. Types One and Two could be "nested" into this level.

The length of time required to build a database is illustrated by research on spiny lobster, *Panulirus argus*, in the Florida Keys. Beginning in the mid-1980s with studies on the ecology of juvenile lobsters and their linkage with shallow coastal nursery habitats, a research team established the basis for replacement of lost sponges with artificial shelters a decade later (Herrnkind et al.

Table 1.3 Organization of the Book to Address Components and Issues of Reef Evaluation

Component	Key Issues	Reference
Reef Goals	To increase, create, implement, manipulate: fishery harvest and production, recreational fishing and diving, conservation and habitat restoration, resource allocation and protection, research	Chapter 1
Assessment Concept	What are the specific objectives for the reef? How will measured characteristics be used to assess the level of success in reaching reef objectives?	Chapter 1
Types of Assessments and Information	One: Description Two: Comparison Three: Interaction and Prediction	Chapters 1–6
Design of Study and Analysis	Summarization: What are the parameter estimates and their confidence intervals? Comparison: Do things vary over time or location? Association: How does one characteristic relate to other characteristics? Prediction: Can knowledge of one set of characteristics predict others?	Chapter 2
Assessment Methods	Physical and Engineering Nutrients and Primary Production Fishes and Macroinvertebrates Economic and Social	Chapter 3 Chapter 4 Chapter 5 Chapter 6
Quality Control	What are the appropriate protocols for data collection and analysis?	Chapters 2–6
Synthesis of Information	Evaluation of objectives Scenarios for implementing assessment	Chapter 7

1999). This research has progressed to the point of establishing population models, built on 15 or more years of prior study.

1.3.4 Reader's Guide to Study Design and Methods

Using this chapter as a point of departure, the remainder of the book discusses reef study design, methods, and implementation to evaluate reef performance. An overview of contents of the chapters is given in Table 1.3 and below:

- General guidance and principles of statistically rigorous study design are given in Chapter 2. They are intended to assist readers in organizing research, including analysis of data. Examples from the literature give the basic practices of statistics a focus on reef systems.
- Specific methods of data collection and study design considerations unique to the disciplines are presented for engineering/physical properties in Chapter 3; nutrients/primary production in Chapter 4; fishes/macroinvertebrates in Chapter 5; and social/economic dimensions in Chapter 6. These chapters discuss the basis for objectives-driven study in the particular field, and identify three types of information that may be obtained according to approach, complexity, and purpose.
- Integration of the approaches to evaluate reef performance is in Chapter 7. It discusses five representative scenarios concerning determination of expected performance of artificial reefs. The five goals addressed are: to promote regional economic development and recreational fishing; to protect valued marine habitat; to develop sustainable artisanal fisheries; to enhance water quality; and to resolve a species life history bottleneck in the context of economic development.

1.4 A BROADER VIEW OF REEF EVALUATION

This book is not about devising new research techniques for gathering data. Rather, it seeks to direct the application of existing techniques to an issue and a set of questions ordinarily not addressed in investigations of artificial reefs, namely, the documentation of projected benefits. By compiling the study techniques in one volume, we seek to create a basis for research worldwide that may yield data that are more directly comparable. The field of research has matured to the point where an examination of consistency among practitioners is recommended. Also, the field has matured to the point that its "toolbox" of methods is compatible with those used in other systems, such as in the economic study of coral reefs.

Whereas the technical methods in individual chapters can produce a gamut of information that includes characterization of basic and detailed aspects of reefs, the context in which they are presented and demonstrated is meant to encourage research of a holistic nature. The aim is to provide feedback for verifying the expectations of reef construction interests. For example, the "reef effectiveness analysis" proposed by Hagino (1991) unites reef planning and study to achieve comprehensive reef assessment. It integrates reef goals, preliminary scientific surveys, "fishing effectiveness analysis," data management and analysis, and evaluation of goals. Alas, such a system is not universally used, even in Japan, but the concept has merit for reef interests globally.

Among centers of information about reefs globally, the European Artificial Reef Research Network recently examined methodologies (Jensen 1997a). Subsequently, in a workshop on research protocols, it reached a number of points of consensus (Jensen 1997b), including the need for research data to be reported in such a fashion as to facilitate comparison between reef programs; standardized 5-year monitoring protocols; development of interdisciplinary studies of the reef system and the inclusion of colleagues from outside the traditionally biological reef scientist base; development of fisheries exploitation strategies; an increase in socioeconomic studies, preferably linking with scientific programs to evaluate results; development of artificial reefs as tools to provide aquaculture and tourist facilities; and understanding and promoting the secondary artificial reef role that coastal defense structures could play. Each recommendation links to the theme of this book.

Resource managers constantly evaluate the potential applications of artificial reef technology. Historical uses of reefs, such as for fishery enhancement, largely were implemented without formal evaluation efforts. Newer applications such as restoration of natural reef ecosystems are just beginning to be implemented and could benefit by measuring attainment of success criteria at the very outset. We endorse the call of Steimle and Meier (1997) for close linkage between the research and management sectors.

Further, we propose that disciplines unite in their efforts to assess reef performance. This is consistent with the growing acceptance of the theme of Ecosystem Management as an organizing framework for environmental studies. Also, there is an opportunity both to emulate and to coordinate with investigations of coral and other natural reef systems. International agreements have led to a global database and monitoring program for coral reefs, with availability of "uniformly gathered data" (McManus and Ablan 1997). By extension, uniformity in data for artificial reefs should be achievable. Comparison of factors such as harvest of extractable resources, tourism impacts, and management of multiple uses could advance understanding of both artificial and natural reef systems.

In little more than a generation, the number of countries in which artificial reefs are deployed has increased dramatically. The size of structures used, and their extent in a given nation, also have increased. To maximize benefits from a growing number of artificial reefs in the world's oceans, we urge that:

- The reef construction and research community maintain accountability to document reef performance for sponsors, funding sources, and users.
- The objectives for a reef be framed so as to allow evaluation of their achievement.
- Each reef is built so as to allow assessment of its performance.
- Competence and consistency of research protocols and personnel is assured through multinational training and collaboration.
- Comparability among datasets is addressed in research design.

1.5 ACKNOWLEDGMENTS

We appreciate the comments from the senior authors of the other chapters in this book, and the review provided by colleagues identified in the Preface. F. Simard, Musée Océanographique, Monaco, kindly assisted with research for this chapter. J. Potter created the drawing on page 1. Photographs were provided by R. Brock, University of Hawaii; J. Halusky, W. Lindberg, and F. Vose, University of Florida; G. Relini, Universita di Genova; P.J. Sanjeeva Raj, Centre for Research on New International Economic Order; and Soon Kil Yi, Korea Ocean Research and Development Institute. Special thanks are due to J. Whitehouse and T. Stivender, University of Florida, for administrative support and typing the manuscript. This chapter was sponsored in part by the National Sea Grant College Program, U.S. Department of Commerce, Grant NA76RG-01020.

REFERENCES

Allemand, D., E. Debernardi, and W. Seaman, Jr. 1999. Artificial reefs for the protection and enhancement of coastal zones in the Principality of Monaco. Pages 151–166. In: A.C. Jensen, K.J. Collins, and A.P. Lockwood, eds. *Artificial Reefs in European Seas.* Kluwer Academic Publishers, Dordrecht, The Netherlands.

Ambrose, R.F. 1994. Mitigating the effects of a coastal power plant on a kelp forest community: rationale and requirements for an artificial reef. *Bulletin of Marine Science* 55(2–3):694–708.

American Fisheries Society. 1997. *Fisheries* (Special Issue on Artificial Reef Management), Vol. 22, No. 4, Bethesda, MD.

Andree, S., ed. 1988. Proceedings, Florida Artificial Reef Summit. Florida Sea Grant College Program, Report 93, Gainesville, FL.

Barshaw, D.E. and E. Spanier. 1994. Anti-predator behaviors of the Mediterranean slipper lobster, *Scyllarides latus. Bulletin of Marine Science* 55(2–3):375–382.

Berger T., J. McGurrin, and R. Stone. 1994. An assessment of coastal artificial reef development in the United States. *Bulletin of Marine Science* 55(2–3):1328.

Bohnsack, J.A., D.L. Johnson, and R.F. Ambrose. 1991. Ecology of artificial reef habitats and fishes. Pages 61–107. In: W. Seaman, Jr. and L.M. Sprague, eds. *Artificial Habitats for Marine and Freshwater Fisheries.* Academic Press, San Diego.

Bombace, G., G. Fabi, L. Fiorentini, and S. Speranza. 1994. Analysis of the efficacy of artificial reefs located in five different areas of the Adriatic Sea. *Bulletin of Marine Science* 55(2–3):559–580.

Bougrova, L.A. and L.Y. Bugrov. 1994. Artificial reefs as fish-cage anchors. *Bulletin of Marine Science* 55(2–3):1122–1136.

Branden, K.L., D.A. Pollard, and H.A. Reimers. 1994. A review of recent artificial reef developments in Australia. *Bulletin of Marine Science* 55(2–3):982–994.

Bulletin of Marine Science. 1994. *Fifth International Conference on Aquatic Habitat Enhancement.* Vol. 55, No. 2 and 3, pp. 265–1359. University of Miami, Miami, FL.

Christian, R., F. Steimle, and R. Stone. 1998. Evolution of marine artificial reef development — a philosophical review of management strategies. *Gulf of Mexico Science* 16(1):32 36.

English, S., C. Wilkinson, and V. Baker, eds. 1997. *Survey Manual for Tropical Marine Resources.* 2nd Ed. Australian Institute of Marine Science, Townsville.

Fabi, G. and L. Fiorentini. 1990. Shellfish culture associated with artificial reefs. FAO Fisheries Report 428:99–107.

Fabi, G. and L. Fiorentini. 1994. Comparison between an artificial reef and a control site in the Adriatic Sea: analysis of four years of monitoring. *Bulletin of Marine Science* 55(2–3):538–558.

Gomoiu, M.-T. 1992. Artificial reefs — means of complex protection and improvement of the coastal marine ecosystems quality. *Studii de hidraulica, ICEM Burcaresti,* 33:315–324.

Grove, R.S. and C.A. Wilson. 1994. Introduction. *Bulletin of Marine Science* 55(2–3):265–267.

Hagino, S. 1991. Fishing effectiveness of the artificial reef in Japan. Pages 119–126. In: M. Nakamura, R.S. Grove, and C.J. Sonu, eds. *Recent Advances in Aquatic Habitat Technology. Proceedings, Japan-U.S. Symposium on Artificial Habitats for Fisheries.* Southern California Edison Co., Environmental Research Report Series 91-RD-19, Rosemead, CA.

Haroun, R.J., M. Gomez, J.J. Hernandez, R. Herreva, D. Montero, T. Moreno, A. Portillo, M.E. Torres, and E. Soler. 1994. Environmental description of an artificial reef site in Gran Canaria (Canary Islands, Spain) prior to reef placement. *Bulletin of Marine Science* 55(2–3):932–938.

Herrnkind, W.F., M.J. Butler, IV, and J.H. Hunt. 1999. A case for shelter replacement in a disturbed spiny lobster nursery in Florida: why basic research had to come first. Pages 421–437. In: L. Benaka, ed. *Fish Habitat: Essential Fish Habitat and Rehabilitation.* American Fisheries Society, Symposium 22, Bethesda, MD.

Jensen, A.C., ed. 1997a. *European Artificial Reef Research. Proceedings, First EARRN Conference,* Ancona, Italy. Southampton Oceanography Centre, Southampton, England. 449 pp.

Jensen, A.C. 1997b. Report of the results of EARRN workshop 1: research protocols. European Artificial Reef Research Network AIR3-CT94-2144. Report to DGXIV of the European Commission, SUDO/TEC97/13. 26 pp.

Jensen, A.C., K.J. Collins, and A.P. Lockwood, eds. 1999. *Artificial Reefs in European Seas.* Kluwer Academic Publishers, Dordrecht, The Netherlands.

Laihonen, P., J. Hanninen, J. Chojnacki, and I. Vuorinen. 1997. Some prospects of nutrient removal with artificial reefs. Pages 85–96. In: A.C. Jensen, ed. *European Artificial Reef Research. Proceedings, First EARRN Conference,* Ancona, Italy. Southampton Oceanography Centre, Southampton, England. 449 pp.

Lindquist, D.G., L.B. Cahoon, I.E. Clavijo, M.H. Posey, S.K. Bolden, L.A. Pike, S.W. Burk, and P.A. Cardullo. 1994. Reef fish stomach contents and prey abundance on reef and sand substrata associated with adjacent artificial and natural reefs in Onslow Bay, North Carolina. *Bulletin of Marine Science* 55(2–3):308–318.

Lozano-Alvarez, E., P. Briones-Fourzan, and F. Negrete-Soto. 1994. An evaluation of concrete block structures as shelter for juvenile Caribbean spiny lobsters, *Panulirus argus. Bulletin of Marine Science* 55(2–3):351–362.

McGlennon, D. and K.L. Branden. 1994. Comparison of catch and recreational anglers fishing on artificial reefs and natural seabed in Gulf St. Vincent, South Australia. *Bulletin of Marine Science* 55(2–3):510–523.

McManus, J.W. and M.C.A. Ablan, eds. 1997. *Reef Base: A Global Database on Coral Reefs and Their Resources.* International Center for Living Aquatic Resources Management, Makati City, Philippines. xi + 194 pp.

Nakamura, M., ed. 1980. *Fisheries Engineering Handbook (Suisan Doboku).* Fisheries Engineering Research Subcommittee, Japan Society of Agricultural Engineering, Tokyo. (In Japanese.)

National Research Council (U.S.). 1996. Linking science and technology to society's environmental goals. National Academy of Sciences, Washington, D.C.

Omar, R.M.N.R., C.E. Kean, S. Wagiman, A.M.M. Hassan, M. Hussein, B.R. Hassan, and C.O.M. Hussiu. 1994. Design and construction of artificial reefs in Malaysia. *Bulletin of Marine Science* 55(2–3):1050–1061.

Ozasa, H., K. Nakase, A. Watanuki, and H. Yamamoto. 1995. Structures accommodating to marine organisms. Pages 406–411. In: H. Sako, ed. *Proceedings, International Conference on Ecological System Enhancement Technology for Aquatic Environments.* Japan International Marine Science and Technology Federation, Tokyo.

Polunin, N.V.C. and C.M. Roberts, eds. 1996. *Reef Fisheries.* Chapman & Hall, London. xviii + 477 pp.

Relini, G. and S. Moretti. 1986. Artificial reef and *Posidonia* bed protection off Loano (Western Ligurian Riviera). FAO Fisheries Report 357: 104–109.

Sako, H., ed. 1995. *Proceedings, ECOSET '95 — International Conference on Ecological System Enhancement Technology for Aquatic Environments.* Japan International Marine Science and Technology Federation, Tokyo.

Sale, P.F., ed. 1991. *The Ecology of Fishes on Coral Reefs.* Academic Press, San Diego. xviii + 754 pp.

Sanjeeva Raj, P.J. 1996. Artificial reefs for a sustainable coastal ecosystem in India, involving fisherfolk participation. *Bulletin of the Central Marine Fisheries Research Institute* 48:1–3.

Schmitt, R.J. and C.W. Osenberg, eds. 1996. *Detecting Ecological Impacts: Concepts and Applications in Coastal Habitats.* Academic Press, San Diego. xx + 401 pp.

Seaman, W., Jr. 1995. Artificial habitats for fish. *Encyclopedia of Environmental Biology.* Academic Press, San Diego. 1:93–104.

Seaman, W., Jr. 1997. Frontiers that increase unity: defining an agenda for European artificial reef research. Pages 241–260. In: A.C. Jensen, ed. *European Artificial Reef Research. Proceedings, First EARRN Conference,* Ancona, Italy. Southampton Oceanography Centre, Southampton, England.

Seaman, W., Jr. and L.M. Sprague, eds. 1991. *Artificial Habitats for Marine and Freshwater Fisheries.* Academic Press, San Diego. xviii + 285 pp.

Simard, F. 1997. Socio-economic aspects of artificial reefs in Japan. Pages 233–240. In: A.C. Jensen, ed. *European Artificial Reef Research. Proceedings, First EARRN Conference,* Ancona, Italy. Southampton Oceanography Centre, Southampton, England.

Steimle, F.W. and M.H. Meier. 1997. What information do artificial reef managers really want from fishery science? *Fisheries* 22(4):6–8.

Stephens, J.S., Jr., P.A. Morris, D.J. Pondella, T.A. Koouce, and G.A. Jordan. 1994. Overview of the dynamics of an urban artificial reef fish assemblage at King Harbor, California, USA, 1974–1991: a recruitment driven system. *Bulletin of Marine Science* 55(2–3):1224–1239.

Stoddart, D.R. and R.E. Johannes, eds. 1978. *Coral Reefs: Research Methods.* Monographs in Oceanography Methods No. 5, UNESCO, Paris.

West, J.E., R.M. Buckley, and D.C. Doty. 1994. Ecology and habitat use of juvenile rockfishes (*Sebastes* spp.) associated with artificial reefs in Puget Sound, Washington. *Bulletin of Marine Science* 55(2–3):344–350.

Yamane, T. 1989. Status and future plans of artificial reef projects in Japan. *Bulletin of Marine Science* 44(2):1038–1040.

CHAPTER 2

Study Design and Data Analysis Issues

**Kenneth M. Portier, Gianna Fabi, and
Paul H. Darius**

CONTENTS

2.1 SUMMARY

Successful reef assessment studies have certain properties in common, such as clearly stated objectives, appropriate measurement techniques, adequate and effective sampling, and powerful statistical analysis. There are no easy formulas to good assessment design, but there are design principles that lead to good designs. To discover these principles, this chapter proposes and discusses 17 questions that should be answered at the design stage of a study. The issues raised by these questions are discussed and illustrated using examples from the artificial reef literature. Conclusions are summarized as a set of 20 principles to use as a study design checklist.

2.2 INTRODUCTION

The issues in designing assessment studies for artificial reefs and statistical analysis of the resulting data are no different from those found in the design of other types of natural resource assessments. At the very least, successful assessment studies have the following:

- A clear statement of the key study objectives, with criteria for when the study objectives have been met.
- Techniques that accurately measure the physical, biological, or economic characteristics needed for the assessment.
- Sampling protocols that define how and how much information must be collected to estimate key reef characteristics, to reliably compare these characteristics or to provide predictions with known precision. The sampling protocol must also indicate how factors producing extraneous variation in the measurements will be managed.
- Estimation protocols that describe how raw data will be transformed into statistical quantities, which can be used to evaluate reef condition or predict future trends.
- Testing protocols that dictate which comparisons will be made, which statistical models and test procedures will be used, methods for assessing the acceptability of assumptions, and how important factors will be incorporated into the analysis.

- Where the assessment involves evaluation of treatments, usually defined as manipulations of various reef parameters, protocols for replication of the treatments and random assignment of treatments to sampling units must be specified.

While these statements concisely describe the properties of a good assessment study, actually developing a study with these properties is a difficult process. There are no easy formulas that, if followed, will result in a good study design. Rather, there are design principles that, if followed, should lead to a good study design. To discover these principles, this chapter explores the questions that arise from a closer examination of the above statements and discusses the statistical issues that result from an even closer examination of the questions.

2.2.1 Chapter Objective

The objectives of this chapter are to (1) discuss the issues arising in the design and statistical analysis of a reef assessment study and (2) recommend principles which should be followed in the design and statistical analysis of any artificial reef assessment study. The first objective is approached through discussion of 17 questions and associated issues which are frequently raised at the design phase of a study. These issues are identified, discussed, and illustrated through references to existing reef literature. The second objective produces 20 design principles which follow from the issues discussion.

This chapter is not intended as a primer on statistical tools as related to reef assessment. The reader is assumed to have a basic understanding of statistical concepts and hypothesis testing or to have access to statistical texts or others trained in statistics. Additional background material on sampling and statistical testing as related to reef assessment is available, but may be difficult to obtain (see Samoilys 1997). Similarly, references on these topics for fisheries (see for example Gulland 1966, 1969; Bazigos 1974; Ricker 1975; Saville 1981), while not recent, nevertheless provide information that can also be used in assessing artificial reefs. Many general statistical design and data analysis texts (Green 1979; Kish 1987; Krebs 1989; Zar 1996) focus on ecology and environmental sciences applications that can be related to reef assessment.

Finally, although the focus of this chapter is on assessment of artificial reefs, most of the principles and methods described are equally applicable to the assessment of natural reefs, with the understanding that research on natural reefs rarely involves the level of direct manipulation found in artificial reef research.

2.2.2 Definitions

Designing a natural resource assessment study is complicated enough without the added confusion of terms that are not precisely defined. In this chapter, technical terms are defined as they are needed in the discussion. The terms and concepts defined in this section are primarily statistical and are just those needed to support further discussion. Wherever possible, definitions reflect the general consensus as defined in the published literature.

Any feature of a reef and/or its associated living assemblages is referred to as a *character* or *characteristic*. A characteristic that can be directly measured, such as sediment transport, the density of fouling organisms, fish abundance, or user satisfaction, is called a *directly measured variable,* or simply a *regular variable*. It is called a variable since its value may change or vary over time and/or over location and/or from individual to individual. Usually more than one variable will be used in the assessment of an artificial reef.

Quantities that are computed from directly measured variables are called *derived variables*. Just like regular variables, derived variables measure characteristics, but these are usually rates or ratios that cannot be directly measured, such as an index of reef stability, the diversity of fish over

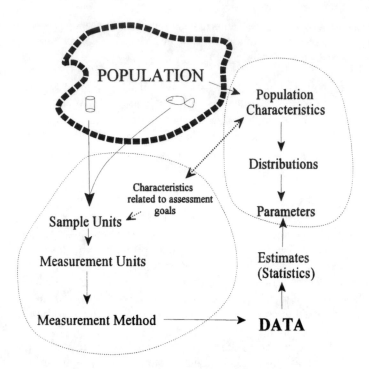

Figure 2.1 Sampling of natural populations requires consideration of population characteristics as parameter-
ized distributions, with the goal of the sampling process to collect data that facilitates estimation of
distribution parameters.

the reef, the richness of the microinvertebrate community, or the cost-to-benefit ratio on the
economics of a reef.

Sampling is the selection of part of something to represent the whole (vaguely defined as the
population, see Figure 2.1). For example, we may collect 30 individuals from a large school of
fish to be representatives of the whole school. Distinct measurements on individuals in the sample
are referred to as *sample values* and the collection of measurements for the whole sample as the
sample set. Sample sets often are referred to simply as the *sample*. This terminology can be
confusing because each individual may also be referred to as a sample. For example, each of the
30 individuals selected from the school of fish may be referred to as a sample or the whole collection
of 30 fish as the sample. For this book, we refer to the whole collection as the sample and each
individual as a sample unit.

A *sample unit* is that specific part, piece, or individual which is collected and/or measured. The
sample unit may be a discrete, identifiable individual, such as a reef, fish, fisherman, or period of
time. The sample unit also may be defined artificially by the tool or method used to collect material
for measurement, such as a core of sediment, a small area of reef, or a volume of water. Sample
units chosen as part of a sample will be measured to provide a value for the assessment variables.
If the sample unit is very large, direct measurement may not be possible. It may be necessary to
subdivide the unit into smaller components that can then be measured. Often, logistics or cost
dictate that only a few of the subunits of the sample unit are measured, with the average being
used as the value for the sample unit. These selected subdivisions are referred to as *measurement
units*. In most cases, the measurement unit will be the same as the sample unit. In other situations,
only part of the sample unit may be measured because the measurement technique only uses a very
small quantity of material.

Statisticians define the *population* of a variable as the set of all possible sample unit measurements. This definition can be quite different from how the term population is used by biologists.
In this chapter, all references to population are to the statistical population. The goal of sampling

is to collect a sample in such a way that the statistical characteristics of the sample unit measurements are as similar as possible to what would be obtained for the same variables measured on every possible unit of the whole population. Since most populations exhibit great variability from site to site, time to time, or sample to sample, achieving a representative sample is not always simple or straightforward.

Every variable has associated with it a *distribution* which describes the relative frequency with which values of the variable will occur in the population. If the variable can have only a limited number of possible values, such as color classification, count of number of individuals, or response on a questionnaire item (e.g., yes/no or preference scale value — strongly agree, agree, neutral, disagree, strongly disagree), the variable is referred to as *discrete* and values will follow a *discrete distribution*. If the variable can take values from an infinite number of possible values, such as weight, salinity, speed, etc., then the variable is referred to as *continuous* and values will follow a *continuous distribution*. A *representative sample* is defined as a sample whose distribution of observed values closely matches the distribution one would obtain if values from the whole population were available.

Estimation is the process whereby we calculate or approximate the value of an unknown quantity, referred to as a *parameter*, which describes an aspect of the associated characteristic's distribution. A *statistic* is simply a *function* of sample values, such as the mean, standard deviation, median, or range. An *estimate* is the specific value of a statistic computed using a set of observed values. Examples of parameters for which estimates are usually obtained include mean water temperature, mean fish weight or length, total fish density, average benthic densities, average wave cycle time, average cost of a fishing trip, survival rate, recruitment rate, or average catch size. The parameters of interest are usually an average, a total, a rate, or a ratio. In comparative studies, the parameter estimates are compared to either a predefined value from a standard or natural population, or to the same statistic computed using values from measurement at another level of one or more *factors* (a different reef site, time, condition, etc.).

2.3 STUDY DESIGN ISSUES

Designing a study is an exercise in balancing the desire to know everything about a system with the reality of limited resources, measurement methodologies, and time (Kish 1987). The objectives for the study reflect our current understanding or lack of understanding of artificial reef processes. The scope of the study will depend on available resources and time.

All effective study designs follow certain basic principles (Green 1979). The best design for a given set of objectives is the one that incorporates all these basic principles while staying within resource constraints. The principles define a common logical flow to the design process for assessment as well as research studies. This flow is evident from the series of questions (Table 2.1) usually asked at the design phase of a study. Each question relates to one or more of the major design components (Figure 2.2). These design components are discussed and illustrated in the sections that follow.

2.3.1 Reef Assessment Concept — Reef Objectives and Success Criteria

The basic aspects of the reef assessment concept were covered in Chapter 1. The reef assessment concept primarily consists of (1) statements of the general and specific objectives for the artificial reef and (2) listings of the criteria to be used in assessing whether the reef has met these objectives. The purpose of an assessment study is to collect the information needed to reliably determine whether the artificial reef has met the objectives assigned to it. To make this determination, the reef objectives must be linked to characteristics which can be measured or, more correctly, with statistics calculated from data collected on measurable characteristics. Success is

Table 2.1   Questions Usually Asked During the Design Phase of an Assessment Study

1. What are the goals and success criteria of the assessment?
2. What type of assessment is needed?
3. What characteristics will be measured?
4. How will samples be collected?
5. Where will samples be collected?
6. When will samples be collected?
7. For descriptive studies, what level of description will be used?
8. For a comparative study, which factors are controlled?
9. What are the hypotheses associated with study factor level comparisons?
10. What is the replicate for a comparative study?
11. How will hypotheses be tested?
12. What level of uncertainty is allowed in test conclusions?
13. How many replicates are needed?
14. For interaction and prediction studies, how will association be measured?
15. For prediction studies, how will predictions be computed?
16. How will data records be handled? Data quality assured?
17. How will data be statistically analyzed?

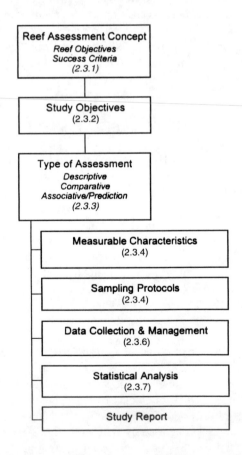

Figure 2.2   Components of a reef assessment study design (Section 2.3.1).

declared when these statistics are determined to meet certain prescribed criteria or fall within some predefined levels.

For example, suppose the general objective for an artificial reef is to improve recreational fishing. Specific objectives might be stated as:

- Increase the *abundance* of sport fish on the reef,
- Increase the *catch success* by sport fishers, or
- Increase the *income* of fishing boat owners in the geographic region nearest the reef.

Success criteria for these specific objectives would be:

- *Average abundance* of sport fish in the area is twice prereef levels,
- *Catch-per-unit effort* by sport fishers is 50% greater than prereef levels, or
- Boat owner *average income* is greater than prereef levels after taking into account any other factors which might have contributed to this income (general wage increases due to inflation, increased numbers of boat owners, etc.).

Depending on the criterion chosen, the subject of our study in this example will be either the sport fish in the area where the reef is established, the fishers, or the boat owners who used the area before and after the reef was installed. The statistics or parameters of interest also are different (see the italicized words in the success criteria statements). Each of these success criteria directs consideration of quite different study designs. For this reason it is important that reef goals and success criteria be discussed and specified as the first step in the study design process. More on this topic appears in Chapter 7.

2.3.2 Study Objectives

In addition to specifying reef objectives, it is important to develop objectives for the study itself by identifying which of the specific objectives of the reef will be examined in a particular assessment study. These study objectives can be narrow, addressing only one of the reef's many objectives. Examples include a postdeployment survey performed to determine if a new reef is where it is supposed to be or a one-time survey to determine if a reef is where it was originally built. On the other hand, study objectives can be quite broad, such as relating reef biotic community processes to reef-associated physical processes. Clearly, the study objectives should address one or more of the specific reef goals. Similarly, success criteria for the study should address some aspect of success for the reef as well.

2.3.3 Type of Assessment

Study objectives dictate the type of assessment to be used. Three levels of assessment were introduced in Chapter 1. These levels describe broad classes of studies ordered by the complexity of study objectives and the amount of sampling and measurement effort used (Figure 2.3). Thus, descriptive studies almost always use fewer resources and address simpler objectives than comparative studies which in turn use fewer resources and address simpler objectives than association/prediction studies.

All studies will involve some type of comparison. For example, an artificial reef constructed to offer opportunities for recreational fishing might use a monitoring-type study to assess whether the reef actually attracts fishers (Rhodes et al. 1994). An artificial reef constructed to replace a natural reef that was accidentally destroyed might be assessed on how its biological communities compare, at specified times in its evolution, to those of a natural reef of the type destroyed (Clark and Edwards 1994). An artificial reef constructed to quantify ecological associations among fish

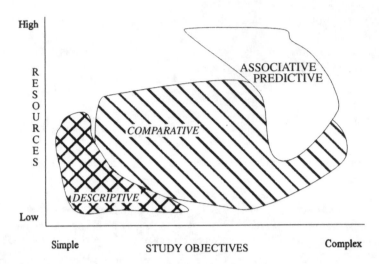

Figure 2.3   Assessment studies are ordered by the complexity of study objectives and the amount of sampling and measurement effort used.

and benthic communities as the reef communities evolve would require careful timing of sampling sessions, extensive sampling across multiple characteristics, and comparisons across time and space, in order to estimate the levels of associations present (Hueckel and Buckley 1989; Ardizzone et al. 1997; Ecklund 1997).

   Classifying an assessment study to a type is useful in identifying which study design questions to ask. For example, if the study type is a descriptive, monitoring, association, or prediction study, the question regarding how many sample units to collect will depend on the precision desired for key parameters of interest (Section 2.3.5.4). On the other hand, if the study is comparative, the number of sample units will depend on how willing one is to make mistakes in the conclusions from statistical hypothesis tests (Section 2.3.5.5.6).

2.3.4   Measurable Characteristics — What to Sample

   A critical stage in the study design process is choosing characteristics to be measured, defining the measurement methods to be used, and relating these characteristics directly to reef objectives. A rich collection of characteristics and measurement methods are available to the study designer. Chapter 3 discusses measurement methods for reef physical characteristics, Chapters 4 and 5 describe methods for reef biological communities, and Chapter 6 discusses measurement of the socioeconomic aspects associated with artificial reefs. Each characteristic can be linked to specific reef or study goals. The study design concepts and statistical analysis methods discussed in this chapter can be combined with any of these measurement techniques in developing a study plan.

   In choosing what to measure, one needs to be aware that measurement is never exact; some uncertainty is always associated with the values obtained. There are two main sources of uncertainty — *measurement error* and *systematic error*. Systematic errors occur as the result of flaws in the data-gathering process, either through the device used to collect the sample material or the protocol used to locate when and where the sampling occurs. Measurement error arises from the practical or technical limitations of the measurement method. For assessment, we want to use measurement methods with sampling protocols which together have the least amount of error. We want to use methods that produce distributions of the observed values that mimic what the true distributions of these values would be if we could examine the whole population. It is important that only methods of demonstrated validity for the stated purpose be used. Methods should be reasonably rugged in the hands of an average user. If the method is accurate and precise, but only when used by an extremely skilled individual, chances of success with unskilled labor will be low.

Consider visual censusing techniques for reef fishes (Samoilys 1997, Chapter 3). These methods typically produce counts that underestimate the actual number of fish present in the area being sampled at the same time overestimating the average biomass. This is because it is difficult to see smaller fish and hence smaller fish are under-represented in the sample as compared to the true population. Systematic errors in the measurement method will produce biased estimates which can vary from situation to situation. For example, Ricker (1975: 142) discusses how data on young-age groups of fish, not fully vulnerable to fishing, can impact the estimate of the recruitment rates for the population. In general, it pays to research each proposed method until the statistical characteristics of the resulting measurements and their impact on target parameter estimates are understood.

Study objectives and uncertainty properties of available measurement techniques determine which characteristics are used in a study. Characteristics not related to reef success criteria have no place in a reef assessment study. Similarly, characteristics which are expensive to measure, have measurements subject to large uncertainties, and/or have measurement methodology that is difficult for all but an expert to perform, may also be excluded from consideration. Any characteristic measured should have a purpose that can relate to study objectives, reef objectives, and success criteria. If this linkage cannot be made during the design phase of the study, the time spent measuring, recording, and analyzing the characteristics will be wasted.

2.3.5 Sampling Protocols

Once the study type and measurement characteristics (what to sample) have been identified, issues relating to sampling protocols must be addressed. These issues are characterized by the questions when to make measurements, where to take measurements, what protocol will be used to select sample units, and how many sample units (measurements) will be taken for each characteristic. The best answers to these questions are obtained by being specific to the reef or reefs under study. In this section, some general guidance on how to answer these questions is provided. Additional recommendations are provided in the discussions of Chapters 3 to 6. To reiterate, each sampling situation is unique and requires that the general approaches illustrated here be adapted for the specific situation.

2.3.5.1 When to Make Measurements

The first sampling protocol decision is the determination of when samples will be collected. Common assessment sampling frequencies include one-time assessments of reef condition, pre- and postdeployment samples, short-term periodic sampling, or long-term periodic sampling, often referred to as *monitoring* (Figure 2.4). Limitations on resources available for sampling are a major determining factor of when and how often sampling will occur. In addition, the changes that occur over time as biotic communities evolve or the artificial media of the reef change are often issues in assessing an artificial reef. More complex studies examine species accrual on new reef habitat and the interactions among species which develop over time, changing the species mix and eventually changing the ecosystem (Pamintuan et al. 1994; Fabi and Fiorentini 1994; Relini et al. 1994; Fabi et al. in press). Biological communities and populations on artificial reefs evolve over time, making the choice of when to sample a difficult but critically important decision.

The success criteria for the reef and the study objectives also drive the choice of sampling protocols. For example, if the study objective is to establish the physical stability of the reef material, a one-time assessment (Figure 2.4a) soon after deployment may be all that is required. To determine whether the establishment of an artificial reef has changed the biotic communities of an area requires

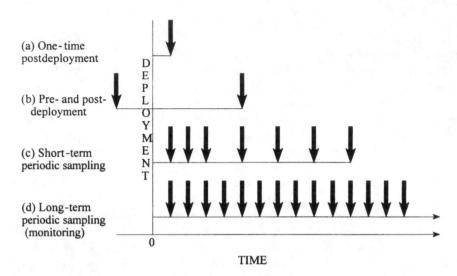

Figure 2.4 Typical frequencies of sampling used in assessment studies.

use of at least two descriptive studies, one performed prior to deployment of the reef construction material and one or more postdeployment surveys (Figure 2.4b) (Lindquist and Pietrafesa 1989; Bombace et al. 1994). The first survey establishes levels for assessment characteristics prior to reef creation and serves as a baseline for comparisons to postdeployment levels. The second time point takes place some months or years postdeployment, the exact time depending on the expected rate of change in the characteristics of direct interest. Usually the same measurement methods and sample selection protocols are used at both time points to allow direct comparison of estimates.

If an objective of the study is to gain an understanding of the timing of key reef processes and/or the time to critical ecological stages, short-term periodic sampling or systematic long-term monitoring is used (Reimers and Branden 1994; Szmant and Forrester 1996). If the assessment concept for a newly evolving reef requires multiple measurements of reef characteristics over time, samples are taken more often during the initial months and years when rapid changes are occurring (Linquist and Pietrafesa 1989; Ardizzone et al. 1989) (Figure 2.4c). Even if sampling will take place only twice a year (Figure 2.4d), the status of characteristics during these two sampling times should be a factor in picking the specific times for sampling (Herrnkind et al. 1997).

2.3.5.2 Where to Take Measurements

The decision of where to take measurements often depends on the characteristic being measured and its spatial variability. For example, when visual counting methods are used to establish fish abundance on a reef, the sampling location is that position above the reef which affords the diver/enumerator the best view of the complete reef (Samoilys 1997, Section 2.4.1). Placement of core sample units to establish composition and biomass for the benthic community surrounding a reef also must take into account expected spatial patterns (see Badalamenti and D'Anna 1997 for a general review).

Spatial aspects of sampling protocols are important for a number of reasons. Many reef variables, especially chemical and biological characteristics, have values that change greatly depending on location and distance from the reef. Large variations are often found even within fairly small areas (Posey et al. 1995). Even reef-associated socioeconomic characteristics such as reef value depend on distances to fishing ports or to neighboring reefs. Because of this, parameter estimates are often very dependent on where measurements are taken. For example, estimates of average water current are different if measurements are taken only near the base of the reef instead of from locations some distance from the reef. If the study objective is to obtain a number that best represents the mean condition of a characteristic in and around the reef, careful consideration is needed in determining where sample units are sited and how measurements will be combined to produce the final estimate.

2.3.5.3 Sample Unit Selection Protocols

A number of general statistical approaches or protocols are used to select individuals or locations to be included in the sample. The broadest classification considers *nonrandom selection* vs. *random selection*. Random selection includes *simple random selection, stratified random selection, cluster selection,* and *systematic selection with random starting locations.* Selection protocols for reef assessment designs usually incorporate some or all of these approaches. A good review of siting issues and selection methods for sampling environmental parameters is provided in Gilbert (1987).

Nonrandom selection of sample locations occurs when sites are chosen for measurement without any recourse to a randomization scheme. Other names for this approach are *purposeful selection, judgmental selection,* and *convenience selection.* A sample selected by a chance mechanism is called a *random selection.* Many researchers prefer to use nonrandom samples since they allow selection of locations or individuals deemed "important," "typical," or "representative." Statisticians, on the other hand, require the use of random selection since it provides assurance that the estimates obtained will have minimum systematic error, hence less bias (at least up to the bias of the measurement device) and, at the same time, allows determination of a measure of confidence in the estimate.

Nonrandom selection often is used in situations where nature itself is assumed to be sufficiently random, deeming additional randomization to be redundant, or when the benefits in cost and convenience outweigh the known biases that result from the selection method (see Gauch 1982). For example, in monitoring a reef, sampling times may be established on a fixed, nonrandom schedule, say every 60 days. Similarly, a visual fish census might always be taken for a reef from the same location, say the east side of a reef, at the same time of day, in order to afford maximum visibility and consistency of counts, to minimize measurement error, and to provide the best data for comparative purposes.

Data collected with a nonrandom selection protocol are usually biased and produce inaccurate estimates, limiting the usefulness of results (Figure 2.5). For example, if the sampling times always

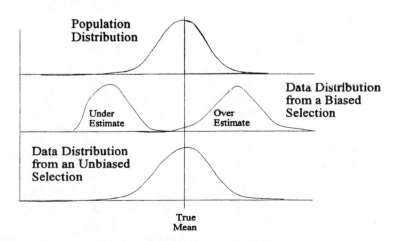

Figure 2.5   Effect of biased selection on the distribution of observed values.

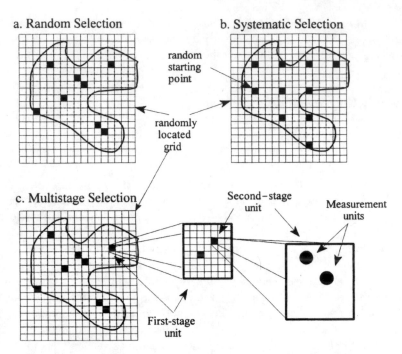

Figure 2.6   Spatial selection schemes for sampling environmental characteristics on a reef.

coincide with high tides, the resulting estimates will only reflect what happens on the reef at high tides. Similarly, by always viewing the reef from the same side, activities and abundances on the far side are not included, with the result that counts and subsequent estimates may over- or under-estimate true values. In this case, if the study objective requires an unbiased and precise estimate of average fish density, a nonrandom selection protocol would be inappropriate. Finally, the precision of estimates from nonrandom selected samples is difficult if not impossible to compute since the underlying statistical distribution for the estimates cannot be determined. This suggests that selection of sample units should not be left completely up to the diver or the individual who is actually doing the sampling. Randomness in sample unit selection should be used unless the study is strictly descriptive and confidence statements on estimates are not required, and/or statistically valid comparisons among populations are not needed. In general, nonrandom selection is not recommended for assessment studies.

Depending on the sample unit, incorporating randomness in unit selection can be easy or difficult to accomplish. Suppose you wish to use a 0.5 m² quadrat as the sample unit for measuring plant biomass on the surface of a reef. Random selection may be accomplished by first overlaying a grid of 0.5 m² cells onto the reef surface. These cells can then be numbered and a random number generator can be used to select which cells are actually measured (Figure 2.6a). An example of a *systematic selection protocol* (Figure 2.6b) would be to measure every fifth cell in a sequential numbering of all possible grid cells. When using a systematic selection protocol, care must be taken to introduce randomness directly into the selection process. This is usually accomplished by selecting a starting cell at random then identifying all other measured cells by their distance from the starting cell. Samples selected using a systematic selection protocol can produce sample measurements which are not strictly independent, adding complexity to the parameter estimation and subsequent data analysis. Modern statistical analysis tools, such as mixed effects linear models (Littell et al. 1996), can accommodate these data dependencies. Systematic selection protocols are efficient when the objective of the sampling is to provide information sufficient to map the characteristic over a geographic area (Andrew and Mapstone 1987), assess the spatial pattern of a set of characteristics, or examine seascape (landscape) characteristics.

When large areas must be sampled with small sample units, a *multistage selection protocol* is recommended. Consider the sampling of bottom sediments for determination of chemical or micro-benthic populations. Typically a small-diameter (say 10 cm) circular core defines the sample unit. These are to be taken from a fairly large area of sea floor surrounding a reef. In multistage selection (Figure 2.6c), a grid of cells, say 1 m² in size, is defined over the entire survey area and each cell is given a unique number. This grid defines what are referred to as the first-stage sample units. A specified number of these units (grid cells) are selected at random and comprise the first-stage sample. Each first-stage sample cell is then further subdivided using a finer grid of, say 0.1 m² cell size, to produce a set of second-stage sample units. One or more of these second-stage units are then selected at random and comprise the second-stage sample. From each second-stage sample unit, one or more cores are collected, usually from locations determined by the diver. This layered division and associated random selection allows incorporation of randomness into the selection process while simultaneously limiting the number of locations that must be sampled. The final selection protocol is extremely effective and efficient.

The cost of processing the material collected from an individual sample unit is often fairly high. Consider, for example, the time needed to perform a complex chemical analysis of a sediment sample or the time needed for a technician to enumerate microbenthos in a grab sample. To reduce this processing cost, samples collected close to each other in space are combined and only one set of measurement values is produced for the combined sample. These combination samples are referred to as *composite samples*. Compositing of samples is a common practice, primarily because it reduces costs while also increasing the area represented by each sample. Information on sample-to-sample variation is lost when using composite samples, therefore, care should be taken to ensure that critical study questions can still be answered with the information provided by the composites. Multiple composite samples should be used in order to provide the variability information needed for statistical comparisons.

Systematic and multistage selection protocols are also useful in assessing the economic and social aspects of an artificial reef. For example, a fish creel survey may dictate that fishers from every third boat returning to the dock be interviewed (systematic selection). In addition, the interviews may occur over a 2-week period, with interview effort evenly distributed over the daylight hours. To assure that all daylight hours are included in the survey, each day may be *stratified* into four or six time periods, with only one or two time periods being used on any one day, but balanced such that each time period is used the same number of times over the 2-week period (multistage selection). Such designs allow for the efficient utilization of resources while providing for good representative coverage of the populations of interest. Stratification is also used to assure complete coverage of daily trends in characteristics such as fish densities, water flow, light levels, and fishing effort. In situations where the area being sampled is quite variable and one has prior information on factors producing this variability, stratification can be used to assure representative coverage, compute more precise overall estimates, and provide additional information for each stratum. While a truly random selection protocol might result in no samples being collected in certain areas, stratified selection will assure that samples are placed in all areas.

More details and examples of the different sample selection protocols can be found in any sampling methodology text. Gilbert (1987) and Green (1979) in particular have good discussions of these techniques for sampling terrestrial environments. Note that discussions of sampling pro-tocols reported in the reef assessment literature often do not provide sufficient detail to determine whether systematic, multistage, or stratification protocols were used.

2.3.5.4 *Number of Sample Units*

Because of its direct effect on the overall cost of an assessment study, the question of how many sample units to measure/observe is one every study designer will have to face. Often, the answer to this question is simple — the number of units observed is the number that can be obtained with the resources, money, time, and personnel available for the study. But this answer misses the true importance of the question, since the number of units measured will be directly proportional to the precision of the final estimates and the resolution of the comparisons performed. Use too few sample units, and the estimates obtained will be relatively imprecise and possibly unusable in assessing whether the reef has reached it goals. Too few units, and the comparisons made will have little discriminatory power, leading to an inability to clearly decide whether differences in characteristics were really observed. The number of units needed to estimate a parameter to a prespecified precision or make comparisons within some predefined probabilities of making an error in the conclusions can be computed for certain simple situations. While it is not possible to cover all probable situations, the approach described below can be expanded to cover many of the more common situations encountered in reef assessment.

We begin by recognizing that the measurement data collected on each characteristic will be summarized at the analysis phase using *statistics,* such as the mean or variance. These values represent estimates of what the true mean or variance would be if the whole population could be sampled. Because a *sample* is not a complete census, the estimates will only approximate the true underlying value. If the measurement method or the sampling protocol produces systematic under- or over-counting or measurement, the resulting estimates will be *biased*. This means that even as the sample size increases, the calculated statistics will not equal the true population values.

Suppose it were possible to repeat a particular sampling selection protocol with given sample size over and over again in the same area at the same time. Each time the protocol was repeated, the sample units measured and, hence, the sample data would be different, and the resulting estimates would also be slightly different. This variation in the estimates is due to a third source of uncertainty in sampling called *sampling* variability, also referred to as *sampling error*. The relative distribution of values of the statistics computed using data from the repeated sample sets is called the *sampling distribution of the estimator*. The spread of this distribution reflects the degree of sampling error associated with the sampling protocol that produced the estimate and the sample size used (Figure 2.7). An estimate having a sampling distribution with small associated sampling error is said to be *precise*.

Precision goals for the values to be estimated should be made at the design phase of a study. It is these precision goals that determine the necessary sample size. The precision goal is composed of two parts: (1) the precision target level and (2) the desired level of confidence that the precision target has reached for our sample. For example, the precision goal for a study might be stated as "the estimate of average fish density should be within 10% of the true density with less than a 1 in 20 chance of being above or below this value." In this case, the target level of precision is specified relative to the true but unknown density. The 1 in 20 chance (0.05 or 5%) statement specifies confidence, in this case stating that if the sampling study were repeated 20 times, it would be acceptable if 1 of the 20 sample means was 10% larger or 10% smaller than the true mean. These two statements are used in combination with the *central limit theorem of statistics* (see Zar 1996) that links estimates of underlying variability in sample values to the sample size needed to achieve the specified level of precision.

The central limit theorem in its simplest form states that the precision of an average, measured by its *standard error*, is the standard deviation of the values divided by the square root of the number of sample values used to make the average (standard deviation of $\bar{x}$ is $\sigma/\sqrt{n}$, where σ is the population variance). The central limit theorem further states that the difference between the sample mean and the true value, divided by the standard error, is distributed as a standard normal deviate, a bell-shaped distribution centered about zero and having variance of one. With this information, the precision statement can be stated as follows:

$$P(|\bar{x} - \mu| > r\mu) < 0.05 \tag{2.1}$$

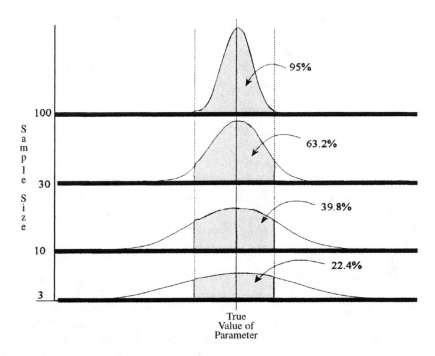

Figure 2.7 The relative precision of an estimate, measured by the spread of the sampling distribution, will increase as the sample size increases.

where μ is the true unknown average, r is the target relative precision (e.g., 0.1 for 10%) and $\bar{x}$ is the sample mean. From the central limit theorem we have:

$$P(|\bar{x} - \mu| > z_{1-\alpha/2}\sigma/\sqrt{n}) = \alpha \qquad (2.2)$$

where $z_{1-\alpha/2}$ is the $100(1 - \alpha/2)$ percentile of a standard normal distribution, n is the sample size, and σ is the true standard deviation computed from all possible measurements on the characteristic of interest. Equating the right-hand terms in the probability statements to each other and solving for the sample size n we get:

$$n = \left(\frac{\sigma\ z_{1-\alpha/2}}{\mu r}\right)^2 = \left(\frac{CV\ z_{1-\alpha/2}}{r}\right)^2 \qquad (2.3)$$

Note that this sample size equation requires specification of the underlying variability in measurements, in this case stated as the coefficient of variation, CV, as well as the target precision, r. For example, assuming a target precision of 10% ($r = 0.10$) of the true mean with 95% ($\alpha = 0.05$) confidence and CV of 30% ($CV = 0.30$), the estimated sample size is given as:

$$n = \left(\frac{CV}{r}z_{1-\alpha/2}\right)^2 = \left(\frac{0.30}{0.10}1.96\right)^2 = 34.57 \approx 35 \qquad (2.4)$$

Variability information often can be obtained from previous studies or as rough estimates generated using data from a small pilot study. The higher the precision target, the more sample units required to achieve a desired level of confidence. Similarly, the higher the confidence desired of our estimate, the more sample units needed to reach the specified precision target. In practice, this sample size calculation must be performed for all characteristics of major importance, and the largest sample size derived is used as the target in the study design. In this way, the study design provides acceptable precision for all characteristics, not simply those with small sampling variation. A large number of more complex sample size formulations have been developed to address situations where the central limit theorem may not provide the best description of the sampling distribution for the estimator. Some of these approaches are provided in statistics texts (see, for example, Gilbert 1987). Specific approaches are covered in detail in Kraemer and Thiemann (1987) and Peterman (1990). Aronson et al. (1994) provides an explicit description of how the sample size for the study was chosen. Sample size issues related to comparisons are covered in the next section.

Another sample size issue relates to the physical size of the sample unit. The size of the sample unit will almost always be dictated by common practice. Chapters 3 to 6 describe sampling techniques and provide guidance on unit size. When a new sampling technique is being developed, some experimentation with size of sample unit is required in order to make an informed recommendation as to the optimal unit size.

2.3.5.5 Issues Specific to Comparative Studies

The issues discussed in the previous section can be encountered in all types of assessments. Assessment study designs, which have as their main objective the comparison of characteristics from a number of situations, have additional aspects that require discussion. These aspects — factor definitions, hypothesis formulation, statistical testing, and replication — together define the conditions necessary for good comparative studies.

2.3.5.5.1 Factors and Mensurative and Manipulative Experiments — When a reef study involves comparison of two or more populations, situations, or groups, the characteristics that differentiate these groups are referred to as *factors,* and the study is referred to as an *experiment.* A factor is simply a reef characteristic that is assumed to affect or in some sense change the average values of other reef characteristics. It is important to realize that there are two kinds of factors to consider. *Manipulative experiments* involve factors representing different situations that are constructed by changes imposed by the experimenter. *Mensurative or observational experiments* involve factors that define comparisons arising from sampling under different naturally occurring conditions.

In a manipulative experiment, the investigator actively sets up the study conditions in such a way that the comparisons directly address the effects of the factors of interest. In Bohnsack et al. (1994), reefs were constructed using one to eight standardized modules. Here the study factor driving the comparisons was reef size. The replicate reefs constructed at each level of size are standardized in all other aspects in order to remove the effect of factors other than size on the characteristics being measured. Bortone et al. (1994) examined two factors simultaneously for their effect on fish assemblage development. One factor was the height of the plastic cone (two levels) used to construct each reef. The other factor was hole size (three levels) used in the manufacture of the cone. The six possible artificial reef types were replicated for the experiment. Fabi et al. (1989) describes an experiment that explores the impact of different reef structures (the factor) on shellfish culture. Experiments also often are used in the preliminary stage of study design to determine which measurement techniques are to be used. See, for example, Thresher and Gunn (1986) for an experiment examining different fish-counting methods.

In a mensurative experiment, the factors of interest do not involve manipulations of conditions, but instead the comparisons arise through sampling under different conditions. Bombace et al. (1994) used a mensurative experiment in comparing reefs at five different areas in the Adriatic Sea. In this case, the location factor identifies the comparisons of interest.

The distinction between a manipulative experiment and a mensurative experiment is found in the amount of control the investigator has in setting up comparisons. In the manipulative experiment, sufficient control is available to create situations that allow direct comparisons across the levels of the factors of interest, while taking into account the effects of other extraneous factors. A manipulative experiment usually starts with locations, situations, or material that are as similar to each other as possible. To these similar units, *treatments,* reflecting the different levels of the factors being studied, are randomly allocated or applied. In the mensurative study, the best the investigator can do is to direct sampling in such a way that the different levels of the factors of interest can be compared, but these comparisons may not be completely free of the effects of other extraneous factors, which cannot be controlled.

Factor effects are estimated using responses measured on samples taken over a range of levels for the factors of interest. It is the difference in average response among sample units at different factor levels that are compared. In some situations, a reef characteristic might be considered a factor, in others, a response. For example, in a manipulative study, fish abundance might be the response of interest, whereas in a different, mensurative-type study of benthic communities, study reefs might be defined by having different fish abundances (the factor). Partitioning characteristics into factors and responses facilitates study design and provides direction to subsequent data analysis. We recommend Underwood (1990) for more detailed discussion on experimental design in assessment and ecology and Box et al. (1978) as a text for experimental designs in general.

2.3.5.5.2 Hypotheses — For each experimental factor in a comparison study, *null and alternative hypotheses* can be formed. A *hypothesis* is simply a statement of what one expects will happen to the characteristics or relationships in question as the levels of the experimental factor change. The null hypothesis states the question as if no change or effect will be observed, that is, the status quo will be maintained. The alternative hypothesis states the question as an affirmation of expected change or effect. Thus, for example, in Bortone et al. (1994) height and void space are the study factors. The associated question, do the height and void space of a reef affect fish assemblage, is translated into the null hypothesis, fish assemblage is not affected by the height and void space of the reef, and the alternative hypothesis, fish assemblage is affected by the height and void space of the reef.

Null and alternative hypotheses are important at the design phase of a comparative study, since they specify exactly which effects will have to be estimated and/or comparisons made. Often, the goal is to reject the null hypothesis in favor of the alternative hypothesis. A hypothesis is deemed *untestable* if data are collected in such a way that multiple factors may be used to describe the differences observed. For a hypothesis to be testable, the only possible explanation for the differences found should be differences in just those factors involved in the hypothesis. Suppose we find differences in fish abundance between two reefs. If one reef is constructed of large and clumped debris and the other reef is constructed of small and dispersed debris, the hypothesis related to the debris size (large vs. small) cannot be tested independently of the hypothesis related to debris distribution (clumped vs. dispersed). In this case, neither the debris size nor the debris distribution hypotheses are testable with these data. It is obvious that with a different type of experiment, hypotheses related to both of these factors could be tested.

In reef assessment studies that involve manipulative or mensurative experiments, only *testable hypotheses* are of interest, since it is only through a test that hypotheses are identified as supported or not. A hypothesis supported by a favorable test is usually accepted as true. When multiple researchers have independently tested and found support for a hypothesis it becomes *accepted fact.* The importance of good hypothesis generation to good research and assessment cannot be over stressed. Most scholarly articles reporting the results of experiments will provide a short list of the hypotheses being examined. An excellent discussion on how hypotheses are linked to experimental design is given in Underwood (1990). A broader review of some of the current null and alternative hypotheses in marine biodiversity, conservation, and management can be found in Bohnsack and Ault (1996).

2.3.5.5.3   Replication — In experiments, the basic situations or factor levels to be compared must be *replicated*. Replicates are two or more situations, locations, or groups having similar factor levels. In a comparative study of, for example, the effect of reef size on fish communities, multiple reefs (the replicates) would be constructed and deployed for each reef size chosen in the comparison (e.g., Bohnsack et al. 1991; Bombace et al. 1994). Every effort would be extended to deploy these reefs in similar environments such that any differences between the reefs are attributed to reef size and nothing else.

For a manipulative experiment, replicates are easy to define, since they relate directly to the factor being manipulated. Defining replicates for a mensurative study becomes much more difficult. The investigator has less control and often the factors of interest are defined by geography — by where samples are taken (e.g., Clark and Edwards 1994; Reimers and Branden 1994; Ecklund 1997). Here, true replicates are sites having similar levels of the factors of interest and far apart enough that measurements can be assumed to be independent. The danger in defining replicates for mensurative studies is in selecting sampling locations which are not independent, in which case the replicates are not true replicates but are *pseudo replicates* (Underwood 1981; Hulbert 1984; Stewart-Oaten et al. 1986). Because pseudo replicates display dependency among characteristics, e.g., counts of fishes are more similar among pseudo replicates than they would be among true replicates, the variability associated with estimates from these data often will be less than what would be expected from data measured on true replicates. The variability among measurements on true replicates is the standard used in all statistical hypothesis testing and all decisions made about the significance of factor effects. Using measurements from pseudo replicates can result in findings of statistical significance for factor effects in situations where there are no actual factor effects, leading to wrong conclusions and decisions.

The differences between a pseudo replicate and a true replicate for a mensurative experiment can be best described using an example. Suppose we are interested in determining how the density (the characteristic) of shellfish changes as a function of the type of substrate (the factor) used to construct the reef. Suppose three types of substrates are to be examined. In a good mensurative experiment, a number of reefs (true replicates) would be constructed of each substrate type. On each reef, a number of 0.25 m^2 quadrats would be selected at random from which shellfish counts would be collected (see Figure 2.8B). The average density would be estimated for each reef. The mean density across all replicate reefs constructed of the same type would be the statistic used to compare substrate types. The variation among reef means would be used in the statistical comparison. If instead of multiple reefs for each substrate type, only one reef were constructed per substrate type, only one mean would be available per reef substrate type (Figure 2.8A). In this case there is no estimate of mean-to-mean variability within a substrate type and no statistic for variability available for the statistical test. If the test is performed using the sample unit-to-sample unit variation within a reef as the statistics for variability, one has substituted true replication variability for sampling variation and has fallen victim to pseudo replication.

What constitutes a true replicate or a pseudo replicate depends on how broadly the results of the study are to be applied. If one only is interested in researching one specific reef in one specific location, measurements on sample units selected from that one site provide the appropriate information for testing, and hence the sample units are true replicates. On the other hand, if the findings from the study are to be applied or extended beyond the specific reef or location, then sampling on one reef only is not adequate. For example in the experiment examining the effect of reef height and void-space size on fish assemblage (Bortone et al. 1994), multiple reefs for each of the six factor combinations were constructed and measured. In this case, true replication involves multiple reefs at each factor combination located far enough apart to be considered independent.

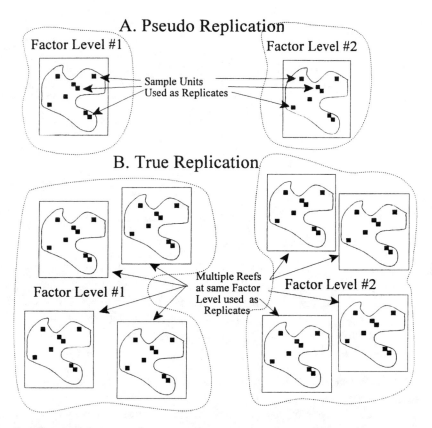

Figure 2.8 True and pseudo replication in mensurative experiments on artificial reefs.

Replication, pseudo replication, and to a certain extent multistage selection protocols are all interrelated concepts. With modern statistical analysis software, some accommodation for dependency among observed values can be made in the statistical analysis model used to test hypotheses (Littell et al. 1996). Nevertheless, any study performed without true replication will have problems presenting credible statistical test results and will in all likelihood not achieve all study objectives.

2.3.5.5.4 Statistical Tests — A *hypothesis test* is performed to determine whether the null hypothesis or alternative hypothesis is supported by the observations. Hypothesis tests are methods for statistically comparing estimators of the characteristics of interest across the levels of the experimental factor. All statistical tests use as the standard for this comparison the sampling distribution of the estimator under the assumption that the null hypothesis is the true description of the situation. The variability associated with this sampling distribution is computed using the fluctuation in estimates from replicate to replicate within each level of the factor being examined, hence, the importance of using true replicates instead of pseudo replicates.

A wide range of statistical methods have been developed to test hypotheses. These range from a simple t-test (or signed ranked test) to comparing means (medians) between two levels of a factor, to mixed effects general linear models, which use complex sampling designs to test hypotheses on the effects of multiple factors observed. Each test is designed to accommodate comparisons of specific types of measurement scales (binary, categorical, ordinal, or continuous), and each makes assumptions which allow the test to be performed. It is important to understand and confirm support for the assumptions that underlie any test proposed for the study design. Statistical methods available for data analysis cannot be easily categorized and described in a few paragraphs. Any of a number

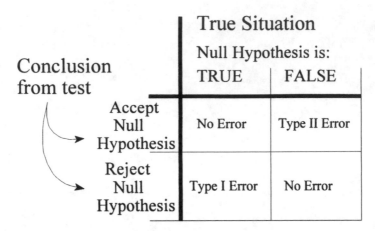

Figure 2.9 Errors in decisions related to hypothesis testing.

of statistical texts (e.g., Box et al. 1978; Snedecor and Cochran 1980; Peterson 1985; Kish 1987; Zar 1996) provide the details for performing most of the tests encountered in assessment studies. Ricker (1975) provides guidance on statistical analysis for fish population studies. Complicated research studies require the guidance of a professional statistician, but in general, careful application of basic statistical methods often produces all of the data analysis needed for decisions related to the assessment.

2.3.5.5.5 Errors in Decisions from Statistical Tests — The statistics used in hypothesis testing are computed from sample data and subject to sampling variation. Because of this, there is always a chance that the conclusions from the hypothesis test will be wrong. No amount of additional sampling, short of a complete census, will produce an error-free decision process. In drawing a conclusion from a statistical test, there are two types of errors one can make (Figure 2.9). A *Type I error* occurs if the null hypothesis (denoted Ho) is rejected when in fact the null hypothesis is true. A *Type II error* occurs when the null hypothesis is accepted, or more correctly not rejected, when in fact the alternative hypothesis is true. The less likely the chances of one or both of these errors occurring, the better the study design.

Common practice defines a good study design as one in which the probability of making a Type I error, denoted by the Greek letter α, is between 0.05 and 0.01. This means that there actually are no differences in the factors being tested (i.e., Ho is true). If the same study (same selection protocols, same environment, etc.) is performed 100 times, somewhere between 5 and 1 of these studies would produce the incorrect conclusion that there is an effect or difference. The *confidence of a test* is defined as $1 - \alpha$ and is often expressed as a percent, $(1 - \alpha)100\%$.

The probability of a Type II error, denoted by the Greek letter β, is closely tied to the number of replicates used for each factor level. Good comparison designs typically have values much less than 0.5, usually around 0.2. This means that if there really is a difference among study factors, and the experiment is repeated 100 times under the same conditions, in 20 of these studies, we would incorrectly conclude that there are no factor differences. The *power of a test* is defined by $1 - \beta$ and also is expressed as a percent, $(1 - \beta)100\%$. The Type II error rate is important in setting the number of replicates needed for a comparative study. In practice, sample sizes are computed for a range of power values associated with a *specific alternative hypothesis*. It is not sufficient to say that there is a difference among the estimates of the factor level effects. Instead, a *minimal difference,* considered important to detect among the estimates at different levels of the experimental factor, must be specified. This *minimal significant difference*, denoted MSD, also can be specified as a fraction of some threshold or overall average value in which case it is referred to as the *minimum relative detectable difference* or MRDD. The value or values used for the MSD or MRDD will depend on the characteristic being considered and the number of levels being compared (Fairweather 1991).

2.3.5.5.6 Number of Replicates

2.3.5.5.6 Number of Replicates — The number of replicates needed for each level of an experimental factor will depend on the Type I and Type II error probabilities chosen and the value of the associated MSD or MRDD (Mapstone 1995). A number of approaches can be used to estimate the number of replicates needed to achieve these criteria (see Gilbert 1987). We describe the simplest approach below by assuming that two groups are being compared on a continuous characteristic.

Define $D = MRDD/CV$ with CV denoting the coefficient of variation expressed as a percent related to replicate-to-replicate variation. Further, let $z_{1-\alpha/2}$ and z_β denote critical values from a table of the standard normal distribution. The sample size needed is the next largest integer greater than n_0 computed as:

$$n_0 = 2\left[\frac{(z_{1-\alpha/2} - z_\beta)}{D}\right]^2 \qquad (2.5)$$

Suppose we are interested in comparing benthic biomass between two types of reefs, and suppose the variation from replicate to replicate in average benthic biomass was estimated to be 30% of the true, but unknown, mean ($CV = 0.30$). Based on findings in the literature, we further set the minimum relative significant difference of importance to MRDD = 0.20 or 20% of the true overall mean. Specifying a slightly low 90% confidence ($\alpha = 0.1$, $z_{1-\alpha/2} = 1.645$) and also slightly low 70% power ($\beta = 0.3$, $z_\beta = 0.525$), the minimum number of replicates per factor level would be $n_0 = 6$ computed as:

$$n_0 = 2\left[\frac{1.645 - (-0.845)}{\dfrac{0.30}{0.20}}\right]^2 = 5.5 \approx 6 \qquad (2.6)$$

This means that to compare average benthic biomass between the two types of reefs would require that six replicates of each type of reef be constructed and sampled.

This simple approach can be used to obtain a rough estimate of the number of replicates needed for most simple comparative studies in reef assessment. For research studies where larger numbers of factors are considered and the study design is more complex, the determination of the number of replication may need to be done by a professional statistician.

2.3.5.6 Issues Specific to Association and Prediction Studies

Studies with association and prediction objectives usually will be much more research oriented than other assessment studies. Association studies aim to examine the relationships among biotic communities and between biotic communities and the surrounding environment. At their foundation, association studies strive to model how the system changes with time and with changing environmental factors and how the biotic community itself changes these environmental factors. Prediction also requires this model.

2.3.5.6.1 Measuring Association

2.3.5.6.1 Measuring Association — *Association analysis* addresses the issue of correlations among species. This may be as simple as quantifying which species are typically found together and which are typically never found together. *Relationship analysis* addresses more complex questions such as "how density changes over time in one species or guild are correlated with density changes over time in other species or guilds?" (e.g., Moffitt et al. 1989; Falace and Bressan 1994). A number of association indices addressing these issues are available from ecological and statistical literature (see Green 1979; Gauch 1982; Pielou 1984; Ludwig and Reynolds 1988). The Pearson product moment correlation coefficient is the standard index used when looking at associations among characteristics having continuous measurements (Steele and Torrie 1980; Sanders et al. 1985). Computing this index on the relative rankings of the observations instead of on their actual values produces Spearman's rank correlation coefficient or Kendall's tau coefficient (Anderson et al. 1989; Greene and Alevizon 1989; Hueckel and Buckley 1989; Buckley and Miller 1994). Measures of independence based on counts of the number of samples where two species occur together vs. where they occur separately result in the use of concordance indices such as the Jaccard (Ludwig and Reynolds 1988; Krebs 1989; Falace and Bressan 1994; Potts and Hulbert 1994) or Sorensen coefficients (Sorensen 1948; Brower et al. 1990; Ardizzone et al. 1989; Pike and Lindquist 1994).

When multiple characteristics are measured on each sample, a multivariate analysis of the correlation structure may be attempted. Multivariate statistical methods have been developed to address a large number of questions. These methods fall into one of three categories. In an *exploratory multivariate analysis* the objective is to tease out association structure in the data. Basic techniques include *principal components analysis* (Pimentel 1979; Patton et al. 1985) and *correspondence analysis* (Hill and Gauch 1980; Bortone et al. 1994; Ardizzone et al. 1997), both procedures classified in statistics as *dimension-reduction techniques*. In a *relationship analysis*, the goal is to relate values from one set of characteristics to values in another set of characteristics. Statistical methods used for this purpose include *canonical correlations analysis, canonical correspondence analysis,* and *factor analysis* (Mardia et al. 1995).

The third type of multivariate analysis is directed toward a testing or inference objective. Statistical methods in this category include a whole suite of *multiple/multivariate regression* (Johnson and Wichern 1992) and *general linear model* (Neder et al. 1996) approaches as well as specialized approaches for categorical responses such as multiple logistic regression (Hosmer and Lemeshow 1989) or *structural equation models* (Hair et al. 1998). While the concepts underlying these types of analyses are straightforward, properly performing the analyses, checking assumptions, and interpreting tests results are not.

2.3.5.6.2 Computing Predictions

2.3.5.6.2 Computing Predictions — Some study objectives aim to develop predictions of future values of a reef characteristic using contemporary values for these characteristics and/or relationships with other easily measured characteristics (Gulland 1969; Beddington and Cook 1983; Edwards and Megrey 1989). Prediction for a target characteristic is accomplished by using measurements of the characteristics over time and space and knowledge of how this characteristic relates to other characteristics that describe the future situation, environment, or circumstances. Prediction is primarily a form of estimation, but one which may take us outside (*extrapolation*) the scope of the existing data.

Prediction can only be done through the use of a mathematical or statistical model. The model is based on a set of explicitly stated or implicitly accepted assumptions. Interactions between the target characteristic (response) and predictor characteristics (sometimes called factors) are stated in formal mathematical equations. The model must be calibrated for the situation under study using observational data to estimate the model parameters. In most situations, for this process to be successful, large amounts of data of reasonable accuracy must be collected, and the relationships must actually exist. If the relationships do not exist, or the strength of the relationship is very weak, no amount of sampling will produce a precise prediction.

Designing a study with a prediction objective is very difficult. It requires collecting data on multiple parameters in such a way that the nature of the interactions or associations among these parameters can be quantified. Knowing something about one set of parameters (e.g., physical setting) will help you predict what will happen for another set of variables (e.g., fish densities). The best advice for these studies is to sample as wide a range of conditions as possible for as long a time frame as possible in an attempt to collect data that capture the strength and nature of the relationships necessary for developing the predicting model.

Current understanding of artificial reef processes is such that prediction objectives, even for research studies, are unrealistic. We know of no studies where prediction has even been attempted at any but the most basic level. As research continues to tease out what happens to biotic and abiotic populations on an artificial reef in response to manipulations of various factors, our ability to predict with any accuracy will also increase.

2.3.6 Data Collection and Management

Every assessment study generates large amounts of data. How these data are managed and reported is critical to the success of the study. If data are not recorded correctly when they are initially taken, it may be impossible to make corrections later, thus complicating the statistical analysis and jeopardizing the whole study. Good data management is not difficult. It consists of (1) clear identification of what measurements must be recorded; (2) limitations on recording of noncritical information; (3) properly designed recording forms; (4) quality control and quality assurance procedures; and (5) processing procedures. The sections that follow provide general guidance on each of these aspects of data collection and management. Unfortunately, the best sources for details on data collections and management approaches are technical reports and study reports, which are often difficult to obtain.

2.3.6.1 Separating Critical from Noncritical Information

Unless one has performed a similar survey before, it is sometimes difficult to know exactly what information is critical to the project's success and what information is simply nice to have but not necessarily critical to success. It is common to want to record everything, but this thinking rapidly becomes unsustainable in the face of limited resources and, in particular, limited time. In addition, recording too much noncritical information can lead to a degradation of the data recording process, which is critical to the study objectives.

A good statement of study objectives helps focus on what data are critical. Similarly, developing a statistical analysis plan even before the data are collected helps identify exactly what information will be needed. When possible, a pilot study should be performed to see whether the data you think should be collected can in fact actually be collected in the field. Many times, data collection processes devised on land are found to be impossible to implement on the open seas.

Finally, documentation of where, when, and how data were collected must be part of permanent study records. When the team who originally collected the data is no longer available, these *meta data* will help others in the future to understand the data and to properly assess whether these historical data meet their needs. In addition, these meta data facilitate comparison and adjustments between studies that use different sampling methodologies and selection protocols.

2.3.6.2 Field Sheets and Recording Forms

Properly designed field sheets or recording forms greatly facilitate the correct and time-efficient recording of critical study data. Such forms help to standardize data collection, establish the necessary level of recording precision for measurements, and resolve any ambiguities in values. Properly designed forms also facilitate transferring data to computer files for subsequent reporting and statistical analysis.

A good recording form has the following features (abstracted from Saville 1981):

- Easy to use, with sufficient space for entering numbers and text as necessary;
- Organized to follow the natural operations of the recording staff; data entry proceeds from top to bottom as the recording staff processes the necessary measurements;
- Explicit in what needs to be recorded, with sufficient labels, titles, and text to remind the recording staff of exactly what goes in which field;
- Laid out in a simple format, allowing an informed user to scan the form quickly and identify obvious errors in recording;
- Requires a minimal amount of information and data;
- Allows for data entry into computer files to be performed directly from the field recording form without the need for transcription to another intermediate document, increasing efficiency and eliminating another source of errors;
- Provides space for recording staff members to comment on any aspects of the data collection process that will add to the understanding of the data being collected; excessive comments should be discouraged, but critical comments can often help explain why observed values deviate from expected values for a particular sample location.

Different data recording forms may be needed for different measurement tasks. The decision of how many forms to use depends on the scope of the data collection process. For example, one form may be used for visual fish counts by divers on the reefs with a separate form used in the lab to record fish length, weight, and other information. Identification numbers for samples collected in the field should be recorded on the lab form at the same time the identification tag is attached to the specimen while still in the field.

Relatively inexpensive, light, and easy-to-use *Global Positioning System* (GPS) units are revolutionizing terrestrial sampling by facilitating the recording of the geographical position of each field observation and sample collected. This geographical position for data is then used in *Geographical Information Systems* (GIS) to analyze and process sampling data. GIS systems have become the standard computing environment for formal data management and data visualization for most environmental studies, since geo-coding study data significantly enhances their value to future researchers and natural resource managers. For aquatic and marine sampling, GPS technology currently cannot provide accurate position for underwater samples. The speed at which technology changes suggests that this is just a short-term limitation. Study designers should keep abreast of this technology and implement it into the data collection scheme as soon as it becomes reliable and cost-effective to do so.

Examples of data forms for large-scale fish sampling and user surveys are given in Bazigos (1974) and Saville (1981). Keep in mind that these forms were created prior to the advent of personal computers, GPS, and GIS, hence these forms must be modified to take geo-coding into account and facilitate transfer to computer files. In addition, for very large studies with long-term repetitive sampling, such as the ICLARM ReefBase Survey (McManus et al. 1997), it is cost-effective to print special data recording forms which can be directly scanned into the computer. The cost of special forms and associated scanning software continues to decrease, and it is conceivable that this will be the preferred method of data recording in the future. Similarly, small handheld computers and/or data recorders in waterproof cases are also available for direct data entry. The initial cost for these devices is still fairly high, but their reusability and ease of reprogramming means that they can be used for multiple studies. The reliability of these two high-technology approaches is currently low, suggesting that a backup plan for manually recording information or physically printing out values as they are collected is still necessary.

2.3.6.3 Quality Control

Good professional practice requires that measurement tasks be documented and executed in such a way that the resulting data can be used with a high degree of confidence. Every study should have a document spelling out those *quality assurance* and *quality control* (QA/QC) procedures that will be used to check on the quality of data as it is collected. QA/QC is a term used to describe those operations which are undertaken to make certain that recorded measurements are actually what were measured. This includes (1) calibrating instruments before and after each sampling session; (2) double-checking written records in the field so that mistakes can be corrected while memory is still fresh; (3) retaking questionable measurements immediately; and (4) taking multiple measurements on the same sample (*subsampling*), then using the value most consistently recorded. When these procedures are in place, the project is said to be "in control." Control is necessary if systematic errors (biases) and to some extent measurement errors are to be kept within tolerable limits. Investigators have to continuously monitor QA/QC procedures to ensure that the best data possible are being recorded for future analysis.

Formal definitions of a quality data set may vary from region to region. For example, in the U.S. many natural resource management and regulation agencies have detailed rules on quality control and quality assurance procedures that must be followed for any of their funded studies. Similar guidance can be found for Food and Agriculture Organization-funded projects. At the same time, many countries are just beginning to conduct formal natural resources studies, and QA/QC guidance has not been prepared. In these cases, study coordinators will have to define and adopt their own procedures. See Geoghegan (1996) for a discussion of QA/QC issues in fisheries science.

2.3.6.4 *Processing Procedures*

Processing field and laboratory data should begin immediately after the completion of data recording activities. This is a multistage task consisting of (1) visual rechecking of completed recording forms by supervisors and experts to identify errors; (2) entry into computer files; (3) cross-checking computer printouts to recording forms to validate the accuracy of the computer entry process; (4) initial computation of frequency distributions and visualization via box plots and histograms to identify extreme values which need to be checked for validity; (5) formal statistical analysis; and (6) creation of tables and graphs for the final study report. Note that steps (1) to (4) can be quite time consuming but are critical for ensuring that the data going into the statistical analysis and the final study report are as "clean" and accurate as possible.

2.3.7 Data Analysis and Reporting

Field and laboratory data only become information when they are evaluated through a statistical analysis. For many, the statistical analysis will consist of the simplest descriptions of the study data, usually provided by a table of sample statistics — the mean, standard deviation, minimum, maximum, median, and range. Graphical portrayals, including box plots, bar charts, histograms, pie charts, scatter plots, etc. will often be used to display and compare frequency distributions for various set of observations. If the study design involves short- or long-term periodic sampling, these same tabular and graphical portrayals are used to illustrate time trends and/or changes in measured characteristics (Chambers et al. 1983; Fabi and Fiorentini 1994).

Beyond simple statistics, a whole array of statistical analysis procedures are available to analyze assessment study data. A number of these procedures, such as statistical hypothesis tests, general linear models and analysis of variance, correlation and association measures, and multivariate and multiple regression analysis, were mentioned in the sections on comparison design and association/prediction issues. In addition to these, one could include formal time series analyses (Box and Jenkins 1976; Priestley 1981; Fabi et al. 1989), categorical data analyses (Agresti 1990; Sokal and Rohlf 1969), and spatial analyses (Cressie 1993) as commonly implemented in geographical information systems. These statistical analyses range from simple to complex. Each statistical procedure carries with it a set of assumptions that must be checked each time the procedure is used.

The most difficult part of a statistical analysis is the correct determination of the appropriate statistical technique to use with a given analysis objective and set of data. Some guidance in this

Table 2.2   Contact Information for a Limited List of Comprehensive Statistical and Mathematical Analysis Packages Commonly Used by Statisticians and Other Scientists

Product	Worldwide Contact
BMDP	BMDP Statistical Software, Inc., 1440 Sepulveda Boulevard, Suite 316, Los Angeles, CA 90025 U.S.A.
GAUSS	Aptech Systems, Inc., 23804 S.E. Kent-Kangley Road, Maple Valley, WA 98038 U.S.A.
GENSTAT and GLIM	NAG (The Numerical Algorithms Group) Ltd, Wilkinson House, Jordan Hill Road, Oxford, OX2 8BR, U.K.
MACSYMA	Macsyma Inc., 20 Academy Street, Arlington, MA 02476-6436 U.S.A.
MAPLE	Waterloo Maple, Inc., 57 Erb Street W., Waterloo, Ontario, Canada N2L 6C2
MATHEMATICA	Wolfram Research, Inc., 100 Trade Center Drive, Champaign, IL 61820-7237 U.S.A.
MATHLAB	The MathWorks, Inc., 24 Prime Park Way, Natick, MA 01760-1500 U.S.A.
NCSS	NCSS Statistical Software, 329 North 1000 East, Kaysville, UT, 84037 U.S.A.
SAS	SAS Institute Inc., SAS Campus Drive, Cary, NC 27513-2414 U.S.A.
SPSS and SYSSTAT	SPSS Inc., 444 N. Michigan Avenue, Chicago, IL 60611 U.S.A.
STATGRAPHICS	Manugistics, Inc., 2115 E. Jefferson Street, Rockville, MD 20852 U.S.A.
STATISTICA	StatSoft, 2300 E. 14th Street, Tulsa, OK 74104 U.S.A.
STRATA	Strata Corporation, 702 University Drive E., College Station, TX 77840 U.S.A
S+ , MATHCAD and AXUM	MathSoft International, Knightway House, Park Street, Bagshot, Surrey GU19 5AQ, U.K.

Note: This list does not include all currently available statistical and mathematical analysis software products. Researchers and reef assessors should perform their own evaluations of the quality and effectiveness of all software available to them for data analysis purposes.

task is available (see, for example, Chatfield 1988; Gilbert 1989), but in general the more statistics one knows and performs, the easier the task gets. The computer aspects of statistics have gotten much easier in recent years with the availability of statistical analysis packages, spreadsheet, and database management systems on personal computers. Similarly, graphical analysis of data can be performed using the graphical tools provided in all of these software packages. A list of the more comprehensive statistical and mathematical analysis packages commonly used by statisticians and other scientists is given in Table 2.2.

2.4 DESIGN PRINCIPLES

The issues discussed in the preceding sections can be briefly summarized as 20 design principles.

1. Start the design process by preparing a clear, written statement of the goals for the reef and the criteria for assessing success.
2. Decide on the study objectives, keeping in mind reef goals, success criteria for the reef, and available resources.
3. Determine which reef characteristics will be measured along with their associated measurement methods. Keep in mind the costs and known biases of each measurement method.
4. Specify the target level of precision and associated confidence for all characteristics critical to meeting the study objective.
5. Calculate the number of samples needed to meet the specified precision and confidence levels for each major characteristic. The sample size for the study will be the maximum of these values.
6. Specify sample selection protocols consistent with assessment criteria and available resources.
7. Specify the timing of sampling consistent with the assessment criteria and available resources.
8. For descriptive studies, specify how measurement will be summarized.
9. For comparative studies, identify those reef characteristics that serve as factors for comparison purposes. Each factor defines a set of experimental conditions.
10. Define the number of replicates to be used for each experimental condition.

11. For comparative studies, state the null hypothesis associated with each key comparison. Identify these comparisons with experimental conditions and determine if the hypothesis is testable.
12. Specify the Type I and Type II error probabilities for all testable hypotheses.
13. Specify the minimally significant difference for comparisons for each of the major hypotheses.
14. Select a statistical method for testing each hypothesis.
15. Calculate the number of replicates needed to achieve confidence levels defined in No. 12 and minimally significant differences in No. 13 for comparisons associated with each hypothesis.
16. Where possible identify the statistical tests to be applied once the data have been collected, and use these tests to examine whether the specified replication and sample sizes will be adequate to observe differences of magnitude considered significant.
17. For association studies, identify the statistical measure of association to be used to relate characteristics of interest.
18. For prediction studies, describe how prediction models will be developed and validated.
19. Prepare data recording forms, a processing plan, and a quality control/quality assurance plan. Field test the data recording forms and make modifications as necessary.
20. Reexamine the final study plan to determine if the study, as designed, appropriately addresses the stated objective. This last review before implementation is important for ensuring that study objectives will be met by the study to be implemented.

Although checking your study design against this list will not guarantee a successful implementation, it provides some assurance that the measurements taken and analysis performed will address the real questions of interest.

REFERENCES

Agresti, A. 1990. *Categorical Data Analysis.* John Wiley & Sons, New York. 558 pp.

Ambrose, R.F. and T.V. Anderson. 1990. Influence of an artificial reef on the surrounding infaunal community. *Marine Biology* 107:41–52.

Anderson, T.W., E.E. DeMartini, and D.A Roberts. 1989. The relationship between habitat structure, body size and distribution of fishes at a temperate artificial reef. *Bulletin of Marine Science* 44(2):681–697.

Andrew, N.L. and B.D. Mapstone. 1987. Sampling and the description of spatial pattern in marine ecology. *Oceanography and Marine Biology Annual Review* 25:39–90.

Ardizzone, G.D., M.F. Gravina, and A. Belluscio. 1989. Temporal development of epibenthic communities on artificial reefs in the central Mediterranean sea. *Bulletin of Marine Science* 44(2):592–608.

Ardizzone, G.D., A. Belluscio, and A. Somaschini. 1997. Fish colonization and feeding habits on a Mediterranean artificial habitat. Pages 265–273. In: L.E. Hawkins and S. Hutchinson, eds. *The Responses of Marine Organisms to Their Environments. Proceedings, 30th European Marine Biology Symposium.* University of Southampton, Southampton, England.

Aronson, R.B., P.J. Edmunds, W.F. Prect, D.W. Swanson, and D.R. Levitan. 1994. Large-scale, long-term monitoring of Caribbean coral reefs: simple, quick, inexpensive techniques. *Atoll Research Bulletin* 421:1–19.

Badalamenti, F. and G. D'Anna. 1997. Monitoring techniques for zoobenthic communities: Influence of the artificial reef on the surrounding infaunal community. Pages 347–358. In: A.C. Jensen, ed. *European Artificial Reef Research. Proceedings, First EARRN Conference,* Ancona, Italy, March 1996. Southampton Oceanography Centre, Southampton, England.

Bazigos, G.P. 1974. The design of fisheries statistical surveys — inland waters. FAO Fisheries Technical Paper 133. 122 pp.

Beddington, J.R. and J.C. Cook. 1983. The potential yield of fish stocks. FAO Fisheries Technical Paper 242. 47 pp.

Bohnsack, J.A. and J.S. Ault. 1996. Management strategies to conserve marine biodiversity. *Oceanography* 9:73–82.

Bohnsack, J.A., D.L. Johnson, and R.F. Ambrose. 1991. Ecology of artificial reef habitats and fishes. Pages 61–107. In: W. Seaman and L.M. Sprague, eds. *Artificial Habitats for Marine and Freshwater Fisheries.* Academic Press, San Diego.

Bohnsack, J.A., D.E. Harper, D.B. McClellan, and M. Hulsbeck. 1994. Effects of reef size on colonization and assemblage structure of fishes at artificial reefs off southeastern Florida. *Bulletin of Marine Science* 55:796–823.

Bombace, G., G. Fabi, L. Fiorentini, and S. Speranza. 1994. Analysis of the efficacy of artificial reefs located in five different areas of the Adriatic Sea. *Bulletin of Marine Science* 55(2–3):559–580.

Bortone, S.A., T. Martin, and C.M. Bundrick. 1994. Factors affecting fish assemblage development on a modular artificial reef in a northern Gulf of Mexico estuary. *Bulletin of Marine Science* 55(2–3):319–332.

Box, G.E.P. and G.M. Jenkins. 1976. *Time Series Analysis: Forecasting and Control.* Holden-Day, San Francisco. 575 pp.

Box, G.E.P., W.G. Hunter, and J.S. Hunter. 1978. *Statistics for Experimenters: An Introduction to Design, Data Analysis, and Model Building.* John Wiley & Sons, New York. 653 pp.

Brower, J.L., J.H. Zar, and C.N. von Ende. 1990. *Field and Laboratory Methods for General Ecology.* 3rd Ed. Wm. C. Brown, Dubuque, IA. 237 pp.

Buckley, T.W. and B.S. Miller. 1994. Feeding habits and yellowfin tuna associated with fish aggregation devices in American Samoa. *Bulletin of Marine Science* 55(2–3):445–459.

Chambers, J.M., W.S. Cleveland, B. Kleiner, and P.A. Tukey. 1983. *Graphical Methods for Data Analysis.* Wadsworth Publishing, Belmont, CA. 395 pp.

Chatfield, C. 1988. *Problem Solving: A Statistician's Guide.* Chapman & Hall, London. 261 pp.

Clark, S. and A.J. Edwards. 1994. Use of artificial reef structures to rehabilitate reef flats degraded by coral mining in the Maldives. *Bulletin of Marine Science* 55:724–744.

Cressie, N.A.C. 1993. *Statistics for Spatial Data: Revised Edition.* John Wiley & Sons, New York. 899 pp.

Davis, N., G.R. VanBlaricom, and P.K. Dayton. 1982. Man-made structures on marine sediments: effects on adjacent benthic communities. *Marine Biology* 70:295–303.

Ecklund, A.M. 1997. The importance of post-settlement predation and reef resources limitation on the structure of reef fish assemblages. Vol. 2, Pages 1139–1142. In: H.A. Lessios and I.G. Macintyre, eds. *Proceedings, 8th International Coral Reef Symposium.* Smithsonian Tropical Research Institute, Balboa, Panama.

Edwards, E.F. and B.A. Megrey, ed. 1989. *Mathematical Analysis of Fish Stock Dynamics.* American Fisheries Society, Bethesda, MD. 214 pp.

Fabi, G. and L. Fiorentini 1994. Comparison of an artificial reef and a control site in the Adriatic Sea. *Bulletin of Marine Science* 55(2–3):538–558.

Fabi, G., L. Fiorentini, and S. Giannini. 1989. Experimental shellfish culture on an artificial reef in the Adriatic Sea. *Bulletin of Marine Science* 44(2):923–933.

Fabi, G., F. Grati, F. Luccarini, and M. Panfili. In press. Indicazioni per la gestione razionale di una barriera artificiale: studio dell'evoluzione del popolamento necto-bentonico. *Biologia Marina Mediterranea.*

Fairweather, P.G. 1991. Statistical power and design requirements for environmental monitoring. *Australian Journal of Marine and Freshwater Research* 42:555–567.

Falace, A. and G. Bressan. 1994. Some observations on periphyton colonization of artificial substrate in the Gulf of Trieste (northern Adriatic Sea). *Bulletin of Marine Science* 55:924–931.

Fricke, A.H., K. Koop, and G. Cliff. 1986. Modification of sediment texture and enhancement of interstitial meiofauna by an artificial reef. *Transactions of the Royal Society of South Africa* 46(1):27–34.

Gauch, J.G., Jr. 1982. *Multivariate Analysis in Community Ecology.* Cambridge University Press, Cambridge. 298 pp.

Gilbert, N.E. 1989. *Biometrical Interpretation — Making Sense of Statistics in Biology.* 2nd Ed. Oxford University Press, Oxford. 146 pp.

Gilbert, R.O. 1987. *Statistical Methods for Environmental Pollution Monitoring.* Van Nostrand Reinhold, New York. 320 pp.

Geoghegan, P. 1996. The management of quality control and quality assurance systems in fisheries science. *Fisheries* 21(8):14–18.

Green, R.H. 1979. *Sampling Design and Statistical Methods for Environmental Biologists.* John Wiley & Sons, New York. 257 pp.

Greene, L.E. and W.S. Alevizon. 1989. Comparative accuracies of visual assessment methods for coral reef fishes. *Bulletin of Marine Science* 44(2):899–912.

Gulland, J.A. 1966. *Manual of Sampling and Statistical Methods for Fisheries Biology: Part 1. Sampling Methods.* FAO Manuals in Fisheries Statistics No. 3. FRs/M3. 87 pp.

Gulland, J.A. 1969. *Manual of Methods for Fish Stock Assessment: Part 1. Fish Population Analysis.* FAO Manuals in Fisheries Statistics No. 4. FRs/M4. (FAO Fisheries Series No. 3). 150 pp.

Hair, J.F., Jr., R.E. Anderson, R.L. Tatham, and W.C. Black. 1998. *Multivariate Data Analysis with Readings,* 5th Ed. Macmillan, New York. 730 pp.

Herrnkind, W.F., M.J. Butler, IV, and J.H. Hunt. 1997. Artificial shelters for early juvenile spiny lobsters: underlying ecological processes and population effects. Pages 12–17. In: *Technical Working Papers from a Symposium on Artificial Reef Development,* Tampa, Florida. Special Report No. 64, Atlantic States Marine Fisheries Commission, Washington, D.C.

Hill, M.O. and H.G. Gauch, Jr. 1980. Detrended correspondence analysis: an improved ordination technique. *Vegetation* 42:47–58.

Hosmer, D.W. and S. Lemeshow. 1989. *Applied Logistic Regression.* John Wiley & Sons, New York. 307 pp.

Hueckel, G.J. and R.M. Buckley. 1989. Predicting fish species on artificial reefs using indicator biota from natural reefs. *Bulletin of Marine Science* 44:873–880.

Hulbert, S.H. 1984. Pseudo replication and the design of ecological field experiments. *Ecological Monographs* 54:187–211.

Jensen, A.C., K.J. Collins, A.P.M. Lockwood, J.J. Mallison, and W.H. Turnpenny. 1994. Colonisation and fishery potential of a coal-ash artificial reef, Poole Bay, United Kingdom. *Bulletin of Marine Science* 55(2–3):308–318.

Johnson R.A. and D.W. Wichern. 1992. *Applied Multivariate Statistical Analysis.* Prentice-Hall, Englewood Cliffs, NJ. 642 pp.

Kish, L. 1987. *Statistical Design for Research.* John Wiley & Sons, New York. 267 pp.

Kraemer, H.C. and S. Thiemann. 1987. *How Many Subjects: Statistical Power Analysis in Research.* Sage Publications, Newbury Park, CA. 120 pp.

Krebs, C.J. 1989. *Ecological Methodology.* Harper & Row, New York. 654 pp.

Linquist, W.J. and L. Pietrafesa. 1989. Current vortices and fish aggregations: the current field and associated fishes around a tugboat wreck in Onslow Bay, North Carolina. *Bulletin of Marine Science* 44(2):533–544.

Lindquist, D.G., L.B. Cahoon, I.E. Clavijo, M.H. Posey, S.K. Bolden, L.A. Pike, S.W. Burk, and P.A. Cardullo. 1994. Reef fish stomach contents and prey abundance on reef and sand substrata associated with adjacent artificial and natural reefs in Onslow Bay, North Carolina. *Bulletin of Marine Science* 55(2–3):308–318.

Littell, R.C., G.A. Milliken, W.W. Stroup, and R.D. Wolfinger. 1996. SAS® System for Linear Models. SAS Institute, Cary, NC. 633 pp.

Ludwig, J.A. and J.F. Reynolds. 1988. *Statistical Ecology: A Primer on Methods and Computing.* John Wiley & Sons, New York. 337 pp.

Mapstone, B.D. 1995. Scalable decision rules for environmental impact studies: effect size, Type I and Type II errors. Pages 67–80. In: R.J. Schmidt and C.W. Osenberg. eds. *Detection of Ecological Impacts: Conceptual Issues and Application in Coastal Marine Habitats.* Academic Press, San Diego.

Mardia, K.V., J.T. Kent, and J.M. Bibby. 1995. *Multivariate Analysis.* Academic Press, London. 518 pp.

McManus, J.W., M.C.A. Ablan, S.G. Vergara, B.M. Vallerjo, L.A.B. Menez, K.P.S. Reyes, M.L.G. Gorospe, and L. Halmarick. 1997. *ReefBase Aquanaut Survey Manual.* International Center for Living Aquatic Resource Management, ICLARM Educational Series 18, Manila. 61 pp,

Moffitt, R.B., J.A. Parrish, and J.J. Polovina. 1989. Community structure, biomass and productivity of deepwater artificial reefs in Hawaii. *Bulletin of Marine Science* 44(2):616–630.

Neder, J., M.H. Kutner, C.J. Nachtsheim, and W. Wasserman. 1996. *Applied Linear Statistical Models.* 4th Ed. Richard D. Irwin, Chicago. 1408 pp.

Nelson, W.G., T. Neff, P. Navratil, and J. Rodda. 1994. Disturbance effects on marine infaunal benthos near stabilized oil-ash reefs: spatial and temporal alteration of impacts. *Bulletin of Marine Science* 55(2–3):1348.

Pamintuan, I.S., P.M. Alino, E.D. Gomez, and R.N. Rollon. 1994. Early successional patterns of invertebrates in artificial reefs established at clear and silty areas in Bolinao, Pangasinan, Northern Philippines. *Bulletin of Marine Science* 55:867–877.

Patton, M.L., R.S. Grove, and R.F. Harman. 1985. What do natural reefs tell us about designing artificial reefs in Southern California? *Bulletin of Marine Science* 37(1):279–298.

Peterman, R. 1990. Statistical power analysis can improve fisheries research and management. *Canadian Journal of Fishery and Aquatic Sciences* 47:2–15.

Peterson, R.G. 1985. *Design and Analysis of Experiments.* Marcel Dekker, New York. 429 pp.

Pielou, E.C. 1984. *The Interpretation of Ecological Data: A Primer on Classification and Ordination.* John Wiley & Sons, New York. 263 pp.

Pike L.A. and D.G. Lindquist. 1994. Feeding ecology on spottail pinfish (*Diplodus Holbrooki*) from an artificial and a natural reef in Onslow Bay, North Carolina. *Bulletin of Marine Science* 55(2–3):363–374.

Pimentel, R.A. 1979. *Morphometrics: The Multivariate Analysis of Biological Data.* Kendall-Hunt, Dubuque, IA. 276 pp.

Posey, M., C. Powell, L. Cahoon, and D. Lindquist. 1995. Top-down vs. bottom-up control of benthic community composition on an intertidal tide flat. *Journal of Experimental Marine Biology and Ecology* 185(1):19–31.

Potts, T.A. and A.W. Hulbert. 1994. Structural influences of artificial and natural habitats on fish aggregations in Onslow Bay, North Carolina. *Bulletin of Marine Science* 55(2–3):609–622.

Priestley, M.B. 1981. *Spectral Analysis and Time Series. Vol. 1: Univariate Series.* Academic Press, New York. 308 pp.

Priestley, M.B. 1981. *Spectral Analysis and Time Series. Vol. 2: Multivariate Series, Prediction and Control.* Academic Press, New York. 736 pp.

Reimers, H. and K. Branden. 1994. Algal colonization of a tire reef — influence of placement date. *Bulletin of Marine Science* 55:460–469.

Relini, G., N. Zamboni, F. Tixi, and G. Torchia. 1994. Patterns of sessile macrobenthic community development on an artificial reef in the Gulf of Genoa (Northwestern Mediterranean). *Bulletin of Marine Science* 55:745–771.

Rhodes, R.J., J.M. Bell, and D. Laio. 1994. Survey of recreational fishing use of South Carolina's marine artificial reefs by private boat anglers. Project No. F-50 Final Report, Office of Fisheries Management, South Carolina Wildlife and Marine Resources Department.

Ricker, W.E. 1975. Computation and interpretation of biological statistics of fish populations. *Bulletin of the Fisheries Research Board of Canada* 191:382 pp.

Samoilys, M.A., ed. 1997. *Manual for Assessing Fish Stocks on Pacific Coral Reefs.* Department of Primary Industries, Townsville, Queensland, Australia. 79 pp.

Samples, K.C. 1989. Assessing recreational and commercial conflicts over artificial fishery habitat use: theory and practice. *Bulletin of Marine Science* 44(2):844–852.

Sanders, R.M., Jr., C.R. Chander, and A.M. Landry, Jr. 1985. Hydrological, dial and lunar factors affecting fishes on artificial reefs off Panama City, Florida. *Bulletin of Marine Science* 37(1):318–328.

Saville, A. 1981. Survey methods of appraising fisheries resources. FAO Fisheries Technical Paper 171. 76 pp.

Snedecor, S.W. and W.G. Cochran. 1980. *Statistical Methods, 7th Ed.* The Iowa University Press, Ames. 507 pp.

Sokal, R.R. and F.J. Rohlf. 1969. *Biometry.* W. H. Freeman, San Francisco. 776 pp.

Sorensen, T. 1948. A method of establishing groups of equal amplitude in plant sociology based on similarity of species content and its application to analysis of the vegetation on Danish commons. *Biologiske Skrifter Kongelige Dansk Videnskabernes Selskab* 5(4):1–34.

Steele, R.G. and H.H. Torrie. 1980. *Principles and Procedures of Statistics.* 2nd Ed. McGraw-Hill, New York. 633 pp.

Stewart-Oaten, A., W.W. Murdoch, and K.R. Parker. 1986. Environmental impact assessment: 'psuedoreplication' in time? *Ecology* 67:929–940.

Szmant, A.M. and A. Forrester. 1996. Water column and sediment nitrogen and phosphorus distribution patterns in the Florida Keys, USA. *Coral Reefs* 15:21–41.

Thresher, R.E. and J.S. Gunn. 1986. Comparative analysis of visual census techniques for highly mobile, reef-associated piscivores (*Carangidae*). *Environment Biology of Fishes* 17(2):93–116.

Underwood, A.J. 1981. Techniques of analysis of variance in experimental marine biology and ecology. *Oceanography and Marine Biology Annual Review* 19:513–605.

Underwood, A.J. 1990. Experiments in ecology and management: their logics, functions and interpretation. *Australian Journal of Ecology* 15:365–389.

Zar, J.H. 1996. *Biostatistical Analysis, 3rd Ed.* Prentice-Hall, Upper Saddle River, NJ. 898 pp.

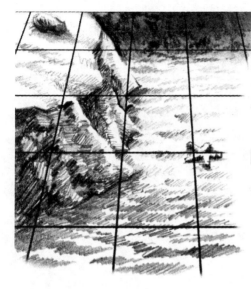

CHAPTER **3**

Physical Characteristics and Engineering at Reef Sites

Y. Peter Sheng

CONTENTS

0-8493-9061-3/00/$0.00+$.50
© 2000 by CRC Press LLC

3.1 SUMMARY

This chapter describes the methods used to evaluate the physical conditions of a reef and its surrounding aquatic environment. The second section presents the primary objectives and defines the "physical characteristics" and "engineering" of reef studies. Section three explains why large-scale as well as local physical variables should be measured at reef sites. In the fourth section, engineering aspects of reef studies are presented in terms of siting considerations, reef stability, effects of physical processes on reef performance, and the effects of the created reef on the environment. The fifth section discusses the three different types of information that need to be gathered in reef studies in order to answer questions of increasing complexity. In the sixth section, assessment methods of the most important physical variables are presented. The seventh section provides several examples of reef monitoring studies. The chapter concludes with recommendations for future reef studies.

3.2 INTRODUCTION

Since physical processes can influence the chemical and biological processes in aquatic environments (Wolanski and Hamner 1988; Sheng 1998, 1999), it is important to incorporate the measurement of physical characteristics into reef monitoring programs. Prior to deployment, it is important to determine the physical characteristics of a potential reef site to assess its suitability. Placing a reef at a site with strong waves and currents could lead to quick burial of the reef. Once a reef is placed at a site, it is essential to monitor the reef and its environment frequently to determine the reef's existence and performance. It is also possible through comparative studies to develop a comprehensive understanding of why a reef functions as intended. Such studies include concurrent monitoring of physical, chemical, and biological attributes of the reef and its adjacent environment. Hence, monitoring of the physical characteristics at reef sites should significantly enhance the ability to build "successful" artificial reefs. Successful reef deployment is, in part, dependent upon the application of sound engineering principles.

3.2.1 Objectives

This chapter addresses the monitoring of physical characteristics, including the reef and the environment, at reef sites. Artificial reefs have been deployed in coastal, estuarine, and fresh waters to enhance fish aggregation and production (Seaman and Sprague 1991). Artificial reefs have been deployed in the nearshore zone in an attempt to dissipate breaking wave energy and to protect beach erosion (Bruno 1993; Dean et al. 1997). Harbor and shore protection works, such as jetties and breakwaters, as well as power plant intake and outfall devices, have been known to enhance plant and fish populations (Helvey and Smith 1985). The objectives of this chapter are to (1) introduce physical characteristics (water circulation, waves, sediment dynamics, and reef structure) that are related to the performance of artificial reefs; and (2) recommend sampling protocols for monitoring physical characteristics at reef sites.

3.2.2 Definitions

In this chapter, *physical characteristics* refers to (1) the physical reef component(s), which could be a concrete module, a derelict or decommissioned vessel, a petroleum platform, or a prefabricated structure of steel or fiberglass; and (2) the physical environment in which the reef is placed, including both large- and small-scale environments.

Large-scale environment refers to the large-scale circulation, wave climate, and sediment dynamics that take place over spatial scales of tens to hundreds of kilometers in the vicinity of the reef. It can also be referred to as the *far-field environment*. The *small-scale environment* (or *near-field environment*) refers to the physical characteristics in the immediate vicinity of the reef — the local current and wave characteristics, temperature, salinity, and suspended and bottom sediments at a reef site.

Engineering in this chapter includes the following aspects: (1) selection of a suitable reef site; (2) designing and monitoring a "stable" reef; (3) understanding the interrelationship between physical characteristics and reef performance; and (4) proper utilization of the knowledge gained in (1) through (3) above to continue to produce successful reefs. These aspects support the general reef goals defined in Chapter 1.

3.3 PHYSICAL COMPONENTS OF REEFS — WHY SHOULD PHYSICAL VARIABLES BE MONITORED?

Artificial reefs are generally placed in shallow coastal (continental shelf) and estuarine waters as well as in lakes and rivers. Artificial reefs are also placed on shallow beaches for erosion control. Reefs placed on the floor of an ocean or estuary are subject to the forcing of currents. Wind, tide, and density gradients can produce slowly varying currents with periods on the order of 1 to 24 h. Waves can induce orbital currents with periods on the order of 1 to more than 10 s. Vertical currents associated with the slowly varying currents are generally small, but short waves can produce significant vertical currents comparable to horizontal currents. Combined, such currents produce stresses on the sea floor and the reef structure. Excessive physical forces may lead to significant erosion of bottom sediments and/or movement of reef structures (Figure 3.1). If sufficient information pertaining to circulation and waves is available for a large region, it is possible to identify "high energy" areas with bottom stresses that are unsuitable for reef placement. Changes in salinity and temperature can lead to changes in water density and, hence, currents and bottom stresses.

Meanwhile, nutrient-rich river plumes entering an estuary or shelf may create a favorable environment for reef deployment, and a reef may alter local circulation and sediment/nutrient dynamics. For example, wakes may be created behind reefs to attract certain fish species, and sediments/nutrients may be brought into the upper water column by resuspension and local upwelling, thereby influencing biological processes (see Chapter 4). Presently, there is a paucity of conclusive data showing exactly how circulation and sediment/nutrient dynamics influence the performance of artificial reefs. Some information on the physical dynamics of reefs can be found in Seaman and Sprague (1991). It is expected that vast field instrumentation improvements in recent years will lead to significantly increased monitoring of the physical environments around reefs.

3.3.1 Reef Structure

Reef structure refers to the physical materials of composition and geometric/geographic distribution of such materials of a reef. As described in Chapter 1, there are significant differences in terms of the shape, dimension, and materials of construction of various types of reefs. Representative examples include cubes, cylinders, etc. and opportunistic materials, including derelict vessels and tires. For successful reef performance, it is essential to ensure that the reef structure will remain in place by (1) resisting the local hydrodynamic forces; (2) not exceeding the ability of the underlying sediments to support the weight of the reef structure; and (3) maintaining the structural integrity of the material. For environmentally sound reef performance, it is also important to ensure that the reef material does not cause any adverse impact on the environment.

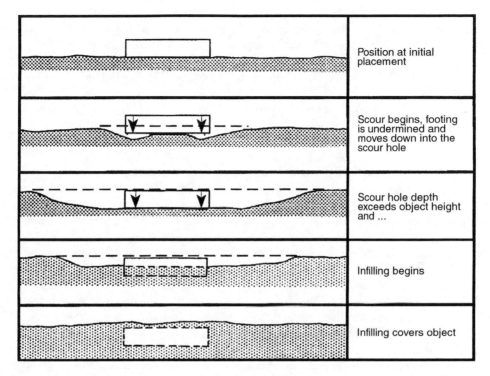

	Position at initial placement
	Scour begins, footing is undermined and moves down into the scour hole
	Scour hole depth exceeds object height and ...
	Infilling begins
	Infilling covers object

Figure 3.1 Possible "burial" mechanism for small structures subject to scouring (adapted from Tian 1994).

3.3.2 Physical Environment

The large-scale environment surrounding a reef could include a large area on a continental shelf or an entire estuary. Examples include the Gulf of Mexico, Japan Sea, Mediterranean Sea, and Chesapeake Bay, U.S. These large-scale environments have distinctively different physical characteristics that will affect the local physical dynamics in the small-scale environments in the vicinity of reefs.

3.3.2.1 Large-Scale Physical Environment

The flora and fauna at a reef site are affected by the local hydrodynamic conditions in the immediate vicinity of the reef, which are in turn influenced by the large-scale physical processes taking place over spatial scales much greater than that comprised by the reef structure. These large-scale processes include the following:

- Tidal circulation,
- Wind-driven circulation,
- Density-driven (baroclinic) circulation, including coastal and estuarine fronts,
- Swell propagated over large distance,
- Locally generated wind waves,
- Tsunami and hurricane-generated currents and waves, and
- Sediment transport processes and bottom sediment characteristics and dynamics.

Each large-scale environment may have a distinctly different circulation system. For example, circulation in the wide (~100 to 200 km) West Florida Shelf in the Gulf of Mexico is influenced by mixed (diurnal plus semidiurnal) tides, wind, fronts, and the Loop Current (Mitchum and Sturges

1982) and is very different from the circulation in the narrow (~5 to 70 km) East Florida Shelf, which is influenced by the Gulf Stream as well as semidiurnal tide and wind (Lee et al. 1988). Circulation in the Pacific coast of Japan is very different from that in the coastal waters of the Japan Sea (Stommel and Yoshida 1972; Ichiye 1984). Circulation in the shallow (10 to 15 m) Taiwan Strait also differs significantly from that in the deep (6600 m) east coast of Taiwan (Nitani 1972; Fan 1984).

The large-scale circulation in a given area generally displays significant seasonal variation. For example, at the Suwannee Regional Reef System (SRRS) in the West Florida Shelf (Lindberg 1996), the winter circulation is characterized by significant vertical mixing, alongshore wind-driven flow, and alongshore and onshore-offshore tidal currents. In contrast, the summer circulation is characterized by baroclinic circulation with significant vertical stratification.

Wave climate also varies significantly with the location of the large-scale environment and season (Hubertz and Brooks 1989). Waves can be generated locally by the large-scale wind field or propagated from a remote area as swell. Waves can generate significant orbital currents in the water column, causing movement of reef units or resuspension of sediments. During hurricanes, typhoons, and tsunamis, strong waves and currents can be generated in the water column. Special consideration must be given to reefs that are to be deployed in areas that are historically prone to frequent hurricanes, typhoons, and tsunamis.

The bottom sediments must be able to support the weight of the reef structures under varying hydrodynamic conditions. It is thus useful to have sufficient data on the composition, size fraction, and strength of the bottom sediments prior to final site selection and development of an artificial reef site. Areas with a large portion of muddy bottom (unconsolidated sediments) should generally be avoided, as differential settlement of reef units (as well as adverse environmental impacts associated with suspended sediments) could result.

3.3.2.2 Small-Scale Physical Environment

The small-scale physical environment in the immediate vicinity of a reef is influenced by the local climate, the large-scale physical processes, and the interaction between the reef structure with its surrounding water and sediment columns. A few examples are provided herein.

3.3.2.2.1 Local Upwelling — The vertical water motion in coastal waters is generally negligible when compared to the horizontal water movement, except in the surf zone. However, a large reef structure in the direction of the flow can create locally significant vertical upwelled current (Otake et al. 1991), or upwelling, which could be of comparable magnitude to the horizontal current. This local upwelling can lead to the transport of sediments and nutrients from the bottom water column to the surface water. Both the significant vertical current and the vertical transport of materials may lead to aggregation of fishes at the reef site (Otake et al. 1991).

3.3.2.2.2 Sediment Resuspension/Scouring — The presence of a large reef structure in a coastal area with significant currents may create a downward flow adjacent to the upstream side of the reef structure. As the downward current approaches the bottom, a horseshoe vortex (Figure 3.2) is formed and may cause resuspension or scouring of sediments around the reef bottom. The resuspended sediments are partly transported to the upper water column (and partly transported to the lee side of the reef, where the currents are weaker) and become deposited there. This process, if repeated over a long time, may lead to instability and even burial of the reef (Tian 1994). Special consideration should be given to minimize the scouring of sediments around reefs. Burial of the reef is also affected by the bearing capacity (Rocker 1985) and the compressibility of sediments (Lamb and Whitman 1969; Tian 1994).

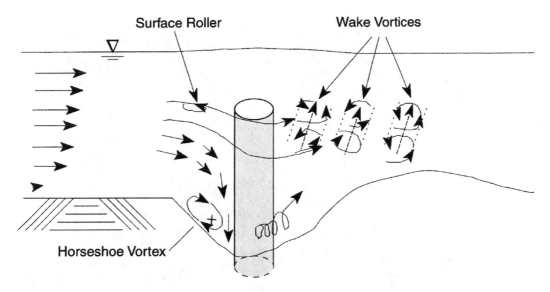

Figure 3.2   Flow patterns and local scour in the vicinity of a cylindrical structure.

3.3.2.2.3   Wake Zone — In shallow waters, the interaction of a reef with a prevailing current usually results in the formation of a wake zone with eddies downstream of the reef. It may attract certain species by providing a shelter, a feeding ground, a spawning ground, a rest area, or a temporary stopover (Takeuchi 1991). On the other hand, the eddies and vortices just outside the wake region contain higher turbulence that may attract certain fish species. The wake zone provides shelter for certain fish species due to its quiescent environment and the deposition of sediments and nutrients typically observed there (Wolanski and Hamner 1988). With sufficient information on the prevailing currents, it may be possible to design a reef structure that will produce a wake zone with user-specified (or desired) dimensions.

3.4 ENGINEERING ASPECTS OF REEFS

Engineering principles of importance in artificial reef planning include: (1) appropriate reef site selection; (2) design of a "stable" reef configuration; (3) utilization of physical processes to enhance reef performance; and (4) minimization of adverse reef impacts on the environment. Application of engineering aspects constitutes the cornerstone of a successful reef.

3.4.1   Siting Considerations

Reefs are generally placed at the bottom of relatively shallow coastal waters with depths between 2 to 100 m. It is reasonable to expect that large-scale ambient physical processes may have a significant effect on reef performance and local fisheries. Thus, as a first step in developing a comprehensive reef plan for a large area such as the West Florida Shelf (eastern Gulf of Mexico) or the Taiwan Strait, it is prudent to seek a quantitative understanding of large-scale physical oceanographic processes (i.e., water circulation driven by tides, wind, and baroclinic/density fields; locally generated wind waves or swells propagated from distance; and sediment distribution and transport) in the ambient aquatic environments, including estuaries and continental shelves.

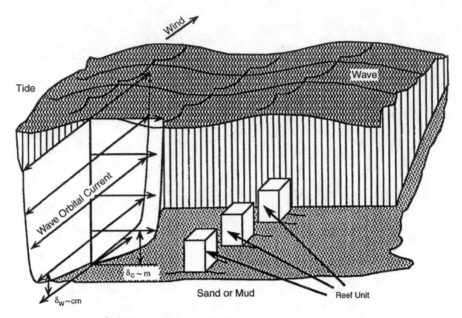

δ_c : thickness of current bottom boundary layer $\sim$ m
δ_w : thickness of wave bottom boundary layer $\sim$ cm

Figure 3.3 A coastal reef system. The reef units are subjected to forcings of slowly varying currents, with a boundary layer (δ_c) on the order of 1 m, and fast wave orbital currents, with a thinner boundary layer (δ_w) on the order of few cm. Wave is assumed to be in the direction of the wind.

3.4.2 Reef Stability

The existence and stability of a reef depend strongly on the ambient physical conditions and the interactions among the flow-sediment-reef system (Figure 3.3). When physical dimensions of reefs are comparable to local water depths, the reefs are expected to cause alteration of the local physical conditions (circulation and sediments) of aquatic environments, which may ultimately affect the reef performance.

In some cases, interaction between a reef and the physical environment may lead to disappearance of the reef. As shown in Figures 3.3, 3.4, and 3.5, the reef may be subjected to the forcing of fast wave-induced orbital currents and slowly varying tidal and wind-driven currents in the form of a skin friction drag (which is proportional to the total wetted area of the reef) plus a form drag (which is proportional to the total projected area of the reef in the flow direction). In the presence of flow separation, form drag is the primary drag force (Sarpkaya and Isaacson 1981). The presence of wave orbital currents produces an inertia force. For the reef to remain stationary, the sum of drag and inertia forces acting on the reef unit must be balanced by the resistive forces of the soil (which depend on the structure and composition of the bottom sediments), while the lift force must be balanced by the difference between a reef's weight and buoyancy force. Thus, lightweight reefs may be washed away at high energy reef sites where the actions of currents and/or waves are strong. A reef placed in shallow water may be moved, buried, or destroyed by strong forces produced by waves and currents associated with the passage of a storm or front.

A reef placed onto a soft muddy bottom may sink through the bottom sediment soon after impact on the sea floor. Even with a hard substrate, local scouring/erosion and accretion of sediments in the vicinity of the reef may lead to partial or complete burial of the reef. Scouring is the result of flow acceleration and separation around the structure and a horseshoe vortex induced by the vertical pressure gradient along the leading edge of the structure. The extent and volume of the

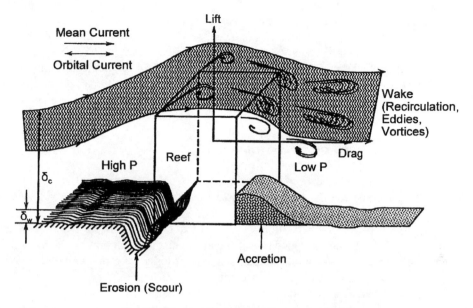

δ_c : Thickness of current bottom boundary layer
δ_w : Thickness of wave bottom boundary layer
P : Pressure

Figure 3.4   Flow-reef-sediment interactions. Downstream of the mean current, a wake region is created behind the reef. Accretion of sediments usually occurs in the wake while erosion usually takes place on the upstream side of the reef. Pressure drop across the reef produces a normal drag force (form drag), which dominates the tangential drag force (skin friction drag). A vertical lift force is balanced by the weight of the reef. In shallow water, waves provide significant modification to the local currents and forces.

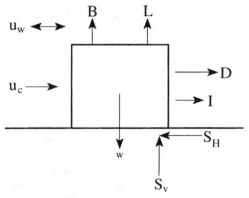

u_c = Bottom Current
u_w = Wave Orbital Current
w = Weight of Reef Unit
B = Buoyancy Force Acting on Reef Unit
L = Lift Force
D = Drag Force (Including Profile Drag and Skin Friction)
I   = Inertial Force due to Wave
S_H = Horizontal Soil Resistance
S_v = Vertical Soil Resistance

Figure 3.5   Forces acting on a reef unit.

wave- or current-induced scour depends on reef shape, size, and location relative to the bottom, the nature of the primary flow, and the sediment parameters. Researchers (e.g., Eadie and Herbich 1987) concluded that the process of structure-induced scour is somewhat different for waves than for steady currents, and the largest scour depths occur with steady currents. The addition of waves to currents will accelerate the rate of scour, but will have little effect on the maximum scour depth.

To ensure that a reef will remain at a site, it is necessary to monitor a reef site before site selection and occasionally after the placement of the reef, so that forces on the reef can be estimated. (See methods discussed later in this chapter.)

3.4.2.1 *Horizontal Forces on Reef Structure*

Horizontal forces on reef structure can be estimated by using existing technology (laboratory experiment, field experiment, and modeling) with varying degrees of sophistication. A few examples follow.

Myatt et al. (1989) performed a 10-month study on two tire reefs off the New Jersey coast to determine the feasibility of using scrapped tires in a fishery enhancement program and the attendant design criteria necessary for the units to remain stable. Test units were numerically scored based on distance moved, and the scores were then compared to various unit parameters to determine common design characteristics of stable units. Daily wave heights at the test areas predicted by the U.S. Weather Service were used in the analysis. The researchers recommended using five unit types for New Jersey reef construction in 18 to 24 m depths and established minimum design criteria for new units. However, the study recommended additional testing for reef deployment in depths shallower than 18 m (~60 ft) where wave effects on reefs are more significant.

Kim et al. (1981) performed a laboratory study of wave forces on submerged fabricated tire reefs to determine design criteria for a reef located in offshore regions with severe wave and current conditions. They found that it is feasible to predict the force on tire reefs, in spite of the three-dimensional shape and elastic responses to the waves, using the Morison equation (Morison et al. 1950) for estimating wave force on a fixed body:

$$F = F_D + F_I = C_D \frac{\rho}{2} A u \,|\, u \,| + C_I \rho V \frac{du}{dt} \tag{3.1}$$

where F_D is the form drag, F_I is the inertial force, F is the total force on the body, u is the flow velocity, $\frac{du}{dt}$ is instantaneous acceleration of the water, ρ is the water density, A is the total projected area of the body in the direction of the flow (see Figure 3.6), V is the volume displaced by the body, C_D is the form drag coefficient, and C_I is the inertia (or mass) coefficient. The drag coefficient depends on the Reynolds number (Re = UD/ν, where U is a representative velocity, D is a representative length scale such as the diameter of the reef, and ν is the kinematic viscosity) and the Keulegan–Carpenter period parameter (KC = $U_{max} T/D$), where U_{max} is the maximum wave orbital velocity, T is the period of the oscillatory wave, and D is the size of the reef along the direction of the prevailing current. For estimation of maximum design force, the following equation should be used:

$$F_{max} = C_f \frac{\rho}{2} A u_{max}^2 \tag{3.2}$$

where u_{max} is the maximum bottom velocity and C_f is the maximum force coefficient (Sarpkaya 1976) that should depend on the shape of the reef structure. In the presence of waves and currents,

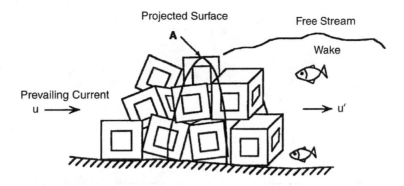

Figure 3.6 Wakes in the lee of reef rocks (adapted from JCFPA 1986).

u_{max} in Equation (3.2) is replaced by $u_o + u_m$, where u_o is the mean current and u_m is the maximum orbital current. Grove et al. (1991) suggested that C_f generally ranges between 2 and 4. In general, C_f can be determined from laboratory experiments for arbitrarily shaped reefs.

For design consideration, wave and current conditions at reef sites should be predicted for a storm return period (e.g., 20 or 50 years) equal to the desired design life of the reef. The maximum bottom current can then be estimated based on linear wave theory and boundary layer theory. The projected area, A, in the direction of the ambient flow can be estimated following the reef mapping procedure presented in Seaman and Sprague 1991. The maximum force coefficient, C_f, can be estimated from either a literature search and/or laboratory experiments. The maximum horizontal force exerted on a reef can then be estimated using Equation (3.2).

It is noted that Equations (3.1 and 3.2) are based on engineering considerations and have been calibrated primarily with laboratory data obtained over simple structures. Therefore, the use of these relationships may contain too much uncertainty for field application in the presence of arbitrarily shaped bodies. For more precise estimation of forces on reefs, sophisticated, turbulent-bottom-boundary-layer models (e.g., Sheng and Villaret 1989) may have to be used.

3.4.2.2 Vertical Forces Between the Reef and the Bottom Sediments

The vertical forces acting on a reef block include its own weight, the vertical resistive force of the sea floor, and the hydrodynamic lift force created by the currents. In some installation procedures, a reef block falls freely through the water column and impacts the sea floor, creating a large physical load (Huang 1994; Tian 1994). The impact may cause the sea floor to settle locally if the force exerted by the reef block on the sea floor exceeds the soil strength. After installation, the reef block will remain at its original impact if sea floor strength is sufficient to support the weight of the reef block and if the weight of the reef block is greater than the hydrodynamic lift force produced by the currents (Tian 1994).

3.4.2.3 Structural and Material Stability

A reef block may be structurally damaged due to the large physical load exerted on the structure when the reef impacts the sea floor. Grove et al. (1991) described a method to compute the impact load on a reef block as it lands on the sea floor by free fall, following the procedures recommended by Japanese planning guides (JCFPA 1989).

Texture and composition of the reef material can also affect reef performance. Concrete, steel, cast iron, rubber, wood, and even coal ash (Kuo et al. 1995) have been used as reef material. Material instability of a reef block may occur due to fouling of reef material resulting from chemical

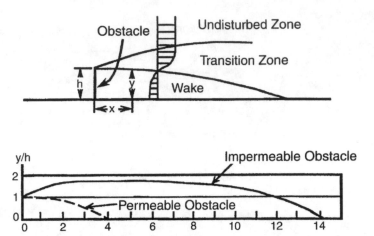

Figure 3.7 Range of disturbance of current due to a two-dimensional obstacle.

interaction between reef material and ocean water. For example, a reef may collapse due to corrosion of steel or cast iron (Nakamura 1980) and disintegration of reef parts connected by nuts and bolts (JCFPA 1984). Recent studies in Taiwan (Kuo et al. 1995) indicate that stabilized coal ash can be used to replace concrete as reef material without harmful effects to the marine environment in which it is deployed.

3.4.3 Effects of Physical Processes on Reef Performance

It is reasonable to expect that interaction between a reef and the local physical environment may affect reef effectiveness. As shown in Figures 3.6 and 3.7, the presence of the reef may lead to the creation of a large wake region with eddies and vortices on the downstream side of the reef. The wake zone may attract fish by providing shelter. Turbulence at the edge of the wake zone may attract certain pelagic species. Upwelled current and sediments/nutrients may benefit certain pelagic species.

It is reasonable to expect the size, shape, and spacing of a reef, as well as flow-reef interactions, to affect the physical conditions (e.g., currents, turbulent eddies/vortices, water density, and sediment erosion/deposition) at the reef site, and thus affect reef effectiveness. However, there presently are insufficient data to suggest a universally valid relationship between fishery or habitat enhancement and the flow-reef-sediment interactions. Comprehensive physical and fishery data at reef sites are needed to facilitate development of cause-effect relationships between flow-reef-sediment interactions and reef effectiveness/performance.

3.4.3.1 *How Should a Reef Be Oriented with Respect to the Flow?*

Lindquist and Pietrafesa (1989) and Baynes and Szmant (1989) reported that interaction between a shipwreck and the local current field can alter the effectiveness of the reef in enhancing fishery assemblages and community structure. However, due to lack of comprehensive data, results of the two studies were not fully consistent as to how to best take advantage of local current conditions.

Extensive studies on the effect of flow-reef orientation have been conducted in Japan. Nakamura (1985) recognized three types of fish species that are attracted to the reef for different reasons (see Chapter 5). Nakamura (1980) reported that when the product of reef thickness/width and ambient current exceeds 100 cm²/s, vortex shedding from the reef occurs and certain fish take refuge in the lee/shadow behind the reef. JCFPA (1986) reported that turbulence reaches 80% of the water column when the reef is 10% of the water depth. Otake et al. (1991) reported that fish schools (measured

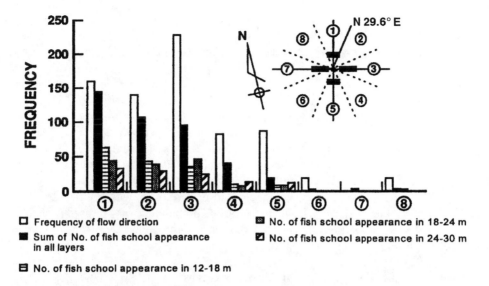

Figure 3.8 Frequency of flow direction vs. number of fish school appearance in the vicinity of an artificial upwelling structure (adapted from Otake et al. 1991). Fish are more likely to swim downstream of the structure where ample turbulence exists.

by echo sounder) swarming around artificial upwelling structures were correlated with the direction and velocity of the impinging tidal current (Figures 3.8 and 3.9). Toda (1991) improved the settling opportunity for larvae by utilizing wakes formed behind an object placed in a current and a wave-induced circulation formed around offshore dikes.

If the prevailing current changes significantly with time, it is more difficult to manipulate the reef orientation to enhance the performance of the reef. In coastal waters where tide, wind, and waves can produce flows of different time scales, it may be difficult to define the "flow" as it may

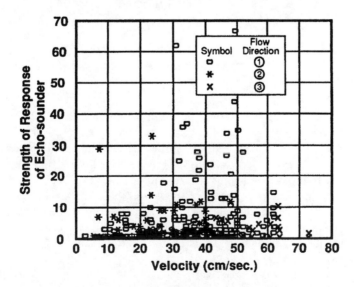

Figure 3.9 Strength of response of echo-sounder (appearance of fish school) vs. current velocity in the vicinity of an artificial upwelling structure (adapted from Otake et al. 1991). Flow directions 1, 2, and 3 correspond to those in Figure 3.8.

mean average or peak tidal/orbital/seasonal flow. Further research is needed to elucidate the effect of flow-reef orientation.

3.4.3.2 *Potential Physical Process Manipulation to Enhance Reef Performance*

Although definitive cause–effect relationships among physical, chemical, and biological processes at reef sites are not presently well understood, one should be aware of the potential to take advantage of such information to someday manipulate the physical processes to enhance reef performance. With vastly improved data gathering technology, it is now possible to gather physical, chemical, and biological data simultaneously at reef sites. With well-designed reef studies, sufficient data may be gathered and used, in conjunction with sophisticated numerical models, as a foundation for developing these cause–effect relationships. Subsequently, with measured physical variables at a reef site and a numerical model, one can attempt to design a reef with a wake zone of desired dimension. Similarly, knowledge of a local upwelling and bottom flow may be used to create desired resuspension and vertical mixing of sediments and nutrients to enhance primary production, and improve reef performance.

3.4.4 Effects of Artificial Reefs on Ambient Environment

In cases where the placement of reef units leads to alteration of the physical (flow and sediment) conditions, it is likely that the local chemical and biological conditions may also be affected. For example, if the placement of the reef leads to significant sediment erosion/resuspension, it is reasonable to expect that nutrients (nitrogen or phosphorus) may also be released into the water column via resuspension and/or diffusion (Simon 1989; Sheng 1993; Sheng et al. 1993), leading to a possible increase in algal concentration and a change in water quality (e.g., dissolved oxygen concentration). To ensure minimal adverse environmental impacts due to reef placement, it is recommended that physical conditions be monitored in conjunction with key chemical and biological parameters at reef sites.

The use of coal combustion solid residues to build artificial reefs appears to cause negligible environmental effects (Kuo et al. 1995; Leung et al. 1995).

3.5 REEF ASSESSMENT AND ANALYSIS

3.5.1 Types of Assessments

A comprehensive monitoring plan should generate sufficient data to satisfy three objectives: (1) to develop a comprehensive database of the large-scale ambient physical processes in order to facilitate the development of science-based reef siting; (2) to validate the existence of reefs; and (3) to measure the effectiveness (in terms of physical interactions) of reefs in meeting the objective of enhancing a fishery or improving habitat and to understand why a reef worked or failed.

As described in the introductory chapter of this book, there are several types of assessments that can be performed concerning physical variables at a reef site. These "types of assessments" are intended as a guideline for developing practical evaluation procedures at reef sites. The type of assessment needed for a reef study depends on the study objectives. The classification of the various types of assessments is not meant to be rigid.

The first type of assessment is descriptive:

1. What are the current and wave conditions at a reef site?
2. How do currents and waves vary over time and space?

3. What are the bottom sediment types and size fractions?
4. How do sediment types and size fractions vary spatially and temporally?

The second type of assessment is for analyses and comparisons:

1. What are the causes for nearfield flow and sediment conditions at a reef site?
2. How do they affect the nutrient distribution and plankton dynamics at the reef site?
3. What are the causes for sediment scouring at the reef site?
4. How does the reef structure affect sediment scouring?
5. How do physical conditions at two distinct reef sites relate to their differences in performance?

The third type of assessment includes studying interactions and predictions:

1. Where is the best reef site in a large coastal area?
2. What are the reef design criteria for enhanced nutrient flux and plankton and baitfish production at an artificial reef area?
3. What are the future physical conditions at a reef site under varying meteorological and hydrological conditions?

Table 3.1 lists the three types of assessments.

3.5.1.1 Type One Assessment — Basic Reef Descriptions

Information needed for a Type One assessment includes data obtained during predeployment monitoring, immediate postdeployment monitoring, and existence/screening monitoring. The questions to be answered are relatively simple:

1. Is this a good site for an artificial reef?
2. Where and how is the reef at the time of its initial deployment?
3. Is the reef still visible?
4. If the reef is not visible, why?
5. What is the spatial distribution and condition of the reef following its initial placement?

Following site selection and reef placement, monitoring should be conducted immediately (within a few days), 6 months later, and every 12 months thereafter.

3.5.1.1.1 Predeployment Monitoring — Individuals or groups wishing to place a reef at a particular site should be responsible for selecting and characterizing a suitable site. The interested party should try to obtain answers to the following typical questions:

1. Is this site a good site?
2. Are the physical characteristics (e.g., currents, wave, temperature, and bottom sediments) at this site suitable for deployment of a particular reef?

Before reef units are deployed at a potential site, it is advisable to collect the following data:

- water depth (m)
- tide (cm)
- current magnitude (cm/sec) and direction (degree) at a minimum of two depths (surface and near-bottom)
- wave height (cm) and period (s) or underwater pressure (psi)

Table 3.1 Types of Assessment for Physical and Engineering Reef Evaluation

Type One — Basic Reef Descriptions

Type One-A — Predeployment Monitoring

Questions to ask:
  Is this site a good site?
  Are the physical characteristics (currents, wave, temperature and bottom sediments) at this site suitable for deployment of a particular reef?

Techniques to use:
  Measurement of water depth, tide, currents at 1 to 2 depths, wave height and period, and collection of bottom sediment samples for analysis of composition, density, and size fraction.

Type One-B — Immediate Postdeployment Monitoring

Questions to ask:
  Where and how is the reef at the time of its deployment?

Techniques to use:
  Diver observation and measurement of reef location and condition using hand-held instruments.

Type One-C — Existence and Screening Monitoring

Questions to ask:
  Is the reef still there?
  If the reef is not there, why?
  Has the reef settled into the bottom, shifted its position, been displaced from its original location, or suffered structural degradation/damage?

Techniques to use:
  Collection of physical data lasting between a few hours to one tidal cycle every 6 to 12 months.

Type Two — Analyses and Comparisons

Questions to ask:
  Is the reef fulfilling its stated objectives?
  If yes, why? If no, why not?
  How does the presence of the reef affect the local and large-scale physical processes?
  How do the physical processes at two different sites differ?
  Why do two reefs at two nearby sites perform differently?

Techniques to use:
  Collection of salinity, temperature, turbidity, and nearfield flow data in addition to data for Type One-C monitoring.

Type Three — Interactions and Predictions

Questions to ask:
  How do the physical processes interact with chemical and biological processes at the reef site?
  Is this site the best site among all possible sites in this region?
  Given the wave and current conditions at a potential site, how can we design a reef to maximize nutrient flux and plankton production to achieve the best possible reef performance?
  How might physical processes have affected the fish aggregation at the reef site?
  How can we select a site so that a reef will not settle or be buried there?

Techniques to use:
  Collection of physical and chemical/biological data concurrently over a longer time period; perform laboratory experiments and numerical modeling to augment the assessment.

- bottom sediment composition and density (g/cm^3)
- presence of existing natural or artificial reef materials.

These data can be used to assess the suitability of the reef site for a reef unit with particular size, shape, and spacing. Analyses must be undertaken to determine if the reef will likely withstand the hydrodynamic forces at the site or not and whether the units might be buried by sediments.

Historically, such predeployment monitoring and analysis have not routinely been carried out for reef projects.

Tide, current, and wave data at a potential site should be collected over sufficiently long time periods prior to deployment by means of long-term moorings of tide gauge, current meter, and wave sensor. Instantaneous data provided by divers do not contain the temporal and spatial variabilities inherent to the physical processes, and are hence of limited value (except for sediment condition and water depth). Temperature and salinity data also can be collected and correlated to current and wave data if possible.

Data collection for siting evaluation may require information and resources that currently do not exist at the local municipal government (e.g., city or county) level. For example, most coastal counties in Florida and other states in the U.S. lack the ability to perform reef monitoring via long-term moorings of current meters; nor do they possess detailed information on regional coastal and estuarine circulation. A cost-effective way to obtain data for siting evaluation is to search for existing information on regional coastal and estuarine circulation from research institutions and to use mathematical models of tide, circulation, and wave to produce the physical data needed for a comprehensive analysis.

For site selection, sufficient data should be collected within a radius of several hundred meters of a site to provide an assessment of spatial heterogeneity. Ideally, one year of data collection is needed. If no data exist, data should be collected over a minimum of 2 to 4 weeks over each different season (e.g., summer and winter and ideally for 1 year) to provide temporal variability. If sufficient data exist, it is still advisable to collect data over at least one tidal cycle to no more than 2 to 4 weeks. (See Table 3.2; details of the methods are discussed in Section 3.6 of this chapter.)

3.5.1.1.2 Immediate Postdeployment Monitoring — To facilitate assessment of reef performance subsequent to its deployment, there is a need to produce a new database of a reef system immediately following its placement. The database of the reef can be expanded with subsequent monitoring data. Should the reef disappear or experience significant change from its original position or condition, monitoring data in the database can be used to determine the reason. The database should consist of precise location and detailed mapping of the reef, in addition to some basic oceanographic measurements. Due to its instantaneous nature, the oceanographic measurement conducted at this time generally consists of diver observation and measurement using hand-held, relatively simple, and less accurate instruments. Such measurement should not replace the more rigorous long-term oceanographic measurement of postdeployment monitoring. Variables and methods for immediate postdeployment monitoring are listed in Tables 3.3 and 3.4. One example of such procedures is generally practiced by the Jacksonville, Florida Scubanauts Reef Research Team (Seaman et al., 1991).

3.5.1.1.3 Existence and Screening Monitoring — Physical variables to be monitored include all those mentioned in the two previous sections, plus reef position and condition. Monitoring should be conducted 6 months after initial deployment and every 12 months thereafter. Duration of the data collection during each monitoring trip may vary between a few hours to one tidal cycle and may be limited by bottom time (depth-dependent) for research divers. Temporal and spatial variability of the oceanographic processes warrant longer term monitoring for more comprehensive analyses. The data to be collected are summarized in Table 3.5.

Extensive data analysis is required with a Type One assessment. For example, if the reef was buried or displaced, it is necessary to perform a detailed analysis to estimate the forces exerted on reef structures and the resistive force of the reef and bottom to identify the cause for burial or displacement. As a result, possible errors in the original siting analysis may be determined. If the reef suffered structural damage or significant marine biological fouling, it is necessary to identify the cause by analyzing data concerning reef material and oceanographic conditions.

Table 3.2 Data to Be Collected for Type One Predeployment Assessments

Parameter	Method	Sampling Duration	Sampling Frequency	Sampling Interval	Horizontal Resolution or # Samples
Location/Position	Loran C GPS (Global Positioning System)	Instantaneous		Once	Covers corners of reef tract area
Water depth	Fathometer	Instantaneous		Once	Sufficient to define 1/4 m contour
Tide	Tide gauge Underwater pressure sensor Numerical model	12 h–4 wk	15 min–2 h	Summer Winter	1
Current	Current meter Numerical model	12 h–4 wk	15 min–2 h	Summer Winter	2 vertical positions
Wave	Wave sensor Underwater pressure sensor Numerical model	12 h–4 wk	1–6 h	Summer Winter	1–2
Visibility	Secchi disk	Instantaneous			4–5
Bottom sediment: Vertical structure	Core sample Probing rod	Instantaneous		Once	5 × 5 grid over the reef tract area
Size distribution	Grab samples and sieves	Instantaneous		Once	
Roughness/Ripples	Camera and video	Instantaneous			

Table 3.3 Data to Be Collected for Type One Immediate Postdeployment Assessments

Protocol	Parameter	Method
Location of placement bench mark	Time Delays (TDs) Latitude/Longitude	Loran C (if GPS not used) GPS or converted from Loran C
Water description	Temperature at surface and bottom Thermocline depth(s) Salinity (surface) Salinity (@ depth) Visibility at surface and bottom Current direction TO	Thermometer −5.0° to 45° Depth gauge and thermometer Hydrometer Refractometer 20 cm dia. Secchi disk, black and white Magnetic oil-filled compass
Bottom description	Sediment depth Sediment sample sieved and weighted Bottom depth Ripple marks	Probing sediment with fiberglass rod Corer for 50 gm sample. Sieves size 6, 20, 40, 100, scale, pan Oil-filled depth gauge or dive computer Visual notes and compass direction across dominant long axis

3.5.1.2 *Type Two Assessment — Analyses and Comparisons*

Type Two assessments attempt to answer more complex questions such as:

1. Is the reef fulfilling its stated objective? If yes, why? If no, why not?
2. How does the presence of the reef affect the local and large-scale physical processes?
3. How do the physical processes at two different sites differ?
4. Why do two reefs at two nearby sites perform differently?

Comprehensive data must be collected to answer these more complex questions.

Table 3.6 provides a list of all variables to be monitored and justification for such monitoring. It is recommended to monitor the following variables in addition to those recommended for Type One Existence and Screening Monitoring:

- salinity (parts per thousand)
- temperature (°C)

Table 3.4 Representative Data to Be Collected for Type One Postdeployment Assessments

Protocol	Parameter	Method
Reef unit, set, or group placement description*	Selection of construction materials (cement, plastic, steel, etc.) Deployment map (single unit reef) Deployment map (multiple unit reef), reef set, or group. Placement date and time Profile Condition of material (breakage)	Visual assessment — Count, measure, and/or photograph Visual description — Photographs, dimensions, orientation of major dimensions Measurements to outer edge of placement in six directions out to 50 m from bench mark Observer records calendar date and local time Measured depth to bottom and highest points of reef units or sets Depth gauge or dive computer Visual inspection
Biological description	Natural live bottom Reef material (any prior fouling?) Flora Fauna	Visual ins., photographs Visual ins., photographs Visual ins., photographs/quadrats Visual ins., photographs/quadrats

* See Chapter 5, Figure 5.4, for an illustration.

Table 3.5　Data for Type One Existence and Screening Assessments

Parameter	Method	Sampling Duration	Sampling Frequency	Sampling Interval	Horizontal Station #	Vertical Station #
Reef Location/Position	Loran C GPS	Instantaneous		At placement, 6 mo later, every 12 mo thereafter	1	
Reef condition	Camera Video Diver observation	Instantaneous		At placement, 6 mo later, every 12 mo thereafter	1	
Water depth	Fathometer	Instantaneous		At placement, 6 mo later, every 12 mo thereafter	1	
Tide	Tide gauge Underwater pressure sensor Model	1–12 h	15 min–2 h	Every 12 mo after placement	1	
Current	Current meter Model	1–12 h	15 min–2 h	Every 12 mo after placement	1	
Wave	Wave sensor Underwater pressure sensor Model	1–12 h	1–6 h	Every 12 mo after placement	1	
Visibility	Secchi disk	Instantaneous		Every 12 mo after placement	1–2	2
Bottom sediment: Vertical structure	Core sample Probing rod	Instantaneous		Every 12 mo after placement	1–2	1–2
Size distribution Roughness/Ripples	Grab samples and sieves Camera and video	Instantaneous			1–2	

Table 3.6   Physical Variables and Reasons for Monitoring Variables at an Artificial Reef

Protocol	Parameter	Reasons for Monitoring
Position	Latitude, Longitude	Reference point for siting consideration and future monitoring
Climate	Wind	Generates current and vertical mixing in water column
	Air temperature	Affects wind stress and heating/cooling of water column
Water column description	Temperature	Affects current, vertical mixing, and possibly fish behavior in water column
	Salinity	Affects current, vertical mixing, and possibly fish behavior in water column
	Visibility	May affect biota, plankton, and oxygen level in lower water column
	Suspended sediment concentration	Affects visibility and nutrient concentration in water column and bottom sediments
	Current	Exerts force on reef and bottom and may affect fish aggregation; causes erosion of sediments/nutrients
	Wave	Causes orbital currents and exerts forces on reef and bottom
	Depth	Determines the influence of current and wave on bottom
	Tide	Causes periodic variation (with 12 or 24 h period) of tidal currents and water depth
Bottom sediments description	Bottom roughness	Affects bottom stress
	Sieve fraction	Affects settling velocity and resistive force of bottom
	Composition	Affects resistive force of bottom and nutrient/dynamics
	Density	Affects resistive force of bottom and nutrient/dynamics
	Biota	Live bottom is not suitable for reef deployment

- turbidity (Nephlometric Turbidity Unit or NTU) or total suspended solids (TSS) concentration (mg/l)
- flow-reef interaction (extent of wake region and/or upwelling current)

Type Two data should be collected twice during the first year, and once every 12 months thereafter. During each monitoring event, duration of data collection should be between one tidal cycle (if the reef appears to be functioning as intended) to 2 weeks (if reef is not functioning as intended) to provide sufficient data for analysis and assessment. Data should be taken from 2 to 3 stations and 1 to 2 depths at the site, depending on the reef and water depth. Except in very shallow water (<1 m), vertical gradients of physical variables (current, temperature, salinity, and suspended sediment concentration) are generally present in the water column. For example, a pycnocline (a thin layer of water with stable density gradient), which prohibits vertical mixing of nutrients and oxygen, often can be found in the water column, particularly during quiescent conditions. Thus, it is important to monitor the vertical structure of physical variables in water columns and to determine the existence and location of the pycnocline.

The physical variables to be measured for a Type Two assessment are summarized in Table 3.7. At this level, a more extensive analysis of the data is required. For example, if fish aggregation was not found near the reef, then it is useful to understand if physical processes might have contributed to it. Due to the complex nature of physical processes and how they affect organisms and habitat, it is advisable to have an interdisciplinary team consisting of physical, chemical, and biological scientists. Examples of Type Two assessments include the case study presented in Section 3.7 of this chapter and those conducted by Baynes and Szmant (1989) and Lindquist and Pietrafesa (1989).

3.5.1.3   Type Three Assessment — Interactions and Predictions

Questions to be answered at this level are very complex and include:

1. How do the physical processes interact with chemical and biological processes at the reef site?
2. Is this site the best site among all possible sites in this coastal or estuarine region?

Table 3.7 Data for Type Two Assessments

Parameter	Method	Sampling Duration	Sampling Frequency	Sampling Interval	Horizontal Station	Vertical Station
Reef location	Loran C GPS	Instantaneous		At placement, 6 mo later, every 12 mo thereafter	1	
Reef condition	Camera Video	Instantaneous		At placement, 6 mo later, every 12 mo thereafter	1	
Water depth	Fathometer	Instantaneous		At placement, 6 mo later, every 12 mo thereafter	1	
Tide	Tide gauge Underwater pressure sensor	1 tidal cycle to 2 wk	15 min–2 h	Every 12 mo after placement	1–2	2
Current	Current meter	1 tidal cycle to 2 wk	15 min–2 h	Every 12 mo after placement	1–2	2
Wave	Wave sensor Underwater pressure sensor	1 tidal cycle to 2 wk	1–6 h	Every 12 mo after placement	1–2	2
Bottom sediment	Grab sample Camera Video	Instantaneous		Every 12 mo after placement	1–2	2
Water temperature	Temperature sensor	1 tidal cycle	15 min–2 h	Every 12 mo after placement	1–2	2
Salinity	Conductivity sensor	1 tidal cycle	15 min–2 h	Every 12 mo after placement	1–2	2
Suspended sediment concentration	Optical back scatter sensor Water sampler	1 tidal cycle	15 min–2 h	Every 12 mo after placement	1–2	2
Wind	Anemometer	1 tidal cycle to 2 wk		Every 12 mo after placement	1–2	1–2
Light intensity	Light meter Secchi disk	1 tidal cycle		Every 12 mo after placement	1–2	2

3. If this site fails, what other sites could be used?
4. Is this reef the most appropriate design for this site?
5. Given the flow and wave conditions at a potential reef site, how can we design a reef to govern nutrient flux and plankton production to achieve the best possible reef performance?
6. Quantitatively, how might physical processes have affected the fish aggregation at the reef site?
7. Is aggregation lower due to a reduced wake zone, resulting from improper orientation of the reef with ambient current?
8. Is fish aggregation lower because of reduced visibility or nutrient concentration in the water column due to reef-induced sediment scouring?

It may be necessary to monitor the physical conditions of multiple sites and over different types of reefs or simultaneously monitor the physical, chemical, and biological conditions at a reef site. At each site, physical variables to be monitored should include all those mentioned for Type Two, plus the wind condition (which is usually measured at adjacent land stations) measured over a longer sampling duration (1 day to 2 weeks) and comparable sampling intervals (every 6 to 12 months) to describe temporal variability. Data should be taken at two to three sites and one to two depths at each site to represent spatial heterogeneity (see Table 3.8). Caution must be exercised when comparing physical data from a developed site to those from an undeveloped site — physical data from the developed site should be collected sufficiently far (e.g., ten times the reef size) from the reef.

To completely answer the questions for a Type Three assessment, it is necessary to utilize field monitoring, laboratory analysis, and numerical modeling. Moreover, Type Three initiatives should gather quantitative information on the interaction among physical, chemical, and biological processes, on both local and regional scales. For example, to enhance the reef performance in an estuary or a shelf, it is advisable to produce a regional database and a model of regional circulation, and a regional comprehensive mapping of suitable reef sites in the entire region. For example, the State of Louisiana has conducted a one-time mapping of the bottom sediment types over its shelf with a spatial resolution of approximately 2 km (approximately 1.2 mi). Accomplishing the goals of a Type Three assessment often requires the effort of an interdisciplinary team. Thus, governments are advised to seek the technical assistance of various universities and qualified firms knowledgeable in physical oceanographic monitoring. In addition, it is recommended to locate and use available numerical models and databases of regional circulation in such an effort.

3.5.2 Sample Designs

Sampling design depends on the level of complexity of the monitoring. As complexity increases, spatial and temporal considerations of sampling (and hence the number of samples) increase.

For Type One monitoring, sampling design varies for immediate postdeployment monitoring and subsequent postdeployment monitoring. The primary purpose of immediate postdeployment monitoring is to conduct detailed reef mapping, but very simple oceanographic measurement. For subsequent postdeployment monitoring, more accurate oceanographic measurement is needed. In general, samples should be taken from the vicinity (e.g., within a radius of a few hundred meters) of a particular reef site, but not from the immediate vicinity (e.g., within 1 m) of the reef structure, to represent the large-scale processes. Water levels generally do not vary significantly over a distance of a few kilometers. Currents, on the other hand, are significantly influenced by local bathymetry and can vary significantly over a distance of 1 km. Inside a typical shallow Florida estuary, there is usually a significant difference in currents between the deep navigation channel and adjacent shallow flats. Downstream of a reef structure, flow can be markedly different from the flow on the upstream side of the structure. However, such detail on the scale of the reef structure is generally not desired for Type One monitoring. Temporal scales of the dominant physical processes will determine sampling duration and frequency. To quantify tidal and wind-driven circulation, data

Table 3.8 Data for Type Three Assessments

Parameter	Method	Sampling Duration	Sampling Frequency	Sampling Interval	Horizontal Station	Vertical Station
Reef location	Loran C GPS	Instantaneous		Every 12 mo	2–3	
Reef condition	Camera Video Diver observation	Instantaneous		Every 12 mo	2–3	
Water depth	Fathometer	Instantaneous		Every 12 mo	2–3	
Tide	Tide gauge Underwater pressure sensor Model	2 wk	15 min–2 h	Every 12 mo	2–3	2
Current	Model current meter	2 wk	15 min–2 h	Every 12 mo	2–3	2
Wave	Wave sensor or underwater pressure sensor Model	2 wk	1–6 h	Every 12 mo	2–3	2
Bottom sediment: Vertical structure	Core sample Probing rod	Instantaneous		Every 12 mo	2–3	2
Size distribution	Grab samples and sieves					
Roughness/Ripples	Camera and video	Instantaneous				
Water temperature	Temperature sensor	2 wk	15 min–2 h	Every 12 mo	2–3	2
Salinity	Conductivity sensor	1 tidal cycle	15 min–2 h	Every 12 mo	2–3	2
Suspended sediment concentration	Optical back scatter sensor Water sampler	1 tidal cycle	15 min–2 h	Every 12 mo	2–3	2
Wind	Wind anemometer	1 tidal cycle to 2 wk		Every 12 mo	2–3	1–2
Visibility	Light meter Secchi disk	1 tidal cycle		Every 12 mo	2–3	2

should be taken over a period of at least 1 tidal cycle to 2 weeks (spring-to-neap variation), and ideally for a period of 1 full year. To quantify wave conditions, data should be collected at a minimum every 3 h over at least 1 day. The bottom sediment structure, which is not expected to change significantly over short time periods, does not need to be sampled frequently. To reduce the amount of data collection, tide tables and numerical models could be used to provide estimation of the water level fluctuation, currents, and waves.

For Type Two monitoring, samples should be collected to represent the large-scale processes as well as the small-scale processes. Thus, more data must be collected to show the horizontal and vertical variability of the physical variables. Duration of the data collection should also be increased. Field work should be conducted more frequently to reflect seasonal and annual patterns. To understand if physical variables are the reason a reef is working or not working, samples may have to be collected from the wake zone behind a reef structure. To quantify the scouring of sediments at a problematic reef site, intensive sampling of suspended and bottom sediments may be required. It is also advised to obtain measurement of turbulence variables such as r.m.s. fluctuating velocity, turbulent kinetic energy, turbulent dissipation, and turbulent mass and salinity fluxes.

For Type Three monitoring, data should be gathered from a number of reef sites, and physical data should be collected simultaneously with chemical and biological data to increase the usefulness of the dataset, thus significantly increasing the sample size. Duration and frequency of the sampling program should be comparable to or more refined than Type Two monitoring. The use of numerical models of physical processes (circulation, wave, bottom-boundary-layer dynamics, and sediment dynamics) to reduce the data collection cost and to guide the design of sampling programs is recommended. For monitoring of reef sites that are adjacent to each other, the monitoring programs should be coordinated.

3.5.2.1 Level of Reliability and Cost

The cost of a physical monitoring program includes: (1) design of sampling program; (2) purchase or leasing of equipment; (3) implementation of a field program; (4) analysis of data; and (5) personnel salaries for the organization performing the work. Obviously, the cost of a monitoring program would increase with the level of complexity. It is difficult to estimate the cost for conducting the various types of monitoring effort, since it may vary significantly with the organization performing the work.

3.5.2.2 Reference Sites

The design of reference sites is an important issue. For example, in order to determine how the presence of a reef may have altered the local upwelling of current and sediment, it is necessary to establish a reference site that is not influenced by the reef structure, but is otherwise similar to the reef site in physical characteristics. The proper distance from a control site to a reference reef will vary case by case.

3.5.3 Processing and Quality Control of Data

To facilitate efficient and accurate storage and retrieval of data, a GIS (Geographical Information System)-based reef database could be developed for a regional coastal and/or estuarine system. Data obtained by various sources would be entered into the reef database and shared by all to facilitate research and management of reefs. Processing of long-term oceanographic data may require the use of such processing packages as MATLAB, SAS, etc. See Chapter 2 for a more detailed discussion on data processing and quality control.

3.6 ASSESSMENT METHODS

3.6.1 General Considerations

When designing a reef study, one must be aware of the questions and problems at hand that will dictate the type of data and assessment methods needed. Some questions may only require "snapshots" of the environment on a few occasions, while others will require continuous *in situ* data over a long time (a month or longer). It is also important to consider the time and spatial scales associated with the various questions and problems when selecting the assessment methods. Once the assessment method is selected, the data obtained only contain certain time and length scales that may limit the usefulness of data.

3.6.2 Data Gathering Methods

Collection of physical and engineering data at a reef site can be carried out by *in situ* deployment of various sensors and/or divers using handheld instruments. Using rather rudimentary instruments, divers can provide instantaneous single-point measurements of position, water depth, current speed, wave, water temperature, reef condition, and bottom sediment condition at a reef site. However, instantaneous data cannot provide sufficient quantitative information (i.e., temporal and spatial variabilities) about the physical oceanographic conditions at a reef site. A comprehensive monitoring plan may require the collection of continuous *in situ* data over a long time, in addition to instantaneous diver-provided data. Long-term (up to one year) measurements of tide, current, wave, temperature, and salinity at a reef site require the deployment of more sophisticated sensors.

A brief review of simplified methods for physical and chemical measurements of seawater can be found in any basic oceanography text or in summary form in Bulloch (1991). More detailed procedures are outlined in Parsons et al. (1984). A review and literature survey of environmental assessment of artificial reefs is found in Bortone and Kimmel (1991). Useful information on diver monitoring can also be found in the *Artificial Reef Research Diver's Handbook* (Halusky 1991). In the following sections, data gathering methods for position, reef structure, wind, air temperature, waves, tide, depth, current, salinity, temperature, light, suspended sediments, and bottom sediments are briefly described. Table 5.1 in Chapter 5 lists variables commonly measured. However, depending on the questions and problems known to exist at a particular reef site, not all the variables listed below must be monitored.

3.6.2.1 Positioning

Horizontal position can be measured by Loran C or Global Positioning System (GPS). Loran C can provide quite accurate relative position, but not absolute position. GPS has advanced rapidly in recent years, and is capable of giving very accurate position (with errors of only a few meters). Differential correction to satellite signal can increase accuracy to sub-meter confidence levels, and is therefore recommended over Loran C recorded position.

3.6.2.2 Reef Structure

Accurate mapping of the configuration of reef materials can be accomplished *in situ* by divers or from a boat using a variety of underwater survey and mapping methods. The method selected will depend on the relative accuracy and precision needed. For example, if an extremely accurate map is needed (accurate to ±1 m over a 100 m² area), one could use methods developed by underwater archaeologists (UNESCO 1972) or use up-to-date imaging sonar systems, including side scan sonar and side scan and section scan sonar. The underwater archaeological methods rely on establishing grids on the bottom, which are costly, and require extensive time, teamwork, and coordination to accomplish. General discussion pertaining to underwater mapping methods is provided in UNESCO (1972), the NOAA Diving Manual (Miller 1979) and Burge (1988). Sidescan Sonar (Tian 1996) and Global Positioning Systems (GPS) are more efficient and economical in providing very accurate mapping and surveying of reefs.

Less accurate methods for sampling reefs, including physical descriptions, water sampling, and postdeployment survey and mapping methods are outlined in Strawbridge et al. (1991). Many of the Jacksonville Scubanauts Reef Research Team (JRRT) methods are the result of 5 years of *in situ* development. Some methods have been field tested and are discussed in Seaman et al. (1991). Many of the JRRT's postdeployment mapping methods use two to three teams of two divers, and can map an area in a 100-m diameter circle, at 24 m depths, during a single dive day.

3.6.2.3 Wind

Wind enhances the transfer of momentum and energy between the atmospheric surface layer and the ocean mixed layer. Wind produces surface currents (which are usually directed slightly to the right in the northern hemisphere and to the left in the southern hemisphere) and surface waves that propagate along the sea surface and produce orbital currents underneath the surface. A steady wind often can produce a setup of the sea surface in the downwind direction that, in turn, can generate a counter-current in the water column. The wind hence produces significant vertical mixing in the water column. During the summer months, the warmer surface water and cooler bottom water produce stable density gradients that prevent the wind mixing from reaching the bottom water. The surface mixed layer is separated from the bottom water by a region of sharp density gradient, i.e., a pycnocline. Persistent wind can also produce Langmuir circulation, which consists of convective cells separated by alternating convergence and divergence zones.

Wind is usually measured by an anemometer at one or two vertical levels within a few meters above the ocean surface. In the absence of a wind anemometer, wind data from nearby meteorological stations can be used.

3.6.2.4 Air Temperature

Air temperature affects the stability of the atmospheric surface layer over the ocean surface. A stable atmospheric surface layer (i.e., warmer air over colder ocean) reduces the transfer of momentum and energy across the air–sea interface, while an unstable atmospheric surface layer (i.e., cooler air over warmer ocean) enhances the transfer of momentum and energy across the air–sea interface.

Air temperature can be measured with an air temperature gauge at one or two vertical levels within a few meters above the ocean surface.

3.6.2.5 Waves

Waves are either caused by local wind or are propagated from distant locations (swell). As the wind blows over the water surface, energy is transmitted to the water and waves are formed. Wave characteristics depend on the wind speed, the distance over which it blows (fetch), and the length of time (duration) that it blows. The sea surface at a particular location may be calm or very irregular, composed of waves traveling in a single direction, or composed of waves traveling in several different directions.

Waves are classified according to the ratio of their length to the water depth as deep water waves, intermediate waves, or shallow water waves. Deep water waves are those whose length is less than twice the water depth. Shallow water waves are those whose length is 25 times the water depth, and intermediate waves lie in between deep and shallow water waves. Waves affect not only the sea surface but also the water column below, since waves generate oscillating currents that extend down into the water. The water particles under a deep water wave tend to move in circular orbits, while those under intermediate and shallow water waves tend to follow elliptical paths.

The magnitude of the water particle (water particle is a parcel of water of a fixed identity) motion is greatest at the surface and decreases with increasing depth. There is little decrease in the case of shallow water waves, since currents at the bottom are almost as strong as those at the surface. Deep water waves, on the other hand, only affect the water column down to a depth approximately equal to one-half the wave length. Thus, while a diver might find it impossible to work in 15 m of water with surface waves greater than a few hundred meters in length, he would not even be aware of waves shorter than 30 m in length. The effects on reef stability would be analogous — deep water waves would not affect reef elements, while intermediate and shallow water waves might actually move the reef elements (and bottom sediments) easily.

Like currents, waves could be observed by divers over short time periods by observing three things at the surface: estimating wave height, wave period, and wave direction. This is relatively easy if only one wave train is present, but may be more difficult if two or more are present. Details of wave observations by divers can be found in Halusky (1991) and Jones (1991). Further descriptions of the three parameters of interest are provided below.

Wave height — If waves are very regular, the height of each successive wave will be nearly identical. If wave heights vary, the "significant wave height" (average height of the highest one-third of the waves) should be estimated.

Wave period — The time it takes successive wave crests to pass a fixed point. When wave period is measured, record the total time it takes n (e.g., 11) wave crests to pass, then divide by $n - 1$, where n is large enough to average out variations in period. Round the result to the nearest second.

Wave direction — The direction from which the waves are coming, not the direction to which they are moving. For example, waves approaching from the northeast would be recorded as 45°. Measure direction to the nearest 5°.

Diver observations can provide crude estimates of wave conditions at a reef site over short time periods. However, peak wave conditions and wave orbital currents generally occur during storms when divers cannot operate. Diver observation may also contain significant bias or error. For quantitative and long-term wave measurement, it is necessary to measure either the underwater pressure using an underwater pressure sensor, or the surface wave using a surface-piercing wave sensor, which records data every 1 to 3 h at a reef site.

3.6.2.6 Tide

Tides continuously raise or lower the sea surface due to the gravitational attraction between the water surface and the Earth-moon-sun system. The vertical tidal movement of the water surface can be as much as several meters in some areas (e.g., the Bay of Fundy, Canada) with one (diurnal) or two peaks (semidiurnal) per day. Equilibrium tide theory, which assumes the ocean surface to be made of a uniform layer of ocean water with no land mass, allows accurate calculation of the period of tidal constituents, which are generally around 12 h (semidiurnal) and diurnal (around 24 h). In reality, however, tides propagate as long waves along the coast, and are hence significantly affected by the local bathymetry and coastline geometry, Coriolis acceleration, and wind. Tidal currents are generally negligible (less than 1 cm/s) in deep offshore water, but can be quite significant and reach more than 1 m/s in coastal and estuarine waters. The to-and-fro motion during a tidal cycle may produce interesting residual (tidally averaged) circulation in coastal and estuarine waters, which can affect the transport of nutrients, suspended sediments, and phytoplankton. In shallow waters, nonlinear interactions between the tidal constituents may produce higher-order constituents with periods around 3, 4, 6, and 8 h. Tidal currents in coastal and estuarine waters can interact with wind-driven currents, wave-induced currents, and internal waves to produce resuspension of sediments and nutrients from the ocean bottom.

Tides are generally measured with staff gauges or underwater pressure sensors. If underwater pressure sensors are used, it is necessary to measure the atmospheric pressure so that the tidal water level can be determined. With sufficient (e.g., 180 days) tide data at a given location, a predictive tide model can be constructed for future estimation.

3.6.2.7 Depth

Depth can be measured with a fathometer. Due to tidal fluctuation of water level, it is important to record the time of measurement and the tidal condition to ensure accurate depth data. Over muddy areas where a layer of unconsolidated sediments often exists, it is also important to ascertain where the "true" bottom is. Depth readings from an echo sounder or fathometer generally signal the top of the flocculent layer.

3.6.2.8 Current

Currents are among the most important factors affecting reef stability and reef performance. Currents past a reef unit can create hydrodynamic forces to shift the reef unit, to scour or deposit sediments, and to reduce the weight-bearing capacity of the bottom.

Currents can be measured by divers using relatively low-cost, handheld current meters over a short period of time. However, these handheld current meters are generally mechanical and are not capable of giving directional information. Moreover, the physical presence of the diver may interfere with the currents that the diver is trying to measure. For reef monitoring and design considerations, it is important to measure, estimate, or assess the maximum current that can occur at a particular site. Maximum currents usually occur during severe storms or seasonal or high tide conditions when divers generally cannot make observations. Thus, for all practical purposes, currents should be measured by self-recording current meters (electromagnetic current meters or an Acoustic Doppler Current Profiler [ADCP]) over sufficiently long time periods. Due to the lower accuracy and short-term usage, low-cost, handheld current meters should only be used to supplement self-recording current meters in collecting current data.

Currents are driven by tide, wind, waves, and density gradients. Various time scales are contained in a long-term current record: wave period (2 to 10 s), inertial period (24 to 26 h), tidal periods (~12 h for semidiurnal tides and ~24 h for diurnal tides), wind event (2 to 14 days), and spring-to-neap period (~2 weeks). To capture the various time scales of physical processes that affect the local currents at a reef site, it is important to obtain measurements of current speed and direction over sufficiently long time periods, ranging from a minimum of 1 tidal cycle to 1 month. Current data could be collected with a sampling period ranging from .5 sec (for wave orbital currents) to 15 min to 1 h (for tidal currents).

Currents vary vertically in the water column. Near the surface, wind and wave mixing creates a thin boundary layer (<1 m) within which the horizontal current decreases logarithmically with increasing depth. Near the bottom (water/sediment interface), a bottom boundary layer (<1 m) exists within which the current decreases logarithmically toward the bottom. In purely tidal situations, currents are generally in the same direction throughout the water column. The presence of wind, however, may set up a "return flow" in the bottom water. Vertical stratification due to salinity and temperature may also create significant vertical variation in horizontal currents. Thus, for reef monitoring, it is important to measure currents at more than one vertical location in the water column. If an ADCP is available, one can use it to measure the vertical profile of horizontal currents at a reef site with a vertical resolution on the order of 5 to 100 cm, depending on the user-selected mode of sampling resolution (Cheng et al. 1997).

For a proper analysis of reef stability, it is necessary to estimate the bottom stress acting on the bottom sediments or reef unit based on measured currents at one or two vertical locations in the water column. The relatively simple Morison-type equation (Equation 3.2) or more sophisticated mathematical models of bottom boundary layer dynamics (e.g., Sheng and Villaret 1989), can be used for computing bottom stress under prescribed current and wave conditions above the bottom boundary layer. A critical parameter for estimating bottom stress is the projected area of the reef structure, which requires accurate mapping of the reef and measurement of ambient currents.

For Type Two and Type Three monitoring, it is essential to obtain current measurements in the immediate vicinity of the reef. Based on current data collected in the vicinity of reef units during a case study at the Suwannee Regional Reef System (see Section 3.7.4), it is evident that local currents are influenced by the reef. For measuring the spatial structure of near-field currents, handheld current meters could be used to augment the time-series data from self-recording current meters. To estimate the forces acting on the reef unit, it is also necessary to measure the frontal area of the reef, i.e., the cross-sectional area of the reef unit in the direction of the flow. Since the flow direction can change significantly over a wave period and a tidal period, estimation of the frontal area is a formidable task, particularly for rather complex reef structures.

3.6.2.9 Salinity

Salinity is a measure of the amount of dissolved solids in water. It is usually expressed in parts per thousand (ppt), with normal seawater having a salinity of 33 to 37 ppt. Salinity usually changes dynamically with time and location. Water density and salinity may be significantly less near freshwater sources such as river mouths, tidal inlets, and offshore springs. In an estuary or in coastal waters, the significant horizontal salinity gradient between the ocean and the river can lead to baroclinic (density-driven) circulation, with seaward surface flow and landward subsurface flow. In the absence of significant vertical mixing, there is usually significant vertical salinity stratification in the water column, with low-salinity surface water and high-salinity subsurface water, and a thin intermediate layer exhibiting an abrupt salinity gradient (halocline) and density gradient (pycnocline). The sharp density gradient in the halocline/pycnocline inhibits the development of turbulent mixing by vertical shearing (velocity gradient), thereby reducing the exchange of dissolved oxygen and other materials between the surface and subsurface waters. To measure the existence of a halocline and a pycnocline, salinity data should be taken with sufficiently fine vertical resolution (~1 m) in the water column.

There is a unique relationship between the density, temperature, and salinity of seawater: knowing any two specifies the third. Salinity is usually measured indirectly by using a sensor to measure the conductivity and then converting it to salinity if the temperature is known. Like water temperature, salinity can be measured by divers or by self-recording instruments. Divers can collect water samples to measure salinity on board a ship or on land by using a thermometer and a hydrometer. A refractometer uses a visual method to determine salinity by measuring the refractive index of the water. Water samples should be collected on the reef as well as in the water column, using instruments manufactured by, for example, Hydrolab or YSI. For long-term salinity measurement at a reef site, self-recording conductivity sensors and temperature sensors (e.g., those manufactured by Sea Bird or Greenspan Technology) should be used.

3.6.2.10 Temperature

The temperature of seawater varies with season, diurnal cycle, location, and depth. Seasonal temperatures can vary by 10°C or more. In deep water, usually a difference between surface and bottom temperatures can be found, while the water column is generally well-mixed in shallow waters. The temperature may drop gradually from surface to bottom or it may drop suddenly, through an interface (called a thermocline or pycnocline) between two distinct layers of water. In some deep waters, there may be a diurnal thermocline superimposed onto a seasonal thermocline. The thermocline inhibits the vertical transfer and mixing of dissolved material across it. Thermoclines move up and down the water column in response to changes in wind and surface heat fluxes.

Water temperatures can be measured as part of a reef siting or monitoring dive using a handheld thermometer. In deep waters, the water temperature should be measured just under the surface, above and below a thermocline, and near the bottom. For reef monitoring, it is recommended to deploy long-term, self-recording current meters, many of which also record temperature.

3.6.2.11 Light

Light is the basis for photosynthesis and hence is the principal determinant of primary production. Visible light constitutes only a small portion of the total spectrum of electromagnetic radiation that reaches the Earth from the sun. For a study on photosynthesis, it is customary to consider only the photosynthetically active radiation (PAR), the sunlight available for photosynthesis by plants, which persists in the 400 to 700 nanometer range (i.e., visible range) of the spectrum. As light travels from the sea surface through the water column, its intensity decreases exponentially as the distance from the ocean surface increases. This "attenuation" occurs due to scattering of light by suspended particles and absorption of light by phytoplankton, particulate matter, dissolved material, and seawater itself. The depth of light penetration determines the depth to which photosynthetic activity can occur. The layer in which photosynthesis takes place, i.e., the photic zone, can extend to 200 m in very clean offshore waters, but may be as shallow as a few meters in coastal waters, where substantial particulate matter may be found due to river input and resuspension from the bottom by currents and waves. The available light at the bottom of the water column significantly affects the benthic community. Plants living on the bottom, e.g., seagrass, must be in the photic zone to survive, and cannot grow in coastal areas with significant light attenuation where less than 20% of incident light reaches the bottom.

Light can be measured in terms of inherent optical properties (e.g., index of refraction and single-scattering albedo) or apparent optical properties (e.g., light attenuation coefficient). In practice, light at a few vertical levels can be measured by light meters or photometers, either as total light or specific wavelengths. Light attenuation coefficient can then be calculated from the light data and related to concurrent data of water quality, suspended sediments, dissolved matter, and phytoplankton. For a preliminary reef study, a Secchi disk (a white-and-black quadrant alternating disk about 30 cm in diameter) is lowered by a line from a boat until it disappears from the viewer to measure the light attenuation depth. This depth (referred to as the Secchi depth) provides preliminary information on water quality, suspended sediments, and biological activity.

3.6.2.12 Suspended Sediments/Color/Turbidity

Turbidity and transparency are both used to describe the clarity of water. Turbidity is a measure of the amount of suspended particles in the water, while transparency is a measure of the ability of water to transmit light. As turbidity increases transparency decreases. These measurements vary with the number, size, and type of suspended particles in the water and with the nature and intensity of the ambient illumination.

One of the simplest ways of measuring water clarity is with a Secchi disk. Oceanographers lower the disk into the water from a ship and record the depth at which the disk is no longer visible. Divers can also employ a Secchi disk, but in a slightly different way by measuring the horizontal distance along the bottom to a point when the disk is no longer visible. Note that divers must be careful not to stir up bottom sediments when this technique is used. If clarity is to be measured on a dive, this should be the first task performed so that diver activity during other tasks will not affect the results (Halusky 1991).

Long-term suspended sediment concentration can be measured by either analyzing water samples in a laboratory or by using self-recording Optical Back Scatter (OBS) sensors or transmissometers in the field. However, it is important to collect water samples to allow both laboratory and field calibrations of the OBS sensor or transmissometer.

One of the problems with the OBS sensor is instrument sensitivity to particle size distribution of the sediment samples. New instruments are available that use a laser diffraction technique and are capable of measuring the particle size distribution as well as the concentration of suspended sediments in the field (Agrawal et al. 1996).

3.6.2.13 Bottom Sediments

Sediment and substrate characteristics are two of the most important factors controlling stability of an artificial reef. If the bottom has a low strength or bearing capacity or if it is susceptible to scouring and/or liquefaction during storms, reef elements will settle.

In general, it is extremely difficult to accurately estimate the strength or bearing capacity of bottom sediments with either *in situ* or laboratory tests. This is because results depend upon the type of test performed and the degree to which the sediment sample is disturbed during sampling and testing. However, some basic correlations exist between the type of sediment, its resistance to penetration, and its bearing capacity.

Bottom sediments can be composed of a variety of types and sizes of particles. Clay, silt, sand, gravel, shell, and rock are those most commonly encountered. In general, coarser and sandy particles are usually found in high-energy areas with strong currents and waves, while finer and muddy sediments are usually found in low-energy areas with weaker currents and waves. The finer particles (clay and some silts) are cohesive, i.e., the particles are bound by electrochemical forces between particles (these could be primary particles, or flocs, which are aggregates of primary particles). The coarser sediments (sand, shell, and gravel) are cohesionless and rely solely upon friction between particles for strength.

Rock may occur in large formations or as fragments suspended on or in other sediments. It sometimes occurs as outcrops on the bottom, but is usually overlain by sediments a few centimeters to 1 m or more in thickness. They possess varying strengths, and in some cases may fracture or wear readily. Gravel is a term used to describe small pieces of rock ranging from several centimeters to a few millimeters in diameter.

Sand particles are smaller than gravel, but larger than silt. The division between sand and silt is from .05 to .74 mm, depending upon the classification system used. Most sand-sized particles occurring offshore are quartz particles or shell fragments. Silt particles are smaller than sand, but larger than .002 to .006 mm in size. Silts can display some cohesive properties, but this is due usually to the presence of small amounts of clay particles. Clay particles are very fine (<.002 to .006 mm) depending upon the classification system used. The most common clay minerals are montmorillonite, kaolinite, and illite. Clays can be very sensitive, losing much of their shear strength when disturbed.

3.6.2.13.1 Sediment Sampling — The best way to sample unconsolidated sediments, both on the surface and beneath, is with a thin-walled core tube (usually clear plastic or PVC pipe, a few centimeters in diameter) following the method described in Blake and Hartge (1996). It is essential that sampling be conducted consistently, while ensuring that all grain sizes present in the sediment are retained in the sample, with minimal disturbance to the sample. If only a surface sample is required, the core tube needs to be about 15 cm long. Partially filling a longer tube is not good practice since this allows the sample to shift in the tube and the sediment structure to be altered during handling. The tube should be pushed fully into the bottom, capped at the top, and then sealed at the bottom with a flat plate or the diver's hand. The tube is then removed, inverted, and capped at the bottom. Procedures for obtaining longer cores are not much different, except it is usually necessary to drive the tube into the bottom. Simple impact corers have been devised to make this task easier.

An important consideration in obtaining samples of unconsolidated sediments is that the location from which they are obtained should be representative of the area being investigated. Pooling multiple samples from the same site will be more representative than a single sample.

3.6.2.13.2 Resistance to Erosion, Bearing Capacity, and Compressibility — Burial of reefs is dependent on the following three properties of the bottom sediment: bearing capacity, compressibility, and resistance to erosion by currents and waves (Tian 1994). The issue of bearing capacity is important during the impact phase of the initial installation only, while the time-dependent nature of the other two properties is crucial. Bearing capacity and compressibility can be determined following marine geotechnical engineering methods (Richards 1967; Poulos 1988). A penetrometer (Jones 1980) or a shear vane apparatus (Dill 1965) could be used. To determine the erodibility of sediments by waves and currents, a laboratory flume (Sheng 1989) or a portable particle entrainment simulator (Tsai and Lick 1986) may be used.

3.6.2.14 Precision and Accuracy of Instruments

Table 3.9 lists the precision and accuracy of some instruments for immediate postdeployment monitoring. Table 3.10 lists more sophisticated instruments for subsequent long-term oceanographic monitoring.

3.7 EXAMPLES

In this section, examples of three different types of reef studies are given, according to the approach used to obtain information as presented in Section 3.5.1.

3.7.1 Type One Studies

Most of the reef studies fall into this category. For example, Lindberg (1996) has monitored the performance of the Suwannee Regional Reef System in the West Florida Shelf in Gulf of Mexico since 1988, with main emphasis on routine fish count but occasional snapshot measurements of physical variables. Shao and Chen (1992) monitored 100 coal ash artificial reefs in 10 m water at Wan-Li, Northern Taiwan for more than 4 years and found that these reefs were as effective as the nearby concrete reefs in attracting fishes and the settlement of benthic organisms. Despite the strong tidal currents (~1.5 m/s) in the 10 m deep water at all of the Wan-Li sites, most of the reef blocks were stable except during typhoons. Kjeilen et al. (1995) reported that the "ODIN Artificial Reef Project" in the North Sea will involve 5-year continuous monitoring at the site of the abandoned oil production platform ODIN, including the following variables: physical integrity of the reef, heavy metal concentration, plankton, benthic community and sediments, fish density, and behavior.

3.7.2 Type Two Studies

Lindberg and Seaman (1991) monitored reefs in the Eastern Gulf of Mexico from 1989 to 1990 to investigate fish abundance and diversity in response to variable dispersion of artificial reef units. Concrete pipes were arranged as six reef sets at 12 m depth and 30 km offshore Florida. Based on monthly fish count by divers, they found that most (but not all) species appeared in greater abundance on the clumped reef sets. Ozasa et al. (1995) monitored the wave conditions and fishes at 24 sites in Japan and established correlations between: (1) wave height and attached organism diversity and biomass; (2) wave height and dominant attached fish species; (3) attached organism biomass and structure shape/material and with respect to dominant current/wave direction; and (4) relationships of attached organism species. Huang (1994) studied the impact load of artificial reefs during placement using laboratory experiments and theoretical modeling. An artificial reef of full scale dimensions was designed, and a model with a length scale of 1/30 was manufactured. The terminal velocity reached by the reef in a wave tank was found to be 0.78 m/s, in complete agreement with the theoretical value. The impact load on a rock bottom was measured to be about

Table 3.9 Precision and Accuracy of Instruments for Immediate Postdeployment Assessments

Protocol	Parameter	Method	Units/Range	Precision	Accuracy	Calibrate
Bench mark (BM) location	Time delays (TDs)	Loran C	Microseconds, ± 50 m	± 0.1 μsec	± 0.1 μsec	Manufacturer standard
	Latitude/Longitude	GPS or Converted from Loran C	Degrees, Minutes, .01 min	± .01 min	± .01 min	Manufacturer standard Loran C is computed from TDs~
Water description	Temperature at surface and bottom	Thermometer −5.0° to 45°	°C	± 1.0°	± 0.5°	With NBS certified thermometer
	Thermocline depth(s)	Depth gauge and thermometer	Meters and °C	± 0.6	± 0.6	With known depth at BM or with dive computer
	Salinity	Hydrometer	‰ parts per thousand	± 1.0	± 0.82	Certified salinity hydrometer set
	Salinity	Refractometer	‰ parts per thousand	± 1.0	± 1.0	Manufacturer standard
	Visibility at surface and bottom	20 cm dia. Secchi disk, black and white	Meters	± 1.0	± 1.0	Standard fiberglass level rod
	Current direction TO	Magnetic oil-filled compass	Direction TO in degrees	± 2.5	± 2.5	Manufacturer standard
Bottom description	Sediment depth	Probing sediment with fiberglass rod	Meter penetration to bedrock	± 0.03	± 0.08	Standard fiberglass level rod
	Sediment sample sieved and weighed	Corer for 50 g sample, sieves size 6, 20, 40, 100, and pan	Sieve fractions weighed in grams	± 0.25 g	± 0.5 g	NBS sieve standards and calibrated lab scales
	Bottom depth	Oil-filled depth gauge or dive computer	Meters	± 0.6	± 0.6	With known depth at BM or against dive computer
	Ripple marks	Visual notes and compass direction across ridges	Degrees from compass	± 2.5°	± 2.5°	Manufacturer standard
Reef unit, set, or group placement description	Selection of construction materials (cement, plastic, steel, etc.)	Visual assessment — count, measure and/or photograph	Material type, number of units, dimensions of units	N/A	N/A	N/A
	Deployment map (single unit reef)	Visual description — photographs, dimensions, orientation of major dimensions	Material type — length, width and height in feet, and degrees magnetic of length	N/A	N/A	N/A

			± 3 m ± 2.5°	± 5 m ± 2.5°	Standard fiberglass measuring tape and oil-filled divers compass
Deployment map (multiple unit reef), reef set or group	Measurements to outer edge of placement in six directions out to 50 m from bench mark	Meters distance and compass degrees of heading away from BM	± 3 m ± 2.5°	± 5 m ± 2.5°	Standard fiberglass measuring tape and oil-filled divers compass
Placement date and time	Observer records calendar date and local time	Year, month, day, local time of day	N/A	N/A	N/A
Profile	Measured depth to bottom and highest points of reef units or sets; depth gauge or dive computer	Meters to bottom; meters to highest points observed	± 0.6	± 0.6	With known depth at BM or against dive computer
Condition of material (breakage)	Visual inspection	Subjective notes	N/A	N/A	N/A
Biological description — Natural live bottom	Visual inspection — photographs	Species list or samples	N/A	Replicates/ reoccupy same station(s)	Replicates/reoccupy same station(s)
Reef material (any prior fouling?)	Visual inspection — photographs	Species list or samples	N/A	Replicates/ reoccupy same station(s)	Replicates/reoccupy same station(s)
Fishes	Visual inspection — photographs	Species list and counts	N/A	Replicates/ reoccupy same station(s)	Replicates/reoccupy same station(s)

Table 3.10 Precision and Accuracy of Selected Instruments for Long-Term Oceanographic Monitoring

Parameter	Method/Range	Units	Sensitivity	Accuracy	Calibration
Temperature	Thermometer −1.0 to 35°	°C	.0001°C	±.01	Factory or other
Conductivity	Flow-through 2-terminal platinum electrode cell 0–6 siemens/meter	siemens/ meter	5×10^{-5}	.001	Factory or other
Current	Electromagnetic meter −305 to 305 cm/s	cm/s	2 cm/s	±3.0	Flow tank
Current	ADCP	cm/s	0.1 cm/s	±0.1	Flow tank
Waterlevel/waves	Transducer 0 to 2 psi	psi	.005	±.03	Factory
Suspended sediments	Infra-red optical back scatter (OBS) 0–500 mg/ℓ	mg/ℓ	1 mg/ℓ	±5 mg/ℓ	*In situ* sampling of suspended solid

9.7 times the reef's weight, while only 3.7 times the weight on a sandy bottom. Kim et al. (1995) experimentally studied the local scour and embedment of an artificial reef by wave action in shallow water. The researchers determined that the local scour depends on the shape of the reef, which significantly influences the local flow. The contact area between the sandy substrate and the bottom of the reef is reduced by the flow developed beneath the reef, which makes the reef unstable and causes it to settle.

3.7.3　Type Three Studies

There have been relatively few Type Three studies. Tian (1996) conducted a comprehensive siting study in the Lieu-Chu Yu Offshore area in Taiwan. Five proposed artificial reef sites were monitored and evaluated through an integrated offshore site investigation technology. The information incorporated in the study included: topography, geomorphology, sediment properties, and sea states. Instruments used included the side scan sonar, echo sounder, GPS, gravity core, ADCP, remotely operated vehicle (ROV), and geomechanical testing apparatus. Sheng et al. (1999) conducted a study at the Suwannee Regional Reef System to understand the interrelationships among the physical, chemical, and biological variables via simultaneous collection of these data at a reef site. Variables measured included wave, current, temperature, salinity, suspended sediment concentration, turbulence, nutrient concentration, phytoplankton, dissolved oxygen, pH, light attenuation, bottom sediments, and fish count. Instruments used included conventional electromagnetic current meters, pressure sensors, OBS sensors, turbulence micro-profiler, Hydrolab, and hydro-acoustic devices for fish monitoring.

3.7.4　Case Study

The objective of the study was to determine the feasibility of gathering long-term physical data at a reef site and to see if and how the presence of the reef modifies the flow in the vicinity of the reef. The results of the study did show that tidal currents were modified near the reef and that waves can be quite significant at the reef. Unfortunately, simultaneous sampling of nutrients and fishes was not carried out; therefore, it was not possible to test the hypothesis that physical processes contributed to increased fish aggregation or production. The case study is presented herein as an example of monitoring protocol, instead of as a model.

3.7.4.1　*Feasibility of Long-Term* In Situ *Gathering of Physical Data*

During February and March 1992, a preliminary monitoring program of physical variables was conducted at a reef tract located at the southern end of the Suwannee Regional Reef System, Levy County, Florida. The purpose was to test the feasibility (and "user-friendliness") of measuring long-term *in situ* physical oceanographic data at a reef site. After the instrument packages were put together in the lab, they were deployed by a four-diver team (with one boat) in one day and retrieved 2 months later from this Gulf of Mexico artificial reef west of Cedar Key, Florida.

Results of the project are briefly summarized here as a case study, giving the reader an example of the considerations and procedures that might be encountered in setting up such a project. (See also examples of reef studies in Chapter 4.)

Two moorings, each consisting of two current meters and one underwater pressure sensor, were placed on the onshore (northeast) and offshore (southwest) sides of a reef block. The distance between the two moorings was 9.3 m (approximately 30 ft). The water depth was 14 m (approximately 45 ft). The bottom current meters and pressure sensor (SEADATA/PACER 635-12 directional wave-tide sensor) were 1.05 m above the bottom at the offshore station but 1.1 m above the bottom at the inshore station. The surface current meters (ENDECO 174 Solid State Current Meter) were at 9.8 m above the bottom. Data collection was initiated on 3 February 1992 and continued for 45 days. The ENDECO meters provided 15-min averaged currents, while the SEADATA sensors furnished average tide data every 2 h, 17-min "burst" data (1 Hz frequency) every 6 h for underwater pressure and current, and instantaneous current readings every 2 h. A sampling of the results follows.

3.7.4.1.1 Surface and Bottom Currents — Surface currents at the inshore and offshore stations are practically identical. Figure 3.10A shows the onshore–offshore current (u) and Figure 3.10B shows the alongshore current (v). The top panel of each figure shows the time series of the 15-min average current that includes the influence of the wind and the tide. The second panel of each figure is the residual current after applying a lowpass third-order Butterworth filter with a 2-day cutoff frequency to the current data in panel 1, thus representing the wind-driven current. The third panel is the tidal current, which is produced by subtracting the second panel from the first panel. The fourth panel shows the energy spectrum.

It is clear that the onshore-offshore current is primarily driven by tide (semidiurnal only) as opposed to wind, while the alongshore current is driven by a mixture of tide (diurnal and semi-diurnal) and wind.

The bottom currents are influenced by the presence of the reef, due to the proximity (within 2.5 m) of the reef block. The bottom currents are generally weaker than the surface currents. Again, u-velocity is primarily tide-driven while the v-velocity is driven by tide and wind. The diurnal influence, however, is significantly reduced when compared with the surface current. The bottom currents on the offshore side are quite different from those on the inshore side, where the diurnal signal is dissipated.

3.7.4.1.2 Tide and Wind Waves — Tides measured at the inshore and offshore stations are basically the same and show a mixture of diurnal and semidiurnal components with the highest tidal range on the order of 1.3 m (~4 ft) during the spring tide.

Since the wind was primarily in the alongshore direction, wind waves at both stations are very similar with a slight difference in direction. Significant wave height reached almost 2 m and wave period reached 11.5 s during passage of cold fronts.

3.7.4.1.3 Bottom Sediments — Two samples of approximately the top 5 cm of sea bottom were taken by hand by divers at each of the two instrument locations next to the artificial reef for a total of four samples weighing approximately 1 kg each.

The samples were dried for over 48 h at 50°C. Then they were split down to ~75 g for sieving through two stacks of six screens each, i.e., a total of 12 meshes of different sizes. Each stack was shaken for 15 min by a Tyler Ro Tap machine. The unused portions of the samples were archived. After the contents of all screens and the last pan were weighed, those portions of the sample that had passed the No. 10 screen were recombined, mixed, and split down to a 2 to 5 g sample for

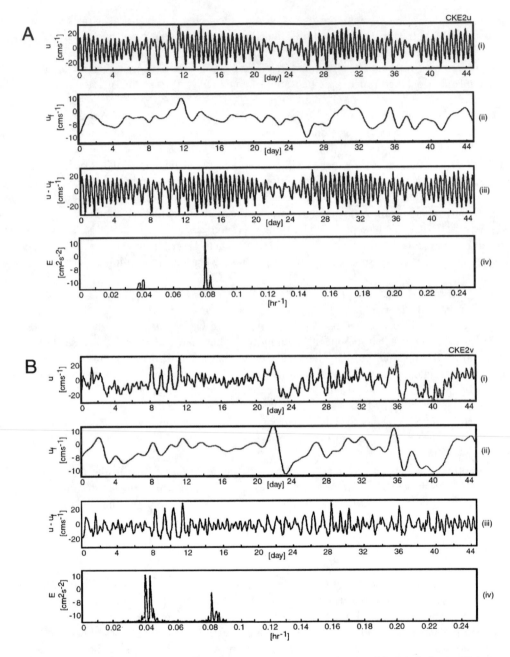

Figure 3.10 A. Measured onshore near-surface current at the inshore station (cke2u) during a 45-day period in February and March 1992. **B.** Measured alongshore near-surface current at the inshore station (cke2v) during a 45-day period in February and March 1992.

carbonate analysis. The resulting sample was weighed and then immersed in 0.5 normal hydrochloric acid and stirred periodically until no bubbles were observed. The sample was then washed with distilled water, dried for over 24 h at 50°C, and then weighed. The carbonate percentage was computed using the initial and final weights of the sample, assuming that all loss was the result of removing all carbonate and that the fraction not passing No. 10 screen was pure carbonate (this fraction consisted of large shell fragments that would have required a longer time to dissolve in acid).

Table 3.11 Representative Format for Data, Sieve Analysis of Sediment

Date: 05/11/92   By: G.C.C.

Soil Sample:	West Bag 2 Soil Sample Weight
Location:	Cedar Keys
Date collected:	03/26/92

Weight of dry soil: 72.1 g   **Weight of dry soil after test:** 72.0 g

Sieve Number U.S. Standard	Sieve Opening mm	Weight of Soil g	%	%	Percent Finer
10	2.00	0.6	0.83	0.83	99.17
20	0.850	3.0	4.17	5.00	95.00
30	0.595	5.7	7.92	12.92	87.08
40	0.42	19.8	27.5	40.42	59.58
50	0.297	26.7	137.08	77.50	22.50
60	0.246	6.95	9.65	87.15	12.85
70	0.210	4.3	5.97	93.12	6.88
80	0.177	3.1	4.30	97.42	2.58
100	0.149	1.3	1.81	99.23	0.77
120	0.125	0.3	0.42	99.65	0.35
140	0.105	0.1	0.14	99.79	0.21
160	0.096	0.05	0.07	99.86	0.14
Pan		0.1	0.14	100.0	
		Total		100.0	

Note: Carbonate content: 5.23%.

Results of the analysis show little variation among the four samples. Results for one sample are summarized in Table 3.11, which illustrates a standard way to organize the laboratory data.

3.7.4.1.4 Summary of Case Study — It is noteworthy that the current data collected on both sides of the reef suggest significant influence of the reef on local circulation. Suspended sediment concentration and turbulence were not measured in this pilot study, but would be important factors in a more thorough project. The case study took place during winter, when frontal systems in the eastern Gulf of Mexico exert significant influence on coastal circulation and hence the water column is generally more or less well mixed. During summer, vertical stratification is expected to be more pronounced due to heating of the water column. Subsequent measurements at the same reef site would likely reveal significant differences between summer and winter circulation.

Data obtained during the case study could be further analyzed to provide estimates of forces exerted on the reef structure during the time period of the case study. With additional data, statistical analysis could be performed to produce estimates of hydrodynamic forces that can be expected during a 10- or 100-year return interval storm.

Wave data also were obtained at the reef site. Although the waves were generally less than 50 cm in amplitude, waves reached almost 2 m in amplitude and 10 s in period during the passage of three fronts. Such waves could reach the bottom and affect bottom current and sediment resuspension and transport.

Visual observation at the reef site suggests that the reef appears to be functioning well physically (after being deployed 9 years) with little settlement or movement and not affected by the monitoring effort. No fish count data were collected to facilitate direct correlation between circulation and fish assemblages.

The successful case study demonstrated that it is feasible to conduct long-term *in situ* physical monitoring at a reef site for the various levels discussed in previous sections at a reasonable cost. A study that includes simultaneous sampling of physical, chemical, and biological variables at the Suwannee Reef has recently been conducted (Sheng et al. 1999).

3.8 FUTURE NEEDS AND DIRECTIONS

3.8.1 Development of a Comprehensive Regional Reef Database

Concurrent with the worldwide increase in artificial reefs, it is useful to consider the development of a regional reef database for future siting studies, reef performance assessment studies, hypothesis testing, and further analysis. A GIS can be used to aid the development of a regional reef database that includes the following information: geometry, bathymetry, waves, currents, wind, tide, bottom sediments, nutrients, and fish species. Data from future monitoring studies can continue to be entered into the database. Numerical models of regional circulation and waves also should be established to provide quick prediction for future use.

3.8.2 Field and Laboratory Experiments

Field and laboratory experiments need to be designed to test hypotheses and enhance our understanding of reef performance. One area that warrants further study is the use of petroleum platforms as artificial reefs (e.g., Kjeilen et al. 1995). Although petroleum platforms in the Gulf of Mexico and the North Sea have served as de facto fish-attracting devices and sites for colonization by plants, invertebrates, and fishes, the use of abandoned petroleum platforms as artificial reefs warrants further study, particularly in terms of structural and material stability, scouring and reef burial, and environmental impact.

Laboratory experiments allow easy manipulation of variables and have been very useful in determining the interrelationships between physical, chemical, and biological variables. Laboratory experiments will continue to be used in the future. However, laboratory experiments may not accurately simulate the field conditions; therefore, results of laboratory experiments must be carefully interpreted, particularly if fishes are used in the experiment as well. It is important to ensure similitude between laboratory models and the prototype by preserving the important dimensionless numbers (e.g., Reynolds number, Keulegan–Carpenter number, and Froude number).

3.8.3 Development of an Artificial Fishery Reef Technology Guide

With increasing interest and investment in artificial fishery reefs, it is timely to consider the development of an artificial fishery reef technology guide. A tentative procedure for the development and maintenance of artificial reefs is shown in Figure 3.11.

3.9 ACKNOWLEDGMENTS

The author thanks W.-M. Tian, C. C. Huang, and K.-T. Shao for providing useful references, Bill Seaman for reviewing versions of this manuscript, Margaret Miller for reviewing the preliminary version, Cynthia Vey for typing the manuscript, and the Florida Sea Grant College Program for providing financial support (U.S. Department of Commerce NOAA Grant NA36 RG-007) for the earlier reef monitoring study, which started my pleasant involvement with artificial reefs.

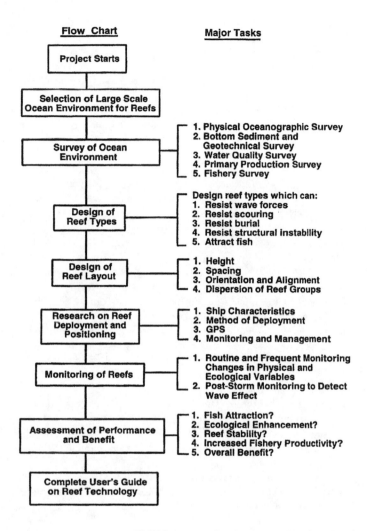

Figure 3.11 Flow chart for developing an artificial fishery reef technology guide (adapted from Huang 1994).

REFERENCES

Agrawal, Y.C., H.C. Pottsmith, J. Lynch, and J. Irish. 1996. Laser instruments for particle size and settling velocity measurements in the coastal zone. *Proceedings, Oceans '96* 1–8.

Baynes, T. and A. Szmant. 1989. Effect of current on the sessile benthic community structure of an artificial reef. *Bulletin of Marine Science* 44(2):545–566.

Blake, G.R. and K.H. Hartge. 1996. Physical and mineralogical methods. Pages 363–375. In: *Methods of Soil Analysis.* Soil Science of America, Madison, WI.

Bortone, S.A. and J.J. Kimmel. 1991. Environmental assessment and monitoring of artificial reefs. Pages 177–236. In: W. Seaman and L. M. Sprague, eds. *Artificial Habitats for Marine and Freshwater Fisheries.* Academic Press, San Diego.

Bruno, M.S. 1993. Laboratory testing of an artificial reef erosion control device. *Coastal Zone '93. Proceedings, Symposium on Coastal and Ocean Management* 2:2147–2154. American Society of Civil Engineers, New York.

Bulloch, D.K. 1991. *The American Littoral Society Handbook for the Marine Naturalist.* Walker Publishing, New York. 165 pp.

Burge, J.W., Jr. 1988. *Basic Underwater Cave Surveying.* The Cave Diving Section of the National Speleological Society, Inc., Branford, FL.

Cheng, R.T., J.W. Gartner, and R.E. Smith. 1997. Bottom boundary layers in San Francisco Bay, California. *Journal of Coastal Research* SI25:49–62.

Dean, R.G., R. Chen, and A.E. Browder. 1997. Full scale monitoring study of a submerged breakwater, Palm Beach, Florida. *Coastal Engineering* 29(3–4):291–315.

Dill, R.F. 1965. A diver-held vane-shear apparatus. *Marine Geology* 3:323–327.

Eadie, R.W. and J.B. Herbich. 1987. Scour about a single cylindrical pile due to combined random waves and a current. Pages 1858–1870. In: *Proceedings, 20th International Conference on Coastal Engineering.* American Society of Civil Engineers, New York.

Fan, K.-L. 1984. The branch of Kuroshio in the Taiwan Strait. Pages 77–82. In: T. Ichiye, ed. *Ocean Dynamics of the Japan and East China Sea.* Elsevier Oceanography Series, Elsevier, New York.

Grove, R.S., C.J. Sonu, and M. Nakamura. 1991. Design and engineering of manufactured habitats for fisheries enhancement. Pages 109–152. In: W. Seaman, Jr. and L.M. Sprague, eds. *Artificial Habitats for Marine and Freshwater Fisheries.* Academic Press, San Diego.

Halusky, J.G. 1991. *Artificial Reef Research Diver's Handbook.* Technical Paper TP-63, Florida Sea Grant College Program, Gainesville. 198 pp.

Helvey, M. and R.W. Smith. 1985. Influence of habitat structure on the fish assemblages associated with two cooling-water intake structures in southern California. *Bulletin of Marine Science* 37:189–199.

Huang, C.C. 1994. The study of the impact load of artificial reefs during placement. Pages D67–78. In: *Proceedings, 16th Conference on Ocean Engineering,* Taiwan.

Hubertz, J.M. and R.M. Brooks. 1989. Gulf of Mexico hindcast wave information. Wave Information Studies on U.S. Coastlines Report 18, Waterways Experiment Station, U.S Army.

Ichiye, T., ed. 1984. *Ocean Dynamics of the Japan and East China Sea.* Elsevier Oceanography Series, Elsevier, New York. 423 pp.

Japan Coastal Fisheries Promotion Association (Zenkoku Engan-Gyogyo Shinko-Kaihatsu Kyokai [in Japanese]) (JCFPA). 1984. Coastal Fisheries Development Program structural design guide (Engan-Gyojo Seibi-Kaihatsu-Jigyo Kozobutsu Sekkei-Shishin). (In Japanese.)

Japan Coastal Fisheries Promotion Association (Zenkoku Engan-Gyogyo Shinko-Kaihatsu Kyokai [in Japanese]) (JCFPA). 1986. Artificial reef fishing grounds construction planning guide (Jinko-Gyosho-Gyojo Zosei Keikaku-Shishin). (In Japanese.)

Japan Coastal Fisheries Promotion Association (Zenkoku Engan-Gyogyo Shinko-Kaihatsu Kyokai [in Japanese]) (JCFPA). 1989. Design examples for Coastal Fisheries Development Program structural design guides. (Engan-Gyojo Seibi-Kaihatsu-Jigyo Kozobutsu-Sekkei Keisan-Rei-Shu). 398 pp. (In Japanese.)

Jones, C.P. 1980. Engineering aspects of artificial reef failures. Notes for Florida Sea Grant's artificial reef research diver training program. NEMAP Fact Sheet 5, Florida Cooperative Extension Service, Gainesville.

Jones, C.P. 1991. Oceanographic data collection and reef mapping. Pages 19–28. In: J.G. Halusky, ed. *Artificial Reef Research Diver's Handbook.* Technical Paper TP-63, Florida Sea Grant College Program, Gainesville.

Kim, J.Q., N. Mitzutani, and K. Iwata. 1995. Experimental study on the local scour and embedment of fish reef by wave action in shallow water depth. Pages 168–173. In: *Proceedings, International Conference on Ecological System Enhancement Technology for Aquatic Environments.* Japan International Marine Science and Technology Federation, Tokyo.

Kim, T.I., C.K. Sollitt, and D.R. Hancock. 1981. Wave forces on submerged artificial reefs fabricated from scrap tires. A final report to The Port of Umpqua Commission and Sea Grant, Report No. RESU-T-81-003, Civil Engineering Department, Oregon State University, Corvallis.

Kjeilen, G., J.P. Aabel, M. Baine, and G. Picken. 1995. Platforms as artificial reefs — advantages and disadvantages, a case study. Pages 513–518. In: *Proceedings, International Conference on Ecological System Enhancement Technology for Aquatic Environments.* Japan International Marine Science and Technology Federation, Tokyo.

Kuo, S., T. Hsu, and K. Shao. 1995. Experiences of coal ash artificial reefs in Taiwan. *Chemistry and Ecology* 10:233–247.

Lamb, T.W. and R.W. Whitman. 1969. *Soil Mechanics.* John Wiley & Sons, New York. 533 pp.

Lee, T.N., J.D. Wang, and J.A. Lorenzzetti. 1988. Two-layer model of summer circulation on the southeast U.S. Continental Shelf. *Journal of Physical Oceanography* 18:591–608.

Leung, A.W.Y., K.F. Leung, K.Y. Lam, and B. Morton. 1995. The deployment of an experimental artificial reef in Hong Kong: objectives and initial results. Pages 131–140. In: *Proceedings, International Conference on Ecological System Enhancement Technology for Aquatic Environments.* Japan International Marine Science and Technology Federation, Tokyo.

Lindberg, W.J. 1996. Fundamental design parameters for artificial reefs: interaction of patch reef spacing and size. Final Report submitted to Florida Department of Environmental Protection, Department of Fisheries and Aquatic Sciences, University of Florida, Gainesville.

Lindberg, W.J. and W. Seaman, Jr. 1991. Design of habitat size and spacing, with special reference to ecological factors. Pages 189–193. In: *Proceedings, Japan–U.S. Symposium on Artificial Habitats for Fisheries.* Southern California Edison Co., Rosemead, CA.

Lindquist, D. and L. Pietrafesa. 1989. Current vortices and fish aggregations: the current field and associated fishes around a tugboat wreck in Onslow Bay, North Carolina. *Bulletin of Marine Science* 44(2):533–544.

Miller, J.W. 1979. *NOAA Diving Manual.* 2nd ed. Stock No. 003-017-00468-6, NOAA, Suppl. of Documents, U.S. Department of Commerce, Washington, D.C.

Mitchum, G.T. and W. Sturges. 1982. Wind-driven currents on the west Florida shelf. *Journal of Physical Oceanography* 12:1310–1317.

Morison, J.R., M.P. O'Brien, J.W. Johnson, and S.A. Schaaf. 1950. The force exerted by surface waves on piles. *Petroleum Transactions.* American Institute of Mining, Metallurgical, and Petroleum Engineers. Page 189.

Myatt, D.O., E.N. Myatt, and W.K. Figley. 1989. New Jersey tire reef stability study. *Bulletin of Marine Science* 44(2):807–817.

Nakamura, M., ed. 1980. *Fisheries Engineering Handbook* (Suisan Doboku). Fisheries Engineering Research Subcommittee, Japan Society of Agricultural Engineering, Tokyo. (In Japanese.)

Nakamura, M. 1985. Evolution of artificial fishing reef concepts in Japan. Bulletin of Marine Science 37:271–278.

Nakamura, M., M. Uuekita, and T. Iino. 1975. Study on the landing impact of a free-falling object in the ocean. *Proceedings, 22nd Annual Japanese Coastal Engineering Conference.* (In Japanese.)

Nitani, H. 1972. Beginning of the Kuroshio. Pages 129–163. In: H. Stommel and K. Yoshida, eds. *Kuroshio, Physical Aspects of the Japan Current.* University of Washington Press, Seattle.

Otake, S., H. Imamura, H. Yamamoto, and K. Kondou. 1991. Physical and biological conditions around an artificial upwelling structure. Pages 299–310. In: *Proceedings, Japan-U.S. Symposium on Artificial Habitats for Fisheries.* Southern California Edison Co., Rosemead, CA.

Ozasa, H., K. Nakase, A. Watanuki, and H. Yamamoto. 1995. Structures accommodating to marine organisms. Pages 406–411. In: *Proceedings, International Conference on Ecological System Enhancement Technology for Aquatic Environments.* Japan International Marine Science and Technology Federation, Tokyo.

Parsons, T., Y. Maita, and C. Lalli. 1984. *A Manual of Chemical and Biological Methods for Seawater Analysis.* Pergamon Press, New York.

Poulos, H.G. 1988. *Marine Geotechnics.* The Academic Division of Unwin Hyman, Ltd., Allen & Unwin, London. 473 pp.

Richards, A.F. 1967. *Marine Geotechnique.* University of Illinois Press, Champaign.

Rocker, K., Jr. 1985. *Handbook of Marine Geotechnical Engineering.* Naval Civil Engineering Laboratory, Port Hueneme, CA. 257 pp.

Sarpkaya, T. 1976. Vortex shedding and resistance in harmonic flow about smooth and rough cylinders at high Reynolds numbers. Report No. NPS-59 SL76021, U.S. Naval Post Graduate School, Monterey, CA.

Sarpkaya, T. and M. Isaacson. 1981. *Mechanics of Wave Forces on Offshore Structures.* Van Nostrand Reinhold, New York.

Seaman, W., Jr. and L.M. Sprague, eds. 1991. *Artificial Habitats for Marine and Freshwater Fisheries.* Academic Press, San Diego. 285 pp.

Seaman, W., Jr., J.G. Halusky, D.W. Pybas, and B. Strawbridge. 1991. Enhanced artificial reef database for Florida: I. State-level reef database demonstration and II. Enhancement and validation of local reef assessment techniques. Sport Fishing Institute, Artificial Reef Development Center, Washington, D.C.

Shao, K. and L. Chen. 1992. Evaluating the effectiveness of the coal ash artificial reefs at Wan-Li, Northern Taiwan. *Journal of the Fisheries Society of Taiwan* 19(4):239–250.

Sheng, Y.P. 1989. Consideration of flow in rotating annuli for sediment erosion and deposition studies. *Journal of Coastal Research* SI5:207–216.

Sheng, Y.P. 1993. Hydrodynamics, sediment transport and their effects on phosphorus dynamics in Lake Okeechobee. Pages 558–571. In: A.J. Mehta, ed. *Nearshore and Estuarine Fine Sediment Transport.* American Geophysical Union, Washington, D.C.

Sheng, Y.P. 1998. Pollutant load reduction models for estuaries. Pages 1–15. In: M. Spaulding, ed. *Estuarine and Coastal Modeling.* American Society of Civil Engineers, New York.

Sheng, Y.P. 1999. Effects of hydrodynamic processes on phosphorus distribution in aquatic ecosystems. Pages 377–402. In: K.R. Reddy, G.A. O'Connor, and C.L. Schelske, eds. *Phosphorus Biogeochemistry in Subtropical Ecosystems.* Lewis Publishers, Boca Raton, FL.

Sheng, Y.P. and C. Villaret. 1989. Modeling the effect of suspended sediment stratification on bottom exchange processes. *Journal of Geophysical Research* 94(C10):14429–14444.

Sheng, Y.P., X. Chen, K.R. Reddy, and M. Fisher. 1993. Resuspension of sediments and nutrients in Tampa Bay. Final Report to Florida Sea Grant College Program, Coastal and Oceanographic Engineering Department, University of Florida, Gainesville.

Sheng, Y.P., E. Phlips, P. Seidle, J. Lee, E. Bredsoe, and T. Groskopf. 1999. Hydrodynamic processes at artificial reefs and effects on plankton and baitfish abundance. Synopsis submitted to Florida Sea Grant College Program, University of Florida, Gainesville.

Simon, N.S. 1989. Nitrogen cycling between sediment and the shallow-water column in the transition zone of the Potomac River and Estuary, II. The role of wind-driven resuspension and adsorbed ammonium. *Estuarine, Coastal and Shelf Science* 28:531–547.

Stommel, H. and K. Yoshida, eds. 1972. *Kuroshio, Physical Aspects of the Japan Current.* University of Washington Press, Seattle. 517 pp.

Strawbridge, E.W., J. Brayton, and M. Barnes. 1991. Underwater methods for the Scubanauts Not-For-Profit, Inc. Reef Research Team. Jacksonville Scubanauts Inc. Reef Research Team, Jacksonville, FL. 31 pp + 9 app.

Takeuchi, T. 1991. Design of artificial reefs in consideration of environmental characteristics. Hokkaido Development Bureau, Sapporo, Hokkaido, Japan.

Tian, W. 1994. Burial mechanism of artificial reefs: a geotechnical point of view. Pages D79–94. In: *Proceedings, 16th Conference on Ocean Engineering.* Republic of China.

Tian, W. 1996. Investigation and evaluation of artificial reef sites: Lieu-Chu Yu offshore area. Pages 878–888. In: *Proceedings, 18th Conference on Ocean Engineering.* Republic of China.

Toda, S. 1991. Habitat enhancement in rocky coast by use of circulation flow. Pages 239–247. In: *Proceedings, Japan–U.S. Symposium on Artificial Habitats for Fisheries.* Southern California Edison Co., Rosemead, CA.

Tsai, C.H. and W. Lick. 1986. A portable device for measuring sediment resuspension. *Journal of Great Lakes Research* 12(4):314–321.

UNESCO. 1972. *Underwater Archeology: A Nascent Discipline.* United Nations Educational, Scientific and Cultural Organization, Paris. 306 pp.

Wolanski, E. and W.M. Hamner. 1988. Topographically controlled fronts in the ocean and their biological influence. *Science* 241:177–181.

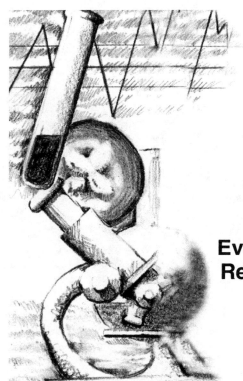

CHAPTER **4**

Evaluation Methods for Trophic Resource Factors — Nutrients, Primary Production, and Associated Assemblages

Margaret W. Miller and Annalisa Falace

CONTENTS

4.1 SUMMARY

This chapter describes the function and assessment of basal trophic resources, including nutrients, primary production, lower level secondary production, and associated assemblages in artificial reef ecosystems. The second section delineates the objectives and some specific definitions of terms used in the chapter. The third section gives background on the functional ecology and biogeographic patterns of trophodynamics of reef ecosystems, both natural and artificial. Sections four and five describe general guidelines and specific methods for assessing nutrients, primary production, and benthic assemblages on artificial reefs. The chapter concludes with a summary of previous research and discussion of future needs in artificial reef assessment with respect to ecosystem function.

4.2 INTRODUCTION

Nutrient and primary productivity regimes have been rarely incorporated into artificial reef monitoring programs because they often require more technical training and expensive equipment than generally available to most reef builders. However, these factors are important aspects of ecosystem function and should be considered in the future. It is especially important to collect nutrient and primary productivity data during the site selection phase of a project to ensure a site's suitability for the success criteria achievement of a given artificial reef project. Two of the major categories of artificial reef objectives are enhancement of fisheries production and mitigation (i.e., creating habitat to compensate for anthropogenic loss or damage to natural ecosystems). In the latter case, duplication of natural reef function should be an explicit objective. In the former case, artificial reef performance is considered in terms of fish production and, in many instances, reef-based food supply for the fishes (Ambrose and Swarbrick 1989; Hueckel and Buckley 1989; Carr and Hixon 1997) is a primary determinant of fisheries production.

4.2.1 Objectives

The objectives of this chapter are to introduce (1) the concepts of nutrients and primary production in the context of how they contribute to reef trophic structure and the performance of artificial reefs; and (2) sampling and analytical methods for assessing the nutrient status, primary production, and associated sessile benthic communities of an artificial reef. The first objective is accomplished largely by consideration of trophodynamics on natural, and, by extension, artificial reefs, including geographical patterns in nutrient and primary production regimes. The chapter then provides a framework for determining the types and methods of assessment for the lower levels of reef trophic webs, including inorganic nutrients, benthic primary production, indicators of water column primary production, and filter-feeding benthic invertebrates (lower level secondary producers). This complex of materials and organisms forms the basis for artificial reef food webs (see Figure 4.1) and constitutes a primary determinant of artificial reef performance.

Even though the focus of this chapter is on assessment of artificial reefs, most of the methods and ecological background are derived from, and are therefore applicable to, natural reef and hard-bottom communities. For additional background on topics in this chapter, useful references include Riley and Chester (1971); Levinton (1982); Nybakken (1982); Parsons et al. (1984a); and Valiela (1995).

4.2.2 Definitions

As pointed out by Bohnsack and Sutherland (1985), imprecise usage of terminology relating to artificial reef "productivity" has yielded much confusion in artificial reef literature. The following definitions are offered in an attempt to minimize such confusion.

The term *nutrients* generally refers to forms of the elements nitrogen (N), phosphorus (P), and silica (Si), which plants need to make basic biomolecules such as proteins, nucleic acids, fats, structural materials, etc. During photosynthesis, the plants combine these inorganic nutrients with carbon (C) obtained from carbon dioxide gas in the atmosphere yielding complex organic materials that animals can use as an energy source (i.e., food). Nutrients can occur in dissolved and particulate forms in aquatic environments; for either of these, there can be inorganic (nitrate, nitrite, and ammonia for nitrogen, and orthophosphate for phosphorus) or organic (e.g., amino and nucleic acids, and phospholipids) forms. Nutrient concentrations are usually expressed in mass per volume (either as milligrams per liter [mg/l] or preferably as micromoles per liter, which is abbreviated as μM).

Primary production is the process or total amount of organic matter produced from carbon dioxide and nutrients via photosynthesis (or via chemosynthesis in some habitats without light). *Primary productivity* is the rate at which primary production is occurring or the rate at which new plant material (biomass) is being produced. Thus, primary production is an absolute variable while primary productivity is a relative variable. Primary productivity can be measured as the mass of oxygen (O_2) generated or carbon dioxide (CO_2) fixed over a given area and time period (e.g., $gC \cdot m^{-2} \cdot$ unit time^{-1}).

Standing crop is the biomass of plant and animal material per area of a given community. It is described in several ways and/or units: as mass of chlorophyll (chl a) per surface area or water volume (mg chl a $\cdot m^{-2}$ or m^{-3}), as wet or dry weight of algal matter per area (g wet or dry weight $\cdot m^{-2}$), and for phytoplankton in millions of cells per volume (10^6 cells $\cdot m^{-3}$). Standing crop may

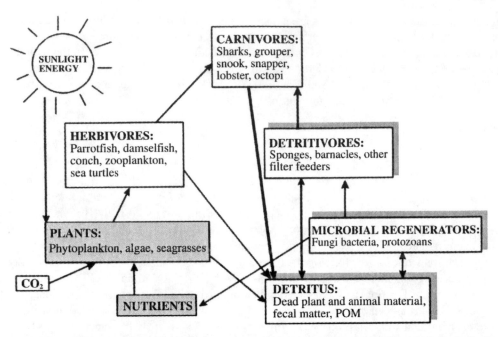

Figure 4.1   Generalized marine food web. Shaded boxes indicate the primary trophic levels that are the target for many of the assessment methods discussed in this chapter. Shadowed boxes indicate compartments relevant to certain lower levels of secondary production (e.g., by sessile filter-feeding ["fouling"] invertebrates) for which similar assessment methods are appropriate. Note that all of these compartments may have both benthic and planktonic members, e.g., there may be detrital feeders (such as isopods) that consume benthic detrital particles as well as filter feeders that consume suspended particulate detritus.

or may not indicate rates of primary production depending on the rate of consumption. That is, areas of high primary productivity may have very low standing crops of plant matter if the rates of consumption are equally high (e.g., tropical coral reef systems).

   Secondary production is the process of converting plant biomass or detritus (dead plant and animal matter) into animal biomass, and *secondary productivity* is the rate at which this happens. There are several levels of secondary production within most aquatic food webs (Figure 4.1). For example, algal turf is converted to herbivore (e.g., a *Mithrax* crab) biomass; the *Mithrax* is eaten by a large wrasse and converted into fish biomass; the wrasse may be eaten by a piscivorous fish (e.g., grouper); and the grouper may be eaten by a shark. This rate of secondary production, often the higher levels of secondary production (e.g., the grouper), is what is often meant by the term "productivity" when used in the artificial reef literature.

   Biofouling community refers to the organisms (including algae, sessile invertebrates, and micro-organisms) that attach to and grow onto the surfaces of a hard structure (ship hulls, pilings, rock reefs, etc.). The macroinvertebrate component is often dominated by solitary and colonial filter feeders such as barnacles, bivalves, ascidians, hydroids, and sponges.

   Trophodynamics is a term used to describe the analysis of food webs, feeding strategies, and energy transfer through ecosystems.

4.3 PRIMARY PRODUCTION AND ASSOCIATED COMPONENTS OF REEFS

   It is known that light, nutrients, and the physical substrate are all strong determinants of primary production in marine systems. However, if the objective of an artificial reef is to increase catches of predatory fishes, why is it important to assess factors relating to primary production and benthic communities of an artificial reef?

4.3.1 Reef Trophodynamics — Why are Nutrients and Primary Production Important?

Whether artificial reefs, in fact, contribute to new production of target fish stocks is still controversial (Bohnsack 1989; Bohnsack et al. 1997). Two mechanisms are most often cited whereby artificial reefs can accomplish increased fisheries production: (1) If shelter is limiting fish production, the additional shelter provided by the reefs may allow more of the resources of a coastal area to flow into fish biomass. The degree to which this mechanism can enhance fish production will be limited by the amount of forage available from the substrates and waters surrounding the artificial reef. (2) If food is limiting fish production, the new primary production and attached benthic secondary production (sessile filter-feeding invertebrates, Fang 1992) fostered by the artificial reef will support a new food web, part of which will end up in fish biomass. Clearly, these factors can only increase fishery production if they are limiting factors to fish population abundance in the region in question. For example, if larval supply to replenish reef fishery stocks is limiting (due to hydrodynamic or other factors), it is unlikely that either the habitat structure nor possibly enhanced primary production of an artificial reef will improve fisheries production.

These mechanisms of habitat enhancement and of enhancement of primary productivity are not mutually exclusive, and in fact, most artificial reefs probably work by a combination of the two (Bohnsack et al. 1991). More interestingly, it has been suggested that in oligotrophic tropical coral reefs there is an interaction of these two factors. Increased structural heterogeneity provides nooks, crannies, and crevices for sediment and organic matter to accumulate and be remineralized (Bray and Miller 1985; Szmant-Froelich 1983, 1984; Szmant et al. 1986). Aside from simply providing surface area for primary producers, this is another mechanism whereby enhancing structural complexity in placement of artificial reefs may increase productivity in oligotrophic systems.

Fishes are heterotrophs and thus need a source of food energy to sustain life, growth, and reproduction. At some point in the trophic chain, fishes are dependent on primary producers (phytoplankton or benthic plants) which, in turn, are dependent on the light and nutrients (complexified in the process of photosynthesis) that are needed for primary production. In the case of benthic herbivores, that dependence on reef primary production is direct. In another case, the primary production supporting a planktivorous fish is occurring in the water column and may be imported from far-off systems via currents. With higher level consumers such as piscivores, the dependence on primary production and the light and nutrients that support it are more removed and diffuse (likely including both benthic and planktonic primary production from both reef-based and off-reef sources).

In the planning of a successful artificial reef project (either for fisheries enhancement or for habitat mitigation), considerable thought must be given to the trophic resources that will support fisheries stocks and/or the ecosystem functioning of the artificial reef system. Light levels are of primary importance to potential primary productivity. Insufficient light availability (due to high turbidity or excessive depth) has inhibited artificial reef primary production and ecosystem function, for example, in the Mediterranean Sea (Falace and Bressan 1994) and Chile (Jara and Céspedes 1994). If a particular type of fish or invertebrate is targeted, its food requirements (and if a carnivore, the food requirements of its preferred prey) should be identified. In some cases, artificial reefs have been shown to increase fish abundance mainly by providing shelter, with reef food resources being of minimal importance to resident fishes (Ecklund 1996). These fishes and mobile invertebrates may be plankton feeders or forage over large areas, including adjacent seagrass or sand-bottom habitats for their food (Randall 1965; Steimle and Ogren 1982; Frazer and Lindberg 1994; Lindquist et al. 1994; Powell and Posey 1995).

If production of such fishes is a targeted objective for an artificial reef project, then availability of sufficient forage base (including area and prey density of adjacent soft-bottom habitats) should be a criterion in artificial reef siting decisions. Indeed, Bortone and Nelson (1995) found that target game fishes at an artificial reef in the northern Gulf of Mexico foraged primarily off-reef and caused significant depletion of the available prey base. In other cases, reef-based food resources are of great importance (Hueckel and Buckley 1989; Johnson et al. 1994; Pike and Lindquist 1994). In either case, assessment of the primary and lower level secondary productivity of reefs and of adjacent areas will help estimate the amount of food available to support the secondary productivity of resident reef fishes.

4.3.1.1 Managing Primary Production

Some researchers have suggested further management intervention to circumvent natural limitations to artificial reef primary and secondary productivity. Spanier et al. (1990), working in low-productivity waters in the southeast Mediterranean Sea, conducted pilot studies demonstrating increased abundance of commercial fish species at an artificial reef when supplemental production was added to the reef in the form of weekly feedings of fish tissue. Clearly, this method is approaching a low level aquaculture activity (sometimes referred to as "marine ranching," Spanier 1989) and beyond the scope of traditional artificial reef projects.

Conversely, artificial reefs may be used to trap excess nutrients and supply food energy to newly formed food chains based on the artificial reefs. A reported objective of some European artificial reef projects (Bombace 1989; Bugrov 1994; Laihonen et al. 1996) has been to utilize excess nutrient supply from wastewater or aquaculture operations by placing artificial reefs in adjacent areas so that fouling organisms (including both filter feeders and plants) can trap the nutrients and excess organic matter, increase benthic primary and secondary production, and possibly prevent water quality problems associated with excess water column productivity. Parchevsky and Rabinovich (1995) calculated that 0.5 to 4 t of nitrogen and 50 to 100 kg of phosphorus might be removed from eutrophic waters of the Black Sea by the seaweeds growing on one hectare of artificial reef every 6 months. Fang (1992) also suggested that artificial reefs can increase fisheries production in regions of high planktonic productivity by allowing for trapping of that productivity by benthic filter feeders and creating a "new" benthic food web to convert that local planktonic production into fish, instead of it being advected out of the local area.

4.3.2 Geographic Patterns in Reef Productivity and Trophodynamics

As for all plant communities, the primary productivity of artificial reefs will be largely determined by two factors: nutrients and light (Parsons et al. 1984a). To some extent, the dominant primary producers and the trophic web based upon them can be predicted, for a given reef, by the light and nutrient regimes, along with other geographic factors such as latitude. This prediction can help to focus assessment efforts toward reef characteristics relevant to the ecological function and the objectives of a given reef.

The nutrient supply to an artificial reef will be a function of local hydrography relative to nutrient sources (coastal runoff, offshore upwelling, and sewage outfalls). Light intensity will be a function of water depth and turbidity. Artificial reefs deployed below 30 m (90 to 100 ft) depth in coastal areas will support minimal primary production (Relini et al. 1994; Valiela 1995) although the depth of specific light penetration will vary according to the turbidity of the site (Valiela 1995). However deep, poorly lit artificial reefs may still develop rich attached filter-feeding invertebrate communities that help funnel planktonic food resources to reef-dwelling fishes by concentrating small particles on which larger fishes could not feed. Shallower reefs in turbid (nutrient rich) estuarine and coastal areas likely also will have low levels of benthic primary production compared to attached benthic secondary production, because sessile invertebrates tend to out-compete algae where water column productivity is high. Thus, benthic (as opposed to planktonic) primary production of artificial reefs will be of greater relative importance to the food web in shallow coastal waters with low to moderate nutrient concentrations.

Marine food webs can be categorized in the following manner: (1) those in which the principal food energy source is benthic plant material generated by photosynthesis within the community (e.g., coral reefs, seagrass beds, and kelp forests); (2) those in which primary production is conducted in the water column with sessile filter-feeding invertebrates (attached benthic secondary production) and planktivorous fishes feeding on plankton as it drifts by (e.g., piling communities and deeper hard-bottom communities); and (3) those in which the basic food source is detrital material (dead plant and animal matter) largely imported from an adjacent community (e.g., deep sea communities and many estuaries and coastal areas receiving terrestrial, marsh, or mangrove detritus).

To a large extent, the delivery of nutrients, the dominant groups of primary producers, and hence, the prevalence of benthic- vs. planktonic-based food webs, are predictable based on geographic patterns in hydrography and ocean productivity (Figure 4.2). Birkeland (1988, 1997) suggests that in tropical coral reefs (generally restricted to areas with low water-column nutrient concentrations) primary production and benthic cover are dominated by animal/plant symbionts, especially reef-building corals, which are efficient internal recyclers of nutrients. In geographic areas with somewhat higher rates of nutrient loading, Birkeland (1988, 1997) predicts that primary production (and hence, benthic community cover) will be dominated by benthic macrophytes (mostly seaweeds). Both of these regimes will be characterized by benthic-based food webs. In geographic areas of very high nutrient loading (including many warm temperate areas), high water column nutrient flux leads to high rates of planktonic primary production. This high planktonic production both obscures light available to benthic plants and provides a rich food source for filter-feeding invertebrates. Thus, benthic communities in these regions will generally have low primary productivity and high abundance of filter-feeding invertebrates (e.g., barnacles, bivalves, and sponges). Of course, all degrees of gradation between these food-web types are found, especially in coastal areas where all three can be important at the same time.

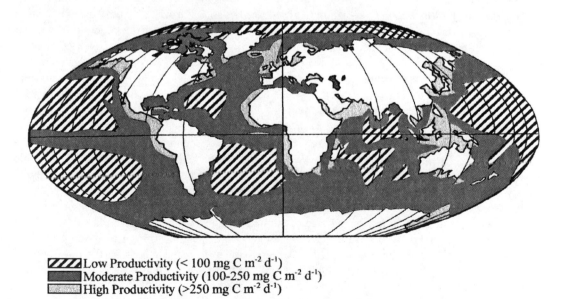

Low Productivity (< 100 mg C m^{-2} d^{-1})
Moderate Productivity (100-250 mg C m^{-2} d^{-1})
High Productivity (>250 mg C m^{-2} d^{-1})

Figure 4.2   Global patterns of planktonic primary productivity. (Adapted from Koblentz-Mishke, O.J., Volkvinsky, V.V., and Kabahova, J.G. In: W.S. Wooster, ed. *Scientific Exploration of the South Pacific*, 185, 1970. With permission. Courtesy of the National Academy of Sciences, Washington, D.C.)

When an artificial structure is constructed, it is important to choose the site carefully with regard to the objectives since the same modules, located in different trophic environments, may lead to development of very different communities and, hence, different characteristics of ecosystem function. In any given site, algal flora is affected by preexisting ecological conditions, which may predestine the area for dominance by certain species or community types. Moreover, it is not possible to build artificial reefs everywhere due to physical constraints such as unstable substrate. Within these constraints, reference to natural reefs in the same region/depth should be a guide to the trophic structure expected to develop in an artificial reef system, and these expectations should guide the monitoring/evaluative program pursued.

4.3.2.1 *Choosing Assessment Parameters*

Table 4.1 summarizes some generalized trophic characteristics and priorities for monitoring of different reef systems. For example, given an artificial reef project with the primary objective to restore or mitigate degradation of natural reef habitat, the appropriate characteristics to assess will depend on the geography and the nature of the reef community expected to develop. In a tropical oligotrophic habitat (e.g., Clark and Edwards 1994), the restoration goal may be primarily defined in terms of restoration of coral populations, and assessment efforts should be focused on coral recruitment, growth, and other environmental characteristics that are important for coral success (including high light and low nutrient concentrations; see next section). In contrast, for a restoration/mitigation artificial reef project in a temperate eutrophic area such as an oyster reef, it would be more important to focus assessment efforts on planktonic resources for filter-feeding oysters, such as water column chlorophyll and particulate organic matter concentrations.

4.3.2.2 *Nutrients and Coral Reefs*

The special case of nutrient regimes of tropical coral reefs bears some discussion (further reviewed in Miller 1998) because there is substantial concern regarding the perceived recent degradation of coral reef systems manifested as a shift from dominance by corals to dominance by macroalgae (the first to second type of community described in Table 4.1). One possible explanation offered by some researchers (and supported by a limited interpretation of the generalized characteristics in Table 4.1) is that increases in anthropogenic nutrient loading have released benthic macrophytes from nutrient limitation allowing them to grow fast and outcompete corals (Smith et al. 1981; Birkeland 1988; Littler et al. 1991; Lapointe 1997). However, experimental nutrient enrichment studies that have sought to test this shift from low to high macroalgal standing stock have failed to document such a shift, at least without simultaneous manipulation of grazing intensity (Hatcher and Larkum 1983; Larkum and Koop 1997; Miller et al. 1999). Other studies offer strong evidence that natural and anthropogenic reductions in grazing regime do consistently induce this phase shift (e.g., Lewis 1986; Hughes et al. 1987; Hughes 1994; Miller and Hay 1998). Thus, while geographical patterns suggest that low ambient water column concentrations are required to maintain healthy coral reef communities (prevent shift to seaweed dominance), the integrity of the complex trophic structure, including extremely high grazing rates, is perhaps more important (Szmant 1997). In the case of artificial reefs in tropical areas designed to emulate or enhance coral reefs, nutrient levels should be monitored to ensure they are not too high, leading to high planktonic primary production and a benthos dominated by filter feeders. However, management for the development of a complex trophic structure ensuring high rates and quality of grazing may be as or more important than minimal nutrient concentrations to successful establishment of a coral community.

In contrast, several studies in temperate waters, where high benthic plant biomass is important to the structural complexity and functioning of the reef ecosystem (e.g., kelp beds), have found that overgrazing is detrimental to community development of primary producers on artificial reefs and suggest that active management to reduce grazing intensity may be beneficial (Carter et al.

Table 4.1 Generalized Trophic Characteristics and Recommended Resource Characteristics to Evaluate for Different Artificial Reef Systems

Characteristic	Reef Community Type		
	Coral-dominated reef (oligotrophic)	Plant-dominated reef (mesotrophic)	Filter-feeder dominated reef (eutrophic)
Predicted geographic distribution	Tropical oligotrophic areas	Temperate areas	Either tropical or temperate coastal areas; often with strong terrestrial influence
Dominant primary producers	Microalgal turfs, Invertebrate/zooxanthellae symbionts (e.g., hermatypic corals*)	Benthic macrophytes (e.g., kelps and other large seaweeds)	Phytoplankton Note: detrital resources are also important
Major nutrient sources	Internal recycling	Advection (e.g., upwelling)	Advection (e.g., upwelling and/or coastal inputs)
Major (lower level) trophic links	High diversity: Herbivorous fishes consuming benthic algae and sometimes off-reef plants (e.g., seagrasses)	Intermediate diversity Some direct herbivory on macrophytes and epiphytic plants, some planktivorous fishes	Sessile invertebrates filter-feeding on planktonic resources (live plankton and detrital particles), fouling invertebrates preyed upon by mobile invertebrates and fishes, herbivory minimal
Relative intensity of predation and herbivory	Very high	Moderate	High
Important resource characteristics to evaluate	Light; water column nutrients (should be low); sediment nutrients; possibly fluxes and internal recycling processes	Light; dissolved water column nutrients (should be moderately high); possibly off-reef prey resources (infauna); growth/productivity of macrophytes	Planktonic resources including water column chlorophyll; suspended organic particulate concentrations
Other management concerns **	Coral colonization — possibly fostered by enhancing grazing intensity	Kelp colonization — possibly fostered by preventing overgrazing	Preventing overgrazing and space dominance by unpalatable species

Note: Generalizations based on hypotheses and results from Birkeland (1988, 1997); Hay (1991); Miller (1998); and others.

* Much of the carbon fixed by corals is converted to skeletal material (calcium carbonate), not organic matter per se and thus never enters the trophic chain even as detritus. Thus, some authors have partitioned the primary production of coral reef systems into "bioconstructional" vs. "trophic" pathways (Done et al. 1996). The bioconstructional pathway may be dominant in coral reef systems, but is not well represented in the other types of systems.

** These issues based on results from studies including Birkeland and Randall (1981); Carter et al. (1985b); Patton et al. (1994).

1985b; Jara and Céspedes 1994; Falace and Bressan 1997). Thus, the trophic structure projected for an artificial reef in a given geographic area also will affect the consideration of more intensive management methods during artificial reef development (Carter et al. 1985b).

4.3.3 Substrate Effects on Artificial Reef Primary Production

Aside from availability of nutrients and light, the community of benthic organisms will be greatly influenced by the nature of the substratum itself. The main physical characteristics of the substratum, influencing the settlement of fouling species are: surface texture, slope, outline (surface shape), color, and light reflection (Relini 1974). Sloping surfaces and optimal surface texture may enhance algal colonization by aiding spore adhesion and reducing sediment accumulation, which is a significant limiting factor for algal colonization in sites with a high regime of sedimentation (Ohgai et al. 1995; Falace and Bressan 1997). Thus, structure shape is considered highly important. Reefs with oblique or almost vertical sides such as pyramids are considered most effective for algae as they provide varying degrees of light, temperature, and other chemical/physical conditions that may be exploited (Bombace 1977, 1981; Akeda et al. 1995). Also, because of the "border effect," surface breaks (e.g., corners and edges) can be critical zones for benthic settlement. To a certain extent, the reef's productivity is a function of the total surface area exposed to the water and available for the settlement and growth of benthic organisms (Riggio 1988; Parchevsky and Rabinovich 1995).

4.4 PURPOSES AND STRATEGIES OF ASSESSMENT

As described in the introductory chapter of this book, there are several levels of complexity at which information concerning nutrients and production of artificial reef systems can be gained. These "Types of Assessment" are intended as a guideline for determining practical strategies for evaluation at reef sites.

4.4.1 Types of Assessments

It is much more difficult to assess the nutrient dynamics of an artificial reef and its environs than it is to determine fish densities because the analytical procedures are more involved, require more expensive laboratory equipment, and require a greater level of technical training. Therefore, nutrient measurements and flux studies are not recommended tasks for routine monitoring or assessment of artificial reefs unless qualified (M.S. and Ph.D. level) geochemists or chemical oceanographers are involved. Studies of plant and biofouling community composition are reasonably straightforward, but those of primary and secondary productivity require more effort. There is a range of complexity of general disciplines of study, within which differing levels of assessment may be pursued.

The first type of nutrient/production assessment is descriptive: what are the concentrations of nutrients at the reef site; what kinds and amounts of plants and fouling invertebrates live on the reef; and how do these factors vary over space and time? The second type of assessment is comparative and process oriented: what are the sources, sinks, and flux rates of nutrients to and from the reef site, how are they incorporated into reef organisms, and what are the pathways of transformation and recycling within the reef system; what are the growth rates of the reef plants and biofouling assemblages; what is the fate of the primary and attached organism production (i.e., rate of transfer into higher food-web groups)? A third type of assessment is integrative: what are the factors that affect rates of primary production; what site or reef design criteria might enhance nutrient flux and primary plus biofouling production of an artificial reef area. These types of assessments and relevant questions and methods related to nutrients, primary production, and associated assemblages are summarized in Table 4.2.

We would like to emphasize that this interpretation of Type One studies differs somewhat from the definition given in Chapter 1. One-time or "snapshot"descriptions (as described in Chapter 1) of these biotic characteristics (especially water column characteristics) are virtually useless in characterizing the nutrient regime of an artificial reef because these factors are extremely temporally variable (e.g., an order of magnitude on the scale of a single day). Generally speaking, benthic organisms display slower dynamics. A "snapshot" sampling of benthic communities might provide much more useful information. A true description of nutrient regime, however, will require sampling over various temporal scales (tidal cycles, seasons, during and after episodic events such as upwelling or storms).

Table 4.2   Types of Information for Nutrients, Primary Production, and Fouling Organisms

Type One — Description

Questions to ask:
What are the concentrations of nutrients (light levels) at the reef site? What kinds and amounts of plants and fouling invertebrates live on the reef? What are the patterns over time and space?

Techniques to use:
Collection of water or sediment samples (analyzed by professional consultants or commercial laboratories); underwater light meters, photographs; algal species (or functional form) checklists; scrapings of algae and invertebrate growth on reef surfaces or settlement plates.

Type Two — Processes

Questions to ask:
What are the sources, sinks, and flux rates of nutrients to and from the reef site? How are they incorporated into reef organisms? What are the growth rates of the reef plant and fouling assemblages? What are the factors that affect rates of primary production?

Techniques to use:
Hydrographic data collection; sediment, reef, and water column nutrient analysis and flux rates; nutrient transformations. Incubation measurements for primary production; settlement panels, with cage manipulations. Phenological studies of algal succession.

Type Three — Integration

Questions to ask:
What reef site criteria may enhance nutrient flux, primary and secondary production of an artificial reef?

Techniques to use:
Incorporation of artificial reef studies from different areas; experimental studies designed to test how different factors influence nutrients and productivity.

Higher levels of assessment usually will provide deeper insight into reef function. Measurements of nutrient concentration or standing stock (snapshot description) are less useful than measurements of nutrient fluxes or the rate of supply and use of nutrients in the system (comparison between inputs and outputs) because concentrations reflect the balance between supply and utilization. For example, nutrient concentrations may be low in spite of high rates of supply if the nutrients are being removed by benthic algae or phytoplankton at a similarly high rate. Measurement of nutrient fluxes involves identifying and quantifying all the potential sources of nutrients to the system, which can be difficult at best and requires expensive ancillary hydrographic work. Furthermore, both the water column and the sediments may be important as nutrient sources. Type One studies will assess patterns in reef characteristics, but Type Two or Three studies are necessary to determine the processes that are causing these patterns. In other words, a Type One study might answer the question, "Is this reef meeting its objective?" (e.g., establishment of a restored kelp bed), but Type Two and Three studies are required to answer the question, "Why?" or perhaps more importantly, "Why not?"

For example, Szmant and Forrester (1996) conducted an excellent Type One study describing temporal and spatial patterns of nutrient standing stocks throughout the Florida Keys. They conducted extensive sampling of water column and sediment nutrients over tidal and seasonal samplings. Despite the high level of effort represented in this study, it represents only a description of pattern. Relini et al. (1994) include Type One descriptive data on water column nutrient concentrations for an artificial reef site in the Mediterranean Sea, including mean, maximum, and minimum concentrations.

Many good ecological studies include multiple types of assessments within a single study. Stimson et al. (1996) sought to determine the factors that determine the standing stock of a tropical green seaweed in Kaneohe Bay, Hawaii. Their study included Type One descriptive data on nutrient levels and other physical characteristics (irradiance and temperature) and on the standing stock of the seaweed in various habitat types. After describing these patterns, Stimson et al. (1996) designed Type Two comparative experiments (measuring algal growth rates under manipulated temperatures and nutrient levels) and Type Three correspondence analyses (seasonal comparisons of algal growth rates *in situ* with seasonal patterns of temperature, nutrient availability and herbivory) to ascertain the relative importance of irradiance, temperature, herbivory, and/or nutrient levels in determining the distribution of the seaweed.

4.4.2 Framework and Approach

Despite the technical challenges, collection of basic nutrient data (concentrations of major inorganic nutrients and chlorophyll) is a highly recommended part of predeployment site assessment measurements even if the study has to be contracted out to professional consultants or samples have to be sent to a commercial laboratory for analysis. These data, together with predeployment assessment of benthic plant and phytoplankton communities (see Section 4.5), can be used to develop a general picture of the nutrients and productivity of the intended reef site. Concerns and protocols for a predeployment nutrient study would be the same as those described in more detail for post-deployment studies. Postdeployment nutrient monitoring would only be recommended for specialized reef projects that require nutrient data as part of a more extensive research program.

There are numerous approaches to measuring plant and fouling biomass (wet or dry weights of individual species or of groupings of functionally similar organisms), species composition (numbers of individuals or percent cover by major species), diversity (number of species per number of individuals, or any of several more sophisticated diversity indices, Peet 1974), and primary productivity (reviewed by Lieth and Whittaker 1975; Holme and McIntyre 1984; Schubert 1984; Littler and Littler 1985). The classic approach to the study of benthic macroalgal communities consists of creating a checklist (inventory of all species of flora present in the studied area). This type of descriptive study should be carried out with a phenological approach during all seasons (time-series data) and over several years to create a temporal description of presence or absence of all species in the species pool at the artificial reef site, which may lend insight into species associations. Descriptions of benthic community dynamics, diversity, and/or productivity can be compared between the artificial reef and nearby natural reefs to evaluate functional equivalence of an artificial reef being used to mitigate for loss of natural reef habitat. Quantitative measurements of benthic plant standing stock and primary productivity under various physical conditions (light levels, temperature, and current flow) can provide predictions of habitat and food resources for fish species of interest over seasonal cycles.

The study of benthic plant communities on an artificial substratum can be carried out by describing the floristic and vegetational components (structural aspects, Type One assessment) or by assessing the role of algae in the marine ecosystem (functional aspects, Type Two or Three assessment). Whenever a study intends to go beyond a simple taxonomic checklist of the species present, it is important to satisfy the following criteria in order to be able to evaluate reef function and possibly discern cause-and-effect relationships:

- Representative and sufficient sampling — The samples need to be representative of the habitat in general. For a representative sampling program a minimum sampling area has to be determined in order to minimize experimental error, and a sufficient number of replicate samples is necessary for meaningful statistical analysis. Chapter 2 deals with these issues of obtaining a representative sample. Some methods are described below for determining an adequate sample size to describe community structure.
- Comparable methodology — Techniques must ensure that biotic and abiotic samples coincide on a spatiotemporal level and that comparability with other studies of systems on both natural and artificial substrates is maintained, both in methodology and data format.

Fulfilling these conditions assures the best possible representation of the environment on which one can test the working hypothesis at the end of the study. The following descriptions, as well as

Table 4.3   Examples of Assessments Described in the Terminology of Chapter 2

Process	Characteristic or Measured Variable(s)	Sample Unit	Derived Variables	Parameter Estimates
Physico-chemical site description	Water column nutrient concentrations	Volume of water (e.g., 10 ml sucked into a syringe)	N:P ratio nutrient flux (mol $\cdot$ l^{-1} $\cdot$ h^{-1})	Mean [Nitrate] Mean [TP]
Primary production	Oxygen concentration	Volume of water (e.g., 0.5 l in an incubation chamber)	Rate of change of [O$_2$] (e.g., μg O$_2$ $\cdot$ l^{-1} $\cdot$ h^{-1})	Mean rate of change of [O$_2$]
Primary production	Standing plant biomass	Area of reef surface (e.g., 0.5 m^2 quadrat)	Rate of change of standing plant biomass (e.g., g 0.5 m^2 $\cdot$ d^{-1})	Mean rate of change of standing plant biomass
Primary production	Individual plant mass	Individual plant	Growth rate (e.g., g $\cdot$ d^{-1})	Mean growth rate
Herbivory	Standing plant biomass	Transplanted sprig of algae	Rate of loss of standing plant biomass (e.g., g $\cdot$ h^{-1})	Mean rate of loss
Benthic community structure	Number of sessile benthic species and abundance of each	Area of reef surface	Species diversity (e.g., H')	Mean diversity Mean barnacle density

the guidelines in Chapter 2, provide means to assure representative, sufficient, and comparable data. Table 4.3 gives some examples from the following measurements and how they relate to statistical terms defined in Chapter 2 to aid in the application of the design principles to the assessment methods that follow.

4.5   ASSESSMENT METHODS

After the reef characteristics to be assessed have been determined based on the specific reef objectives and biogeographical considerations (see Table 4.1), the methodologies must be determined. The following general considerations should provide for a robust, efficient, and scientifically valid assessment plan. Specific methods and measurements are described in Section 4.5.2.

4.5.1   General Considerations

4.5.1.1   Sampling Considerations for Nutrients

There are three major nutrient sources for artificial reef communities: (1) water column nutrients from waters flowing by the artificial reefs; (2) sediment nutrients fluxing out of the benthos surrounding the reefs; and (3) nutrients within the reefs themselves, originating from nitrogen fixation or, predominantly, from microbial regeneration of detrital material (animal feces and plant detritus) entrapped within the reefs. While this third nutrient source may be extremely important, especially in tropical oligotrophic reefs (see Table 4.1), the methods for this level of assessment (i.e., processes of nutrient regeneration) are significantly more involved and are beyond the scope of this discussion. Only the first two are considered here.

4.5.1.1.1   Water Column Constituents: Nutrients, Chlorophyll, and Particulate Organic Matter (POM) — Water column nutrient, chlorophyll, and POM concentrations vary both temporally and spatially (e.g., Andrews and Muller 1983). Temporal variations may be more or less regular, such as those associated with tides and seasons, or associated with episodic events such as storms. Spatial variation is generally related to proximity to sources, such as land (runoff), rivers, sewage outfalls, sediment reservoirs, or oceanic nutrient sources, and the hydrography of the area. It is important to quantify these temporal and spatial variabilities for the concentration measurements to be of any use. This requires sampling at different scales of time and space. Daily sampling over complete annual cycles is ideal but totally impractical.

Resources and logistics will to a large degree place limits on the frequency and replication of sampling. The first step is to determine how much field-time can be afforded (e.g., days per year). The next step is to try to rank the processes thought to have a major control on nutrient distributions, and use this ranking to schedule sampling days. For example, if tides are expected to be important, then sampling should be scheduled for high and low tides, and possibly during spring and neap tides. Sampling stations should be selected taking into account suspected or known nutrient sources and current patterns that affect the redistribution of the nutrients. The number of stations sampled, as well as the number of depths per station and replication per station, will usually be determined or limited by logistics and resources (e.g., the size of the area to be studied or the number of water samples that can be analyzed). It is beyond the scope of this chapter to provide one standard sampling scheme that can be applied to all projects. Suffice it to say that some *a priori* knowledge is needed, as well as a good dose of common sense.

Physical parameters such as current speed, direction, and duration are needed to estimate flux rates and will help to characterize both the source and effects of the measured nutrient load. (See Chapter 3 for associated methods.)

4.5.1.1.2   Sediments — Sampling considerations are different for sediment nutrients. Spatial variability is usually much greater than temporal, although there may be seasonal trends related to variation in the rates of organic material input to the sediments. Because of this reduced temporal variability and the fact that analyses of sediment samples are more laborious (compared with water samples), they are usually done only two to four times per year. But more replication is needed per sampling (three to six cores per station, depending on sediment heterogeneity determined for each site). These stations could be placed randomly among the reef units or with a systematic sampling scheme (see Chapter 2, Figure 2.4). Sediment cores are analyzed for both pore water (readily available to plants) and total nutrients (nutrient reservoir) to estimate the contribution of benthic recycling of nutrients to the water column and reef interstices.

Another way to study sediment nutrient dynamics is to measure fluxes of nutrients out of the sediment with an *in situ* flux chamber. Specially designed chambers are inserted into the sediment, enclosing a portion of the substrate and overlying water. Water samples are withdrawn from the chamber periodically for the measurement of nutrient and oxygen concentrations, and from changes in these over time, nutrient efflux rates are calculated. Similar chambers also can be used for measurement of respiration and photosynthesis rates on hard substrates (see Section 4.5.2.5.3).

For either type of nutrient source, a pilot study is recommended to get an idea of the concentration ranges that will need to be accommodated (analytical techniques can vary depending on whether high or low nutrient ranges are being measured) and to get estimates of the temporal/spatial variabilities (these will determine the necessary replication in sampling stations and sampling frequency).

4.5.1.2   Sampling Considerations for Primary Producers and Biofouling Community

As with nutrients, spatial and temporal variation in algal and invertebrate biomass (standing stock), diversity, and productivity must be considered to keep them from obscuring patterns of interest. Successional stage and seasonal patterns in temperature, photoperiod, and light intensity (well documented even for tropical waters; Harris 1986; Larkum et al. 1989) yield variation in species abundance and primary production. Spatially, plant and biofouling community composition and productivity can vary greatly due to orientation with regard to currents, amount of light (depth and orientation of substrate), and proximity to nutrient sources. Many investigators use randomization techniques for selecting sampling stations, but then have a difficult time explaining the high variance (scatter) in the data. When microhabitat variability is great, then a stratified sampling approach may be preferable. For example, horizontal and vertical surfaces or windward and leeward sides of a reef should be treated separately as far as selecting sampling stations, and there should be an adequate number of samples within each distinct habitat type.

Finally, there are many potential plant groups contributing to the primary production of any system (microalgae, algal turfs, and macroalgae), and the methodologies and sample units used for their study must be suited to the size class and spatial distribution pattern of the group. In most cases, several different methods need to be applied in the same study in order to adequately characterize the various groups. For example, photographic techniques are adequate for estimating the abundance of larger macroalgae, but not of microalgal films. For the latter, chlorophyll measurement of scrapings is a better method of estimating biomass (see next section).

Methods for characterizing species composition and abundance are fairly straightforward, but require skill in identifying algal and invertebrate species, which can be very difficult and extremely time consuming. For most purposes, limiting the identification to higher taxonomic groups (genus, family, or class) or functional form classifications (e.g., Littler and Littler 1980; Steneck and Diether 1994) is sufficient. Indeed, if the benthic community composition is not of interest per se, production can be estimated from the biomass of the entire algal community without any taxonomic division at all (e.g., methods of Falace et al. 1998). This allows more effort to be dedicated to analyzing a larger number of replicates.

The amount of replication necessary to adequately characterize the primary producer and biofouling community composition and biomass will be a function of the degree of heterogeneity in the community and of the size of the sample unit (e.g., quadrat). If species composition is of interest in the assessment, the more diverse and heterogeneous the system or the smaller the quadrat, the larger the number of replicates needed (which amounts to a larger area of substrate sampled). One technique used to evaluate whether the number of replicates, for a given sample unit size, is adequate to estimate species composition is to construct a species vs. area curve (Holme and McIntyre 1984). This is done by plotting the cumulative area of the individual samples on the X-axis vs. the cumulative number of species (or functional forms) identified on the Y-axis. The resulting curve should increase initially as replicates are added, but level off (asymptote) at the point of sufficient sampling. After the asymptote has been reached, there is no benefit to analyzing more samples. A similar curve can be generated to assess adequacy of replication for measuring biomass or percent cover. In this case, the variance or standard deviation is recalculated as each new replicate is added; these values are plotted on the Y-axis against the number of replicates included in the estimate (the X-axis). Here, the variance should be higher initially and then decrease as more replicates are added. Adequate replication is demonstrated by a leveling off of the curve before the last replicate is added.

As with the nutrients, a pilot (Type One) study is recommended, using quick survey techniques such as line transects or photographic methods that utilize a range of sample unit size (i.e., area covered by photograph or length of line transect), to provide a broad overview of the types and quantities of plant and fouling organisms that occur on the study reef. This information can be used to select the specific methodologies and sample units to be used in the monitoring or assessment studies. For example, a pilot study by Littler et al. (1987) determined that systematic samples at fixed intervals along a transect did not differ significantly from samples that were taken at mechanically randomized placements (see Chapter 2). Based on numerous background studies, these authors also recommend a 0.15 m^2 rectangular quadrat for sampling of tropical reef algal communities (Littler et al. 1987). Literature surveys can be used to learn as much as possible about seasonal patterns of growth and distribution of the dominant organisms, in order to design temporal sampling schedules. General references for sampling considerations and sample analysis of benthic organisms include Boesch (1977); Holme and McIntyre (1984); Littler and Littler (1985); and Andrew and Mapstone (1987).

4.5.2 Techniques

This section summarizes sampling and measurement methods for nutrient, primary producer, and biofouling assemblage studies. Some of this material is based on technical literature, but much of it is drawn from scientific experiences and from unpublished information provided by experts. The intent is to provide guidelines, while pointing the way to handbooks and more exhaustive information that fills many more pages than available in this book.

4.5.2.1 Water Column Constituents — Sample Collection and Preparation

In practice, a water sample is divided. One portion is analyzed for dissolved inorganic nitrogen (DIN), dissolved inorganic phosphorus (DIP), and sometimes Si; a second portion is digested for analysis of total N and P (the amounts of dissolved organic N and P in each fraction are then calculated by subtracting the amount of inorganic N or P from the total N or P values), and particulate matter retained on the filters (from a known volume) can be used to determine concentrations of POM and chlorophyll in the sample. Figure 4.3 shows an example of how a water sample would be subdivided for these analyses. In this example, the portion of water used for the dissolved inorganic nutrient analyses was filtered before analysis. Aside from collecting suspended particles for analysis, filtration of water for analysis has two purposes: (1) to remove microorganisms (bacteria and phytoplankton) that can cause changes in the nutrient content of the water sample; and (2) to remove particulates that cause error in the calorimetric analyses. Filtration is always recommended for turbid water samples and those suspected of having high levels of biological activity. However, water samples from low nutrient/low productivity waters are best analyzed without filtration, since filtration increases the risk of contaminating the sample.

As DIN and DIP can be rapidly incorporated into phytoplankton biomass, water column chlorophyll content is generally also measured in conjunction with nutrient sampling. Chlorophyll data can provide a framework to interpret nutrient results (e.g., how much of the total organic matter is live vs. detrital). Water column chlorophyll is measured by extracting with acetone the chlorophyll from phytoplankton cells collected by filtering a known volume of water. Procedures for chlorophyll analysis also are standardized and described in Parsons et al. (1984b).

Similarly, it may be beneficial to quantify the concentration of POM in the water column as an estimate of food availability to filter feeders. A precombusted, preweighed filter containing particles from a known volume of water can be combusted in a muffle furnace to determine the ash-free dry weight (i.e., the organic content) of the particulate matter.

To effectively conduct nutrient analyses on water samples: (1) collect the samples without contaminating them with exogenous nutrients; and (2) maintain the samples without degradation until they can be analyzed. Thus, all sampling equipment and sample storage bottles must be precleaned with acid and rinsed thoroughly with high quality (>16 MΩ) deionized water. Also, ammonia-based cleaners, soft drinks that contain phosphoric acid, fingers (human skin exudes ammonia), cigarette smoke, engines fumes, etc. can contaminate samples. Extreme care in preventing contamination is especially critical when sampling in tropical oligotrophic waters.

After collection, the concern becomes one of preventing postsampling changes in the concentrations of nutrients within the sample bottles. Phytoplankton within the water can take up the

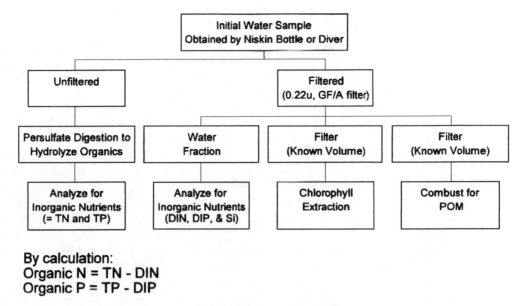

By calculation:
Organic N = TN - DIN
Organic P = TP - DIP

Figure 4.3   Flow chart showing the suggested partitioning of a water sample for multiple nutrient analyses. Some investigators also perform the persulfate digestion on filtered water in order to separate particulate organic N and P from dissolved organic N and P. N = nitrogen; P = phosphorus; DIN = dissolved inorganic N; DIP = dissolved inorganic P; Si = silica; TN = total N; TP = total P.

nutrients, while microorganisms and zooplankton can excrete nutrients. The higher the concentration of organisms or detrital particles within the water sample, the more severe is the problem. Optimally, the samples should be subsampled and filtered soon after collection (but see caution above for clear, low nutrient waters), and the chemical analyses should be performed within hours of collection. In practice this is often difficult, and samples must be kept for some time before analysis. In this case, the sample bottles should be kept on ice in the dark until return to the laboratory. Filtration, if needed, should be done as soon as possible. If samples must be kept for any length of time before chemical analysis, they should be frozen. Other commonly used, but less acceptable, techniques of preservation are acidification with hydrochloric acid (below pH 2) or addition of mercuric chloride. Analyses for total nutrients (TN and TP) are less affected by the degradation problem, since in essence, this technique does not discriminate between the various forms of N and P. Filters with particulates for chlorophyll analyses should be immediately wrapped tightly in aluminum foil to exclude air (to minimize photo-oxidative degradation of the pigments) and kept frozen until analysis.

Water samples can be collected with remote samplers (e.g., Niskin bottles) deployed on a line such that several depths can be sampled at the same time. Recommended sampling depths are dependent on water depth, and degree of water column stratification. If water depth is less than 3 m, generally only one depth is sampled; when more than 3 m, samples are taken 1 m above the bottom and 1 m below the surface. If water depths exceed 20 m (~65 ft), intermediate depths should be sampled as well. Sample containers (smaller precleaned polyethylene bottles) should be rinsed three times with water from the Niskin bottle before filling. Alternatively, water samples can be collected by divers, who can use air from their regulators to purge and rinse the sample bottles three times before filling. Depth of sample collection, sampling time, tidal phase, water clarity and color, temperature, salinity, and sea state should be recorded or measured for each sample (see

Chapter 3). Methods for pretreating sample bottles are described in Parsons et al. (1984b) and APHA (1989).

4.5.2.2 Sediments — Sample Collection and Preparation

Sediment samples can be collected by divers using short (20 to 30 cm) corers made from 7 to 8 cm diameter aluminum core pipe or plastic core liner, or taken with a remote corer or grab. Within safe diving depth, divers offer the best option for precise sample site selection, especially near an artificial reef. Once inserted into the substrate, corers should be capped and carefully dug out so as not to disturb the contents of the corer. A stopper should be inserted into the bottom of the corer before it is pulled out of the sediment column. It is important to keep the core upright, especially for coarse sediments, and to make sure both ends of the corer are well stoppered (to prevent leaking of pore waters). One method to minimize leaking is to keep the cores totally submerged in a bucket of seawater until they are analyzed. Depending on the study objectives and design, either pore water or total sediment, or both, will be analyzed for nutrients. A description of all of the forms of nutrients that have been analyzed in sediment geochemistry studies (pore water, sorbed, dissolved organic, particulate organic, mineral forms, etc.) can be found in Berner (1980).

For pore water sampling, sediment is extruded from the corers 1 to 2 cm at a time, collected (sliced off), and the interstitial waters collected by centrifugation, filtration, or "squeezing" (Mudroch and MacKnight 1991). If the sediments are collected from an anoxic area, pore water extraction should be done in a glove bag under nitrogen or argon gas (to prevent oxidation of N forms and precipitation of P forms). Pore waters generally have high nutrient concentrations (10 to 100 times higher than water column) and thus fewer problems with storage before analysis. Most need to be diluted before chemical analysis, both to increase the volume of water for analysis and because their concentrations are too high for most analytical methods.

For total nutrient (N or P) content, sediments are extruded as above, known volumes of sediment are wet weighed and dried at 100°C to constant weight (the difference between wet and dry weights is used to calculate sediment porosity, a value needed to estimate nutrient efflux rates). Dried sediments can then be ground and used for measurements of total N and P as described below. Excellent reviews of concerns and approaches for studying sediment nutrient concentrations and fluxes can be found in Berner (1980); Mudroch and MacKnight (1991); and Agemian (1997).

4.5.2.3 Chemical Analyses

Nutrient studies minimally measure for the concentrations of the dissolved inorganic forms of N (DIN), P (DIP, also known as SRP: soluble reactive phosphorus), and Si, and optimally for all forms described earlier (see Section 4.2.2). The types of analyses needed or recommended, however, will depend on particular circumstances, and many studies only measure forms of N and P. There are specific standardized colorimetric analyses (in which reagents added to a water sample cause the development of a colored substance) for the individual dissolved inorganic nutrients that can be done manually or by autoanalyzer. Numerous manuals, including USEPA (1983); Parsons et al. (1984b); CMEA (1987, Russia); NIH (1987–88, India); APHA (1989); and Crompton (1992), provide standardized recipes for all the commonly measured nutrient analyses including nitrite, nitrate, ammonium, orthophosphate, and reactive silica. Several brands of automated chemistry analyzers are available commercially, and they provide standardized methodology with their instrumentation. Although these are "cookbook" methods for the most part, to perform them properly requires a background or specific training in analytical chemistry.

The dissolved organic nutrient forms usually are determined by treating water samples with a strong oxidant to break down (digest) organic molecules to release their N and P components. They are then analyzed using the same techniques for the dissolved inorganic forms (Parsons et al. 1984b).

Care should be taken to select techniques appropriate for the concentration range at hand. For example, many of the techniques acceptable for freshwater (such as the standard USEPA methods) are too insensitive to be used in subtropical marine waters where concentrations are low. A research quality spectrophotometer is necessary to measure absorption values, and for large numbers of samples, an automated system is advisable. For total N or P analyses, there are several wet digestion techniques available, some based on chemical oxidation by persulfate or by sulfuric acid (Kjeldahl), or by UV oxidation. The persulfate and UV techniques are generally less efficient at breaking down particulate organics than dissolved organic compounds; the Kjeldahl technique is effective but slow and cumbersome. In samples with a lot of refractory detrital material, the particulate fraction is often analyzed by elemental analysis of particulates filtered from known volumes of water sample. Several brands of elemental analyzers are available, but they are expensive and require experience and work to keep running properly. An important requirement for working with low nutrient concentrations is a source of high quality deionized water to use in making up reagents and standards. For best accuracy with an autoanalyzer, a source of low nutrient seawater (e.g., Sargasso Sea) is needed to make up standards.

4.5.2.3.1 Quality Control — The general importance of quality control was discussed in Chapter 2. The major concerns of quality control and quality assurance procedures for nutrient measurements relate to the problems discussed above: avoidance of contamination and degradation, and at the analytical end, reproducibility of results. Records should be kept of the blanks and standard factors for each run. Deviations from the norm for either should be investigated since they may indicate bad deionized water, reagents, or standards. Sample bottles and glassware used for nutrient sample storage or analysis should be dedicated to this purpose alone. Efforts should be made to ascertain that the techniques are performing properly and to conduct intercalibrations with other experienced laboratories. If samples are sent to a commercial laboratory for analysis, unidentified (blind) standards should be inserted in with the unknowns, and all samples should be run in duplicate or triplicate.

4.5.2.4 Plant and Fouling Invertebrate Biomass Measurements

Biomass of plants or fouling invertebrates can be determined by scraping off all attached organisms from a known surface area of substrate (natural or artificial) and measuring the volume of organisms, wet weight, dry weight, and/or ash-free dry weight (AFDW, usually estimated as loss on ignition). The advantage of AFDW is that it estimates only organic matter, as bioconstructional materials such as shells or tubes do not combust, but remain as ash. Otherwise, separating organic tissues from shells and tubes by hand may be difficult to impossible. The analyses can be done on algal and animal samples sorted by species, or on groups of species sorted into structural or functional groups.

The biomass of plants also can be estimated by measuring the amount of chlorophyll in weighed subsamples of the scrapings. This is especially useful for measuring the biomass of diverse microalgal films and turf communities. Chlorophyll analyses are done by grinding known amounts of plant material (e.g., a weighed subsample of algae) in 90% acetone and then reading the extract's absorbance or fluorescence, depending on concentration (see methods in Parsons et al. 1984b). The grinding and extraction should be done in the dark to prevent photodegradation of the pigments.

Specifically, Falace et al. (1998) have developed and rigorously evaluated a standardized protocol for determining biomass of benthic algal communities. Consistently applied, it could help resolve problems of inconsistency and incomparability in data on artificial reef primary production. This protocol emphasizes the importance of repeated washing of sediments from the algal tissues, especially understory turf communities, before biomass determination since sediments can bias algal dry weight determinations by 50 to 82%. For example, six repeated washings were necessary to remove sediment from algal samples collected from coastal habitats in the Gulf of Trieste (northern Adriatic Sea). Falace et al. (1998) also suggest that algal tissue samples may be ground (homogenized, 30 min for 4 to 32 g fresh weight samples) before accurate wet weight, dry weight, and ash-free dry weight determinations. Homogenization does not significantly alter these biomass estimates and diminishes the variation between aliquots partitioned from the same sample for biomass and extraction for chlorophyll or other biochemical assays (e.g., protein or polysaccharide content). However, if chlorophyll determinations are to be made, this entire protocol (rinsing and homogenizing) must be performed at a low temperature (4°C) in order to avoid degradation of the pigments. If facilities are not available for extensive processing at low temperature, it may be best to freeze the subsamples for chlorophyll separately and grind them in small amounts on ice while processing the samples for biomass at room temperature.

Quantitative recovery of areal samples (either scraped or plucked) underwater can be difficult if there is any current or surge. An alternative approach is to use settlement plates made of a material similar to that of the artificial reef (see Reimers and Branden 1994 for an excellent example conducted in Australia). The plates are attached to the reef at the beginning of the study (soon after the reef is deployed) in a way that is easy to retrieve (e.g., bolted). They are then recovered at designated sampling times, placed in sealed plastic bags underwater, and brought to the surface for examination under a dissecting microscope (to quantify small organisms) and/or scraping for biomass or chlorophyll determination. If this approach is used, then care must be taken to attach enough settlement plates for adequate replication in all the preidentified microhabitats for the duration of the project. While new settlement plates can be attached later in the study, the growth on them will be at a different successional stage than that on the original plates (and reef substrate). This approach can be used to assess explicitly the effect of timing of placement of artificial reefs on the colonization and successional processes (Reimers and Branden 1994). Settlement plates are a particularly powerful tool for experimental hypothesis testing since they can be so easily manipulated (e.g., caged and placed in chambers for direct primary productivity and respiration measurements; see Case Study, Section 4.6.2 and Figure 4.4).

Photographic methods also have been used to estimate changes in percent cover of benthic organisms (Bohnsack 1979; Littler and Littler 1985). In this approach, 35 mm photographic slides taken of the substrate community are overlaid with clear acetate templates marked with 100 randomly selected dots. Organisms under each dot are identified, and the data are used to estimate percent cover by each species. Such methods are useful for dominant and large organisms, but will of course underestimate small or understory components of the community.

Figure 4.4 Caged and uncaged settlement plates to assess benthic plant production on an artificial reef unit used in the Palm Beach, Florida study (Szmant 1993). Photograph courtesy of A.M. Ecklund.

4.5.2.5 Primary Productivity Measurements

Three types of primary productivity measurements should be considered in artificial reef monitoring studies — on the reef surfaces themselves, on surrounding benthos, and in the water column bathing the reef system. Approaches to measuring primary productivity vary according to degree of difficulty and amount of training needed, as well as the level of complexity of information the technique yields. Different methodologies include biomass change measurements, free water methods, and incubation experiments. Selection of methods to be used should be based on project objectives and resources (personnel and funding).

4.5.2.5.1 Biomass Change — Technically, the simplest method of determining primary (or fouling organism) production is to measure plant (or animal) biomass changes over time on either cleared substrate or artificial experimental panels attached to the substrate. This approach only provides an estimate of standing stock or net production after grazing. If consumption rates are high, as is the case in essentially all tropical coral reef systems (Hay 1991) and many other systems as well (e.g., Carter et al. 1985b; Patton et al. 1994), this method will greatly underestimate the true amount of benthic production. To estimate this loss to grazing and determine what Steneck and Diether (1994) call the "productivity potential" of the given environment, cages can be used to exclude most grazers from access to the substrate. Cages are attached over replicate substrates such as settlement plates, and the amount and type of biomass accumulated on caged substrates is compared to that on exposed substrates (Figure 4.4).

Unfortunately, since cages can introduce many types of artifacts (e.g., reduced light and water flow, providing refuges from predation for small mesograzers, such as small crustaceans or gastropods, with the unintended effect of increased grazing intensity inside of cages), great care is required in the design of a scientifically rigorous caging study (Hulberg and Oliver 1980; Steele 1996; Connell 1997). However, a few general practices will minimize the possibility of interpreting artifactual effects of the cage as effects of consumers. These include using cage control treatments (partial cages that allow entry to grazers, but impose similar restrictions of light incidence and water flow as the full cages) and a mesh size large enough to admit small predatory fishes such as wrasses and blennies to prevent accumulation of mesograzers in full cages (Lewis 1986). Another problem in implementing this approach is determining the duration of the growth interval. If it is too short, there may not be enough growth to get a good estimate, and if too long, self-crowding and competitive interactions between organisms may affect the growth rates in a manner difficult to explain. The correct duration of exposure at a given site will obviously depend on the colonization and growth rates of the species involved and must be "played by ear" or, alternatively, multiple samplings can be undertaken over a range of exposure times.

4.5.2.5.2 Free Water Methods — Free water methods measure primary productivity of whole communities or assemblages by measuring oxygen evolution or carbon dioxide uptake in a water parcel as the water flows between the upstream and downstream ends of the study system. This technique measures the entire community metabolism, benthic and planktonic, and is limited in use to highly productive, shallow, well-mixed waters. An advantage of this technique is that it is a measurement under natural conditions, with all parts of the system included. The disadvantages, however, are that there is a lack of biotic specificity, and there is little control over the system and its status. Furthermore, it can be difficult to determine exactly the path the parcel of water takes; usually a large dye spot or a small drogue are used to follow the water. Odum and Odum (1955) used this technique, and many others since have repeated and refined it (Marsh and Smith 1978). A unidirectional flow is necessary, so that the water mass is flowing from one station to another. In areas with restricted circulation, one station can be used with the same water mass resampled over time (Kinsey 1978). A floating instrument package that takes continuous measurements as it follows the water (marked with dyes or drogues) has also been used (Chalker et al. 1985).

Water samples for oxygen or carbon dioxide measurement are taken at time zero at the upstream end and at various time or distance intervals while tracking the dye or drogue. Oxygen concentration is measured by Winkler titration (Parsons et al. 1984b) or with an oxygen electrode (several brands available). Winkler titration is inexpensive but time consuming, whereas use of an oxygen probe is less bulky and faster, but requires care in temperature control and calibration of the electrode. Both methods can produce reliable and accurate results with minimal training.

The oxygen evolved over time or between stations can be converted to carbon fixed (gross production), if the photosynthetic quotient (the ratio between oxygen production and carbon dioxide fixation) is known (it is usually assumed to be 1.0 to 1.1). Carbon dioxide can be estimated indirectly by measuring pH if there is minimal calcification in the system (i.e., assuming no change in total alkalinity), since respiration causes a decrease in pH and photosynthesis causes an increase in pH. This approach is explained in detail in Smith and Kinsey (1978). Although the pH method of measuring carbon dioxide has the potential of being less accurate because of the numerous assumptions on which the calculations are based, its main advantage over oxygen measurements is that there is no need to know the metabolic quotient, and the potential for gas exchange with the atmosphere is less than with oxygen. Other more accurate techniques are available for the measurement of carbon dioxide, but they require expensive instrumentation and highly skilled technicians.

Light measurements must be made throughout the experiment, preferably with a quantum sensor, because short-term productivity rates are strongly dependent on light intensity at the time of measurement. These types of experiments should be done on sunny days when light is above saturation and between 10 a.m. and 2 p.m. when photosynthesis is at its maximum rate. Night measurements are also required for estimates of 24 h (daily) production.

4.5.2.5.3 Incubation Measurements — Incubation experiments are conceptually similar to the ones described in Section 4.5.1.1.2, but aimed at the organismal level. They involve putting algae or pieces of macrophytes in bottles and measuring oxygen evolution, or radiolabeled C-14 incorporation, by a known amount of biomass (or area of substrate, which combined with biomass determinations described above, can be converted to biomass). Larger incubation chambers have been used to enclose small areas of benthos inhabited by assemblages of organisms. Henderson (1981) and Hopkinson et al. (1991) have conducted incubation experiments of reef-flat benthos using chambers sealed to *in situ* natural reef substrate and also for communities growing in flow-through systems in the laboratory. Oxygen production and consumption have been measured on a freshwater artificial reef by means of light and dark plexiglass enclosures (Prince 1976; Prince et al. 1985). Artificial reef studies are especially suited since settlement plates of similar age and material as the artificial reef (as described above for quantification of biomass) can be placed in incubation chambers much more easily than enclosing natural reef substrate (Figure 4.5). A potential problem with this approach is limited water flow within the chambers with consequent reduced access to carbon dioxide. Variations on the method have attempted to address these problems by providing stirring within each chamber.

Several of these bottle/chamber experiments can be done during the course of a day, using light (net photosynthesis), dark (respiration), and control (only water in them) bottles. Water samples are taken at the beginning and end of each incubation and sometimes at intermediate times. The most sophisticated chambers include oxygen electrodes in each chamber so as to get continuous oxygen measurements during the incubation. As with the field studies, it is important to collect light data throughout the incubation. When oxygen is measured, net photosynthesis is calculated as the production (increase) of oxygen in the light bottle, corrected for any increase in the control; respiration is calculated as the depletion of oxygen in the dark bottle, also corrected for the control; and gross photosynthesis is calculated as net photosynthesis to which is added the amount of oxygen

Figure 4.5 Diver sampling from incubation chambers being used to measure primary productivity (i.e., oxygen production) of the benthic plant assemblage on settlement plates in the Palm Beach, Florida study (Szmant 1993). The same chambers were completely covered with black tape for incubations to determine respiration rates (i.e., oxygen consumption) of the assemblage. Photograph courtesy of A.M Ecklund.

consumed by respiration. Rates are normalized to the surface area or biomass within the chambers and expressed per some unit of time (hour or day). The oxygen can be measured by oxygen probe or Winkler titration as described above.

Radiolabeled C-14 can be introduced to the same types of bottles or chambers as are used for oxygen flux. Organisms are harvested after 6 to 24 h and their radioactivity measured. Total carbon dioxide must be measured or estimated in order to convert the radioactivity to total C fixed. Although this technique is the most sensitive available for use in low productivity water, it is more expensive and laborious and requires access to more sophisticated equipment. In most places, it also requires special government permits for using radioisotopes, and thus it is unlikely that this technique can be used except by research scientists licensed for such work.

Sometimes multiple approaches have been used in the same study to evaluate primary productivity. For example, Ferreira and Ramos (1989) combined incubations and biomass information to determine annual productivity for three species of macroalgae; Hatcher and Larkum (1983) scraped algae off previously cleaned coral heads and measured dry weights, AFDW, and chlorophyll concentrations of algae recolonizing the substrate.

4.5.2.5.4 Quality Control — Voucher specimens of all algal and invertebrate species identified in the study should be maintained and, if possible, their identity verified by taxonomic experts. The adequacy of replication should be assessed early in the study, even though in many cases the number of replicates that can be taken may be limited by time and personnel resources. It is important to include sufficient control chambers in primary productivity incubation studies to be able to correct for any oxygen generation or consumption by organisms in the water enclosed within the chambers. It is also important to determine that oxygen and pH meters are properly maintained and calibrated before making measurements, and records should be kept of calibration conditions (temperature and salinity for oxygen, temperature for pH). If a fluorometer is used for the chlorophyll measurements, it should be calibrated at least once per year against a chlorophyll standard and spectrophotometer. Light meters should be returned to the factory annually for recalibration.

4.5.2.5.5 Fate of Primary Production and Fouling Organisms — It also may be of interest to assess the fate of primary production (or fouling organisms) within the artificial reef in order to understand processes of transfer to higher trophic levels (harvestable fisheries) or to make functional comparisons to natural reef communities. There are various methods that can be used to assay the rate of consumption of primary producers and/or fouling organisms. Macrophyte or invertebrate tissues (possibly collected from other habitats) can be preweighed, exposed to artificial reef consumers for a set length of time, and then reweighed to obtain an estimate of the rate of consumption of biomass for comparison to natural reef areas. Often, different types of grazers (fishes vs. sea urchins) leave distinctive scars and thus the relative impact of different grazers can be assessed (Hay 1984; McClanahan et al. 1994). Also, underwater video or *in situ* observations can yield estimates of the consumption rates by grazing fishes of small turf or film communities that are not directly manipulable (No. bites · area^{-1} · time^{-1}). Lastly, by comparing the biomass accumulation inside and outside of grazer-exclusion cages (as discussed above for estimating primary production), an estimate of the cumulative consumption rate on an areal basis can be obtained.

4.6 EXAMPLES

Unfortunately, there is a paucity of scientific, peer-reviewed literature on functional aspects of artificial reefs such as nutrient regimes and lower trophic factors. Here we summarize the literature on nutrients and primary production on artificial (and some natural) reefs and the more extensive literature on benthic assemblage structure and succession on artificial reefs. Lastly, we describe in more detail an unpublished case study, which attempted to assess how nutrient regime and primary productivity influence benthic assemblage structure.

4.6.1 Previous Work on Reef Trophic Resources and Primary Production

Unlike studies of ecological function, studies from all parts of the world describing the benthic assemblage structure (colonization and succession) on artificial reefs abound. In addition to simple description of successional sequence (e.g., Carter et al. 1985a; Palmer-Zwahlen and Aseltine 1994), some of these studies evaluate the influence of various environmental factors, such as seasonal patterns (Bailey-Brock 1989; Jara and Céspedes 1994; Reimers and Branden 1994), current flow (Baynes and Szmant 1989), grazing (Hixon and Brostoff 1985; Relini et al. 1994), and turbidity (Pamintuan et al. 1994), on colonization and successional processes of artificial reefs. There are also several published studies that utilize biomass accumulation to evaluate the suitability of various materials for artificial reef construction (Hatcher 1995; Gilliam et al. 1995; Ohgai et al. 1995).

Very little information exists in the literature on functional aspects of nutrient dynamics and primary productivity of artificial reefs. Fang (1992) presented a theoretical model of artificial reef communities that aimed to evaluate artificial reef performance by relating fish productivity to densities of phytoplankton and epibenthic filter feeders. Rice et al. (1989) described the standing crop and primary production of giant kelp, *Macrocystis pyrifera*, transplanted to breakwaters at a harbor in order to evaluate the success of a habitat mitigation project, and found up to 50% of the kelp production was grazed by fish and invertebrates from the breakwater habitat. Falace and Bressan (1994) quantified the accumulation of algal turf cover to evaluate the influence of temperature and turbidity on primary production of an artificial reef.

There is also extensive literature based on studies of natural reef systems such as coral reefs (reviewed in several chapters of Dubinsky 1990; Hatcher 1988, 1990, 1997) and soft- and hard-bottom communities (Levinton 1982; Rumohr et al. 1987; Thompson et al. 1987; Zieman and Zieman 1989; Hopkinson et al. 1991) that can serve as background for how nutrient and productivity processes may function on and around artificial reefs. Much of this research has been summarized in the earlier sections of this chapter. Most nutrient studies have the purpose of either measuring the sources and rates of flow of nutrients through the system studied or of measuring the rates of transfer of nutrients between different compartments within the system (e.g., Tribble et al. 1988;

D'Elia and Wiebe 1990; Erez 1990). Productivity studies generally focus on measuring amounts of biomass (standing crop) of various primary producer groups, measuring the rates of their primary production, or examining the processes that affect rates of primary productivity (e.g., nutrient supply and grazing; Larkum 1983; Berner 1990; Erez 1990). The process of conversion of primary to secondary production should be an important aspect of evaluating artificial reef "success" when reef goals are focused on fishery production, but is rarely addressed. Even for these sophisticated studies of natural reef systems, few have related the nutrient dynamics and primary production to secondary production (e.g., Odum and Odum 1955; Atkinson and Grigg 1984; Grigg et al. 1984).

4.6.2 Case Study

One Type Two/Three (see Section 4.1) artificial reef study (Szmant 1993) will be described in some detail since it (1) is one of the few artificial reef studies that has attempted to quantify functional aspects of nutrient cycling, primary production, and the correlation of primary productivity with fish stocks; and (2) demonstrates some of the difficulties and pitfalls that can hamper even carefully planned research studies. This study (Szmant 1993) was conducted in the Atlantic Ocean off Palm Beach County in southeast Florida, and was designed to test the hypothesis that high artificial reef structural complexity would foster increased nutrient regeneration and, hence, increased primary productivity and increased fish abundance in oligotrophic coral reef systems. In order to distinguish the effect of structural complexity fostering increased fish abundance via shelter vs. the hypothesized mechanisms of enhanced primary production, an experiment was conducted using artificial reef unit treatments with different levels of structural complexity (hollow or filled with broken cinder blocks) and treatments with differing levels of nutrient or food resources (units that were coated with an antifouling paint to prevent the development of biofouling community and units to which the experimenters added a slow release fertilizer to enrich the nutrient status). Many of the methods described above were used to assess water column and sediment nutrients, water column and benthic chlorophyll, biomass and percent cover of the biofouling community on settlement plates, caged biomass and cover (Figure 4.4), incubation measurements of primary productivity and respiration (Figure 4.5), and extensive sampling of the fish assemblages (see Chapter 5) in all of the reef treatments and at additional control sites (bare sand area with simple a-frame wire structures to suspend the settlement plates).

Despite the careful design of this study, substantial difficulties were encountered in interpreting the results. The first surprising result was the dominance of benthic assemblages by sessile filter-feeding invertebrates, with very low amounts of plant cover and biomass. Indeed, substantial amounts of macroalgae appeared only on uncaged treatments, where grazing fishes consumed competing invertebrates. This is contrary to the expected outcome for tropical coral reef systems (where we would expect more algae in cages where they are protected from herbivorous fishes, see Table 4.1) and indicated difficulty in interpreting results as testing hypotheses regarding nutrient and primary productivity in tropical coral reef systems. In fact, this was not a tropical coral reef system at all.

Results on nutrient regime included no differences in sediment nutrient concentration throughout the study. Water sampling, however, indicated that after more than a year following deployment, interstitial waters of the filled reefs had higher inorganic nitrogen concentrations than both ambient waters and the hollow reefs. Though no difference was found in phosphate concentrations, this result provided limited support for the hypothesis that high artificial reef structural complexity will increase nutrient regeneration.

This possible difference in nutrient regeneration, however, did not appear to translate into higher primary productivity. That is, enriched treatments (either experimentally fertilized or the filled reefs with seemingly higher regeneration) did not show higher primary productivity. Further, the link to secondary production was particularly unclear, as fish studies indicated that habitat structure of the artificial reefs was more important to fish enhancement than any food resources the reefs might provide. The high structural complexity of the filled reefs (even on antifouled reefs that essentially lacked trophic resources) did result in increased fish abundance.

While very limited evidence was found that increased structural complexity of artificial reefs may enhance nutrient regeneration, there was no link of enhanced nutrients with enhanced primary productivity of reefs. The author concluded that the experiment probably was not conducted in an appropriate place. This site was not an oligotrophic tropical site, as evidenced by the dominance of filter-feeding invertebrates that developed on the artificial substrates and a dominance of plank-tivorous guilds in the colonizing fish assemblage (see Table 4.1). Unfortunately, in this case, where nutrients and primary productivity were explicitly examined, planktonic resources were the primary trophic bases for the community that developed, and thus we would not expect nutrient regeneration and reef primary productivity to be important. The importance of structural complexity in providing fish shelter was, however, confirmed. Another factor cited by the author, which will no doubt play some role in constraining all artificial reef projects, was the dictation of the perhaps inappropriate site by governmental permitting and other logistical considerations. Management factors other than hydrological and ecological considerations will always influence artificial reef siting.

4.7 FUTURE NEEDS AND DIRECTIONS

As noted above, there is a severe paucity of published, peer-reviewed literature regarding functional aspects of artificial reef ecosystems such as nutrient cycling, primary production, and conversion of primary to secondary production. This lack of rigorous and accessible information is a severe hindrance to improving our understanding of the aspects of design and the physical and biotic environments that contribute to successful artificial reefs.

Whereas most artificial reef projects in the past had objectives primarily related to fisheries enhancement, in the future, greater emphasis will be placed on artificial reefs with restoration and mitigation objectives. In the former case, it is perhaps logical that monitoring of how many of what species of fishes are present at an artificial reef over time may be the primary criterion for evaluation of success. In the case of restoration and mitigation, the specific success criteria may be less clear (i.e., how does one judge the replication of a complex natural system?) and an area requiring greater attention in environmental management in the near future. It seems logical, though, that the restoration of ecosystem function should be a major criterion for assessing success of restoration projects. It is often assumed that the restoration of ecosystem structure (species richness, relative abundance, and age structure) implies that the underlying aspects of ecosystem function (nutrient cycles and recruit-ment dynamics) are healthy. Bell et al. (1993) examined this assumption for restored seagrass beds and concluded that there was a distinct lack of correlation between the structure of the restored seagrass itself and the functional aspects (specifically, fulfilling habitat requirements including trophic resources for animal residents) of the restored seagrass habitat. This assumption remains untested in reef systems, and, given that the restoration of the natural or unimpacted (and especially diverse) reef community structure is often unattainable on management-relevant time scales, increased effort at direct assessment of artificial reef ecosystem function is required. Given the complexity of the many functional assessment methods described in this chapter, there is also a recognized need for simpler, more applicable methods and success criteria for assessing ecosystem function (Zedler 1996).

4.8 ACKNOWLEDGMENTS

MWM gratefully acknowledges the advice, mentoring, and tremendous contribution to this manuscript by Dr. Alina Szmant (Szmant 1992). Funding for the preparation of this chapter was provided by the U.S. National Oceanic and Atmospheric Administration.

AF would like to thank Prof. Laura Talarico for her helpful assistance.

REFERENCES

Agemian, H. 1997. Determination of nutrients in aquatic sediments. Pages 175–227 In: A. Mudroch, J.M. Azcue, and P. Mudroch, eds. *Manual of Physico-chemical Analysis of Aquatic Sediments.* Lewis Publishers, Boca Raton, FL.

Akeda, S., K. Yano, A. Nagano, and I. Nakauchi. 1995. Improvement works of fishing port taken with care to artificial formation of seaweed beds. Pages 394–399. In: *Proceedings, International Conference on Ecological Systems Enhancement Technology for Aquatic Environments.* Japan International Marine Science and Technology Federation, Tokyo.

Ambrose, R.F. and S.L. Swarbrick. 1989. Comparison of fish assemblages on artificial and natural reefs off the coast of southern California. *Bulletin of Marine Science* 44:718–733.

Andrew, N.L. and B.D. Mapstone. 1987. Sampling and the description of spatial pattern in marine ecology. *Oceanography and Marine Biology Annual Reviews* 25:39–90.

Andrews, C. and H. Muller. 1983. Space-time variability of nutrients in a lagoonal patch reef. *Limnology and Oceanography* 28:215–227.

APHA. 1989. *Standard Methods for the Examination of Water and Wastewater.* 17th ed. American Public Health Administration, Washington, D.C. 1268 pp.

Atkinson, M.J. and R.W. Grigg. 1984. Model of a coral reef ecosystem. II. Gross and net benthic primary production of French Frigate Shoals, Hawaii. *Coral Reefs* 3:13–22.

Bailey-Brock, J.H. 1989. Fouling community development on an artificial reef in Hawaiian waters. *Bulletin of Marine Science* 44:580–591.

Baynes, T.W. and A.M. Szmant. 1989. Effect of current on the sessile benthic community structure of an artificial reef. *Bulletin of Marine Science* 44:545–566.

Bell, S.S., L.A.J. Clements, and J. Kurdziel. 1993. Production in natural and restored seagrasses: a case study of a macrobenthic polychaete. *Ecological Applications* 3:610–621.

Berner, R.A. 1980. *Early Diagenesis: A Theoretical Approach.* Princeton University Press, Princeton, NJ.

Berner, T. 1990. Coral reef algae. Pages 253–264. In: Z. Dubinsky, ed. *Ecosystems of the World,* Vol. 25: *Coral Reefs.* Elsevier, New York.

Birkeland, C. 1988. Geographical comparisons of coral reef community processes. Vol. 1, Pages 211–220. In: J.H. Choat, D. Barnes, M.A. Borowitzka, J.C. Coll, P.J. Davies, P. Flood, B.G. Hatcher, D. Hopley, P.A. Hutchings, D. Kinsey, G.R. Orme, M. Pinchon, P.F. Sale, P. Sammarco, C.C. Wallace, C. Wilkinson, E. Wolanski, and O. Bellwood, eds. *Proceedings, 6th International Coral Reef Symposium.* Symposium Executive Committee, Townsville, Australia.

Birkeland, C., ed. 1997. Geographic differences in ecological processes on coral reefs. Pages 273–286. In: *Life and Death of Coral Reefs.* Chapman & Hall, New York.

Birkeland, C. and R.H. Randall. 1981. Facilitation of coral recruitment by echinoid excavation. Vol. 1, Pages 695–698. In: E.D. Gomez, C.E. Birkeland, R.W. Buddemeier, R.E. Johannes, J.A. Marsh, Jr., and R.T. Tsuda, eds. *Proceedings, 4th International Coral Reef Symposium.* Marine Sciences Center, University of Philippines, Quezon City.

Boesch, D.F. 1977. Application of numerical classification in ecological investigations of water pollution. EPA-600/3-77-003, Corvallis Environmental Research Laboratory, Office of Research and Development, USEPA, Corvallis, Oregon.

Bohnsack, J.A. 1979. Photographic quantitative sampling studies of hard-bottom benthic communities. *Bulletin of Marine Science* 29:242–252.

Bohnsack, J.A. 1989. Are high densities of fishes at artificial reefs the result of habitat limitation or behavioral preference? *Bulletin of Marine Science* 44:631–645.

Bohnsack, J.A. and D.L. Sutherland. 1985. Artificial reef research: a review with recommendations for future priorities. *Bulletin of Marine Science* 37:11–39.

Bohnsack, J.A., D.L. Johnson, and R.F. Ambrose. 1991. Ecology of artificial reef habitats and fishes. Pages 61–107. In: W. Seaman and L. Sprague, eds. *Artificial Habitats for Marine and Freshwater Fisheries.* Academic Press, San Diego.

Bohnsack, J.A., A.M. Ecklund, and A.M. Szmant. 1997. Artificial reef research: is there more than the attraction-production issue? *Fisheries* 22:14–16.

Bombace, G. 1977. Aspetti teorici e sperimentali concernenti le barriere artificiali. Pages 29–42. In: *Proceedings, Atti IX Congresso Società Italiana Biologia Marina,* Ischia, Italy.

Bombace, G. 1981. Note on experiments in artificial reefs in Italy. *Conseil General (de) Peche Maritime Etude et Revues* 58:309–324.

Bombace, G. 1989. Artificial reefs in the Mediterranean Sea. *Bulletin of Marine Science* 44:1023–1032.

Bortone, S. and B.D. Nelson. 1995. Food habits and forage limits of artificial reef fishes in the Northern Gulf of Mexico. Pages 215–220. In: *Proceedings, International Conference on Ecological System Enhancement Technology for Aquatic Environments.* Japan International Marine Science and Technology Federation, Tokyo.

Bray, R.N. and A.C. Miller. 1985. Planktivorous fishes: their potential as nutrient importers to artificial reefs. Abstract only. *Bulletin of Marine Science* 37:396.

Bugrov, L.Y. 1994. Fish-farming cages and artificial reefs: complex for waste technology. *Bulletin of Marine Science* 55:1332.

Carr, M.H. and M.A. Hixon. 1997. Artificial reefs: the importance of comparisons with natural reefs. *Fisheries* 22:28–33.

Carter, J.W., A.L. Carpenter, M.S. Foster, and W.N. Jessee. 1985a. Benthic succession on an artificial reef designed to support a kelp-reef community. *Bulletin of Marine Science* 37:86–113.

Carter, J.W., W.N. Jessee, M.S. Foster, and A.L. Carpenter. 1985b. Management of artificial reefs designed to support natural communities. *Bulletin of Marine Science* 37:114–128.

Chalker, B.E., K. Carr, and E. Gill. 1985. Measurement of primary production and calcification *in situ* on coral reefs using electrode methods. Vol. 6, Pages 167–172. In: C. Gabrie and M. Harmelin-Vivien, eds. *Proceedings, 5th International Coral Reef Congress,* Tahiti. Atenne Museum National D'Historie Naturelle et de L'Ecole Pratiquede Hautes Etudes, Moorea, French Polynesia.

Clark, S. and A.J. Edwards. 1994. Use of artificial reef structures to rehabilitate reef flats degraded by coral mining in the Maldives. *Bulletin of Marine Science* 55:724–744.

CMEA. 1987. *Unified Methods for Water Quality Examination, Part 1: Methods of Chemical Analysis.* 4th Ed., Council of Mutual Economic Assistance, Moscow, 1244 pp. (In Russian.)

Connell, S.D. 1997. Exclusion of predatory fish on a coral reef: the anticipation, pre-emption, and evaluation of some caging artifacts. *Journal of Experimental Marine Biology and Ecology* 213:181–198.

Crompton, T.R. 1992. *Comprehensive Water Analysis Vol. 1: Natural Waters.* Elsevier Applied Science, London.

D'Elia, C.F. and W.J. Wiebe. 1990. Biogeochemical nutrient cycles in coral-reef ecosystems. Pages 49–74. In: Z. Dubinsky, ed. *Ecosystems of the World, Vol. 25: Coral Reefs.* Elsevier, New York.

Done, T.J., J.C. Ogden, W.J. Wiebe, and B.R. Rosen. 1996. Biodiversity and ecosystem function of coral reefs. Pages 393–429. In: Mooney, H.A., J.H. Cushman, E. Medina, O.E. Sala, and E.-D. Schulze, eds., *Functional Roles of Biodiversity, A Global Perspecitve.* John Wiley & Sons, Chichester.

Dubinsky, Z., ed. 1990. *Ecosystems of the World, Vol. 25: Coral Reefs.* Elsevier, New York.

Ecklund, A.M. 1996. The effects of post-settlement predation and resource limitation on reef fish assemblages. Ph.D. dissertation, University of Miami, Coral Gables, FL.

Erez, J. 1990. On the importance of food sources in coral-reef ecosystems. Pages 411–418. In: Z. Dubinsky, ed. *Ecosystems of the World, Vol. 25: Coral Reefs.* Elsevier, New York.

Falace, A. and G. Bressan. 1994. Some observations on periphyton colonization of artificial substrata in the Gulf of Trieste (N. Adriatic Sea). *Bulletin of Marine Science* 55:924–931.

Falace, A. and G. Bressan. 1997. Adapting an artificial reef to biological requirements. In: L.E. Hawkins, S. Hutchinson, and A. Jensen, eds. *Proceedings, 30th European Marine Biology Symposium.* Southampton Oceanography Centre, Southampton, England, September 1995.

Falace A., G. Maranzana, G. Bressan, and L. Talarico. 1998. Approach to a quantitative evaluation of benthic algal communities. Pages 108–119. In: A.C. Jensen, ed. *Final Report and Recommendation,* European Artificial Reef Research Network (EARRN). Report to European Commission, Contract No. AIR-CT94-2144.

Fang, L.S. 1992. A theoretical approach of estimating the productivity of artificial reef. *Acta Zoologica Taiwanica* 3:5–10.

Ferreira, J.G. and L. Ramos. 1989. A model for the estimation of annual production rates of macrophyte algae. *Aquatic Botany* 33:53–70.

Frazer, T.K. and W.J. Lindberg. 1994. Refuge spacing similarly affects reef-associated species from three phyla. *Bulletin of Marine Science* 55:388–400.

Gilliam, D.S., K. Banks, and R.E. Spieler. 1995. Evaluation of a novel material for artificial reef construction. Pages 345–350. In: *Proceedings, International Conference on Ecological Systems Enhancement Technology for Aquatic Environments.* Japan International Marine Science and Technology Federation, Tokyo.

Grigg, R.W., J.J. Polovina, and M.J. Atkinson. 1984. Model of a coral reef ecosystem. III. Resource limitation, community regulation, fisheries yields, and resource management. *Coral Reefs* 3:23–29.

Harris,G.P. 1986. *Phytoplankton Ecology: Structure, Function, and Fluctuation.* Chapman & Hall. New York.

Hatcher, A. 1995. Trends in the sessile epibiotic biomass of an artificial reef. Pages 125–130. In: *Proceedings, International Conference on Ecological Systems Enhancement Technology for Aquatic Environments.* Japan International Marine Science and Technology Federation, Tokyo.

Hatcher, B.G. 1988. The primary productivity of coral reefs: a beggar's banquet. *Trends in Ecology and Evolution* 3:106–111.

Hatcher, B.G. 1990. Coral reef primary productivity: a hierarchy of pattern and process. *Trends in Ecology and Evolution* 5:149–155.

Hatcher, B.G. 1997. Organic production and decomposition. Pages 140–174. In: C. Birkeland, ed. *Life and Death of Coral Reefs.* Chapman & Hall, New York.

Hatcher, B.G. and A.W.D. Larkum. 1983. An experimental analysis of factors controlling the standing crop of the epilithic algal community on a coral reef. *Journal of Experimental Marine Biology and Ecology* 69:61–84.

Hay, M.E. 1984. Patterns of fish and urchin grazing on Caribbean coral reefs: are previous results typical? *Ecology* 65:446–454.

Hay, M.E. 1991. Herbivorous fishes and adaptations of their prey. Pages 96–119. In: P.F. Sale, ed. *Ecology of Fishes on Coral Reefs.* Academic Press, San Diego.

Henderson, R.S. 1981. *In situ* and microcosm studies of diel metabolism of reef flat communities. Vol. 1, Pages 679–686. In: E.D. Gomez, C.E. Birkeland, R.W. Buddemeier, R.E. Johannes, J.A. Marsh, Jr., and R.T. Tsuda, eds. *Proceedings, 4th International Coral Reef Symposium.* Marine Sciences Center, University of Philippines, Quezon City.

Hixon, M.A. and W.N. Brostoff. 1985. Substrate characteristics, fish grazing, and epibenthic reef assemblages off Hawaii. *Bulletin of Marine Science* 37:200–213.

Holme, N.A. and A.D. McIntyre. 1984. *Methods for the Study of Marine Benthos.* Blackwell Scientific Publications, Oxford.

Hopkinson, C.S., Jr., R.D. Fallon, B.O. Jansson, and J.P. Schubauer. 1991. Community metabolism and nutrient cycling at Gray's Reef, a hard bottom habitat in the Georgia Bight. *Marine Ecology Progress Series* 73:105–120.

Hueckel, G.J. and R.M. Buckley. 1989. Predicting fish species on artificial reefs using indicator biota from natural reefs. *Bulletin of Marine Science* 44:873–880.

Hughes, T.P. 1994. Catastrophes, phase shifts, and large-scale degradation of a Caribbean coral reef. *Science* 265:1547–1551.

Hughes, T.P., D.C. Reed, and M.J. Boyle. 1987. Herbivory on coral reefs: community structure following mass mortalities of sea urchins. *Journal of Experimental Marine Biology and Ecology* 113:39–59.

Hulberg, L.W. and J.S. Oliver. 1980. Caging manipulations in marine soft-bottom communities: importance of animal interactions or sedimentary habitat modifications. *Canadian Journal of Fisheries and Aquatic Science* 37:1130–1139.

Jara, F. and R. Céspedes. 1994. An experimental evaluation of habitat enhancement on homogeneous marine bottoms in southern Chile. *Bulletin of Marine Science* 55:295–307.

Johnson, T.D., A.M. Barnett, E.E. DeMartini, L.L. Craft, R.F. Ambrose, and L.J. Purcell. 1994. Fish production and habitat utilization on a southern California artificial reef. *Bulletin of Marine Science* 55:709–723.

Kinsey, D.W. 1978. Productivity and calcification estimates using slack-water periods and filed enclosures. Pages 439–468. In: D.R. Stoddart and R.E. Johannes, eds. *Coral Reefs: Research Methods.* Monographs in Oceanography Methods No. 5, UNESCO, Paris.

Koblentz-Mishke, O.J., V.V. Volkovinsky, and J.G. Kabanova. 1970. Plankton primary production of the world ocean. Pages 183–193. In: W.S. Wooster, ed. *Scientific Exploration of the South Pacific.* National Academy of Sciences, Washington, D.C.

Laihonen, P., J. Hanninen, J. Chojnacki, and I. Vuorinen. 1996. Some prospects of nutrient removal with artificial reefs. Pages 85–96. In: A.C. Jensen, ed. *Proceedings, First European Artificial Reef Research Network Conference.* Southampton Oceanography Centre, Southampton, England.

Lapointe, B.E. 1989. Caribbean coral reefs: are they becoming algal reefs? *Sea Frontiers* 35:82–91.

Lapointe, B.E. 1997. Nutrient thresholds for bottom-up control of macroalgal blooms on coral reefs in Jamaica and southeast Florida. *Limnology and Oceanography* 42:1119–1131.

Larkum, A.W.D. 1983. The primary productivity of plant communities on coral reefs. Pages 221–230. In: D.J. Barnes, ed. *Perspectives on Coral Reefs.* Published for AIMS by Brian Clouston Publishers, Manuka, Australia.

Larkum, A.W.D. and K. Koop. 1997. ENCORE: algal productivity and possible paradigm shifts. Vol. 1, Pages 881–884. In: H.A. Lessios and I.G. Macintyre, eds. *Proceedings, 8th International Coral Reef Symposium.* Smithsonian Tropical Research Institute, Balboa, Panama.

Larkum, A.W.D., A.J. McComb, and S.A. Shepherd. 1989. *Biology of Seagrasses: A Treatise on the Biology of Seagrasses with Special Reference to the Australian Region.* Elsevier, New York.

Levinton, J.S. 1982. *Marine Ecology.* Prentice-Hall, Englewood Cliffs, NJ.

Lewis, S.M. 1986. The role of herbivorous fishes in the organization of a Caribbean reef community. *Ecological Monographs* 56:183–200.

Lieth, H. and R.H. Whittaker. 1975. *Primary Productivity of the Biosphere.* Springer-Verlag, New York.

Lindquist, D.G., L.B. Cahoon, I.E. Clavijo, M.H. Posey, S.K. Bolden, L.A. Pike, S.W. Burk, and P.A. Cardullo. 1994. Reef fish stomach contents and prey abundance on reef and sand substrata associated with adjacent artificial and natural reefs in Onslow Bay, North Carolina. *Bulletin of Marine Science* 55:308–318.

Littler, M.M. and D.S. Littler. 1980. The evolution of thallus form and survival strategies in benthic marine macroalgae: field and laboratory tests of a functional form model. *American Naturalist* 116:25–44.

Littler, M.M. and D.S. Littler. 1985. *Handbook of Phycological Methods — Ecological Field Methods: Macroalgae.* Cambridge University Press, Cambridge.

Littler, M.M., D.S. Littler, J.N. Norris, and K.E. Bucher. 1987. Recolonization of algal communities following the grounding of the freighter *Wellwood* on Molasses Reef, Key Largo National Marine Sanctuary. NOAA Technical Memorandum, NOS MEMD 15. 32 pp.

Littler, M.M., D.S. Littler, and E.A. Titlyanov. 1991. Comparisons of N- and P-limited productivity between high granitic islands versus low carbonate atolls in the Seychelles archipelago: a test of the relative dominance paradigm. *Coral Reefs* 10:199–209.

Marsh, J.A. and S.V. Smith. 1978. Productivity measurements in flowing water. Pages 361–378. In: D.R. Stoddart and R.E. Johannes, eds. *Coral Reefs: Research Methods.* Monographs in Oceanography Methods No. 5, UNESCO, Paris.

McClanahan, T.R., M. Nugues, and S. Mwachireya. 1994. Fish and sea urchin herbivory and competition in Kenyan coral reef lagoons: the role of reef management. *Journal of Experimental Marine Biology and Ecology* 184:237–254.

Miller, M.W. 1998. Coral/seaweed competition and the control of reef community structure within and between latitudes. *Oceanography and Marine Biology: An Annual Review* 36:65–96.

Miller, M.W. and M.E. Hay. 1998. Effects of fish predation and seaweed competition on the survival and growth of corals. *Oecologia* 113:231–238.

Miller, M.W., M.E. Hay, S.L. Miller, D. Malone, E.E. Sotka, and A.M. Szmant. 1999. Effects of nutrients versus herbivores on reef algae: a new method for manipulating nutrients on coral reefs. *Limnology and Oceanography* 44:1847–1861.

Mudroch, A. and S.D. MacKnight, eds. 1991. *CRC Handbook of Techniques for Aquatic Sediments Sampling.* CRC Press, Boca Raton, FL.

NIH. 1987–88. *Physico-Chemical Analysis of Water and Wastewater.* National Institute of Hydrology, Roorkee — 247667(UP), India.

Nybakken, J.W. 1982. *Marine Biology: An Ecological Approach.* Harper & Row, New York.

Odum, H.T. and E.P. Odum. 1955. Trophic structure and productivity of a windward coral reef community on Eniwetok Atoll. *Ecological Monographs* 25:291–320.

Ohgai, M., N. Murase, H. Kakimoto, and M. Noda. 1995. The growth and survival of *Sargassum patens* on andesite and granite substrata used on the formation of seaweed beds. Pages 470–475. In: *Proceedings, International Conference on Ecological Systems Enhancement Technology for Aquatic Environments.* Japan International Marine Science and Technology Federation, Tokyo.

Palmer-Zwahlen, M.L. and D.A. Aseltine. 1994. Successional development of the turf community on a quarry rock artificial reef. *Bulletin of Marine Science* 55:902–923.

Pamintuan, I.S., P.M. Alino, E.D. Gomez, and R.N. Rollon. 1994. Early successional patterns of invertebrates in artificial reefs established at clear and silty areas in Bolinao, Pangasinan, Northern Philippines. *Bulletin of Marine Science* 55:867–877.

Parchevshy, V.P. and M.A. Rabinovich. 1995. Influence of habitat enhancement on yield and biomass renewal of seaweeds in eutrophic coastal waters of the Black Sea. Pages 459–463. In: *Proceedings, International Conference on Ecological Systems Enhancement Technology for Aquatic Environments.* Japan International Marine Science and Technology Federation, Tokyo.

Parsons, T.R., M. Takahashi, and B. Hargrave. 1984a. *Biological Oceanographic Processes.* 3rd Ed. Pergamon Press, New York.

Parsons, T.R., Y. Maita, and C. Lalli. 1984b. *A Manual of Chemical and Biological Methods for Seawater Analysis.* Pergamon Press, New York.

Patton, M.L., C.F. Valle, and R.S. Grove. 1994. Effects of bottom relief and fish grazing on the density of the giant kelp, *Macrocystis. Bulletin of Marine Science* 55:631–644.

Peet, R.K. 1974. The measurement of species diversity. *Annual Review of Ecology and Systematics* 5:285–307.

Pike, L.A. and D.G. Lindquist. 1994. Feeding ecology of spottail pinfish (*Diplodus holbrooki*) from an artificial and natural reef in Onslow Bay, North Carolina. *Bulletin of Marine Science* 55:363–374.

Powell, C. and M. Posey. 1995. Evidence of trophic linkages between intertidal oyster reefs and their adjacent sandflat communities. Abstract. In: J.P. Grassle, A. Kelsey, E. Oates, P.V. Snelgrove, eds. *Proceedings, 23rd Benthic Ecology Meetings.* Institute of Marine and Coastal Sciences, Rutgers University, New Brunswick, NJ.

Prince, E.D. 1976. The biological effects of artificial reefs in Smith Mountain Lake, Virginia. Ph.D. dissertation, Virginia Polytechnic Institute and State University, Blacksburg.

Prince, E.D., O.E. Mauggham, and P. Brouha. 1985. Summary and update of the Smith Mountain Lake artificial reef project. Pages 401–430. In: F.M. D'Itri, ed. *Artificial Reefs Marine and Freshwater Application.* Lewis Publishers, Chelsea, MI.

Randall, J.E. 1965. Grazing effects of seagrasses by herbivorous reef fishes in the West Indies. *Ecology* 46:255–260.

Reimers, H. and K. Branden. 1994. Algal colonization of a tire reef- influence of placement date. *Bulletin of Marine Science* 55: 460–469.

Relini, G. 1974. La colonizzazione dei substrati duri in mare. *Memorie Biologia Marina e Oceanografia,* Numero Singolo, 4(4–6):201–261.

Relini, G., N. Zamboni, F. Tixi, and G. Torchia. 1994. Patterns of sessile macrobenthos community development on an artificial reef in the Gulf of Genoa (northwestern Mediterranean). *Bulletin of Marine Science* 55:745–771.

Rice, D.W., T.A. Dean, F.R. Jacobsen, and A.M. Barnett. 1989. Transplanting of giant kelp *Macrocystis pyrifera* in Los Angeles Harbor and productivity of the kelp population. *Bulletin of Marine Science* 44:1070.

Riggio, S. 1988. I ripopolamenti in mare. Pages 223–250. In: *Proceedings, Atti IV Convegno Siciliano Ecologia,* Porto Palo di Capo Passero, Italy.

Riley, J.P. and R. Chester. 1971. *Introduction to Marine Chemistry.* Academic Press, New York.

Rumohr, J., E. Walger, and B. Zeitschel. 1987. Seawater-sediment interactions in coastal waters: an interdisciplinary approach. *Lecture Notes on Coastal and Estuarine Studies,* No. 13. Springer-Verlag, New York.

Schubert, L.E. 1984. *Algae as Ecological Indicators.* Academic Press, London.

Smith, S.V. and D.W. Kinsey. 1978. Calcification and organic carbon metabolism as indicated by carbon dioxide. Pages 469–484. In: D.R. Stoddart and R.E. Johannes, eds. *Coral Reefs: Research Methods.* Monographs in Oceanography Methods No. 5, UNESCO, Paris.

Smith, S.V., W.J. Kimmerer, E.A. Laws, R.E. Brock, and T.W. Walsh. 1981. Kaneohe Bay sewage diversion experiment: perspectives on ecosystem responses to nutritional perturbation. *Pacific Science* 35:279–395.

Spanier, E. 1989. How to increase the fisheries yield in low productive marine environments. Vol. 1, Pages 297–301. *Proceedings, Oceans '89: The Global Ocean.* Institute of Electrical and Electronics Engineers, New York.

Spanier, E., M. Tom, S. Pisanty, and G. Almog-Shtayer. 1990. Artificial reefs in the low productive marine environments of the southeastern Mediterranean. *Marine Ecology (Pubblicazioni della Stazione Zoologica di Napoli)* 11:61–75.

Steele, M.A. 1996. Effects of predators on reef fishes: separating cage artifacts from effects of predation. *Journal of Experimental Marine Biology and Ecology* 198:249–267.

Steimle, F.W., Jr. and L. Ogren. 1982. Food of fish collected on artificial reefs in the New York Bight and off Charleston, South Carolina. *Marine Fishery Review* 44:49–52.

Steneck, R.S. and M.N. Diether. 1994. A functional group approach to the structure of algal-dominated communities. *Oikos* 69:476–498.

Stimson, J., S. Larned, and K. McDermid. 1996. Seasonal growth of the coral reef macroalga *Dictyosphaeria cavernosa* (Forskål) Børgesen and the effects of nutrient availability, temperature, and herbivory on growth rate. *Journal of Experimental Marine Biology and Ecology* 196:53–77.

Szmant, A.M. 1992. Reef data: nutrients, primary productivity, and fouling. Pages 113–139. In: W. Seaman, Jr., ed. *Environmental and Fishery Performance of Florida Artificial Reef Habitats.* Florida SeaGrant College Program, Gainesville.

Szmant, A.M. 1993. Nutrient cycling and the optimum productivity of shallow-water artificial reefs. Final Report 91-104, Florida Sea Grant College Program, Gainesville. 22 pp.

Szmant, A.M. 1997. Nutrient effects on coral reefs: a hypothesis on the importance of topographic and trophic complexity to reef nutrient dynamics. Vol. 2, Pages 1527–1532. In: H.A. Lessios and I.G. Macintyre, eds. *Proceedings, 8th International Coral Reef Symposium.* Smithsonian Tropical Research Institute, Balboa, Panama.

Szmant, A.M. and A. Forrester. 1996. Water column and sediment nitrogen and phosphorus distribution patterns in the Florida Keys, USA. *Coral Reefs* 15:21–41.

Szmant, A.M., L.M. FitzGerald, and V.I. Hensley. 1986. Nitrogen fluxes in fore-reef sediments *Eos, Transactions, American Geophysical Union* 67:997.

Szmant-Froelich, A. 1983. Functional aspects of nutrient cycling on coral reefs. NOAA Symposium Series on Undersea Research, National Undersea Research Program, Rockville, MD 1:133–139.

Szmant-Froelich, A. 1984. The role of herbivorous fish in the recycling of nitrogenous nutrients on coral reefs. NOAA Hydrolab Final Report, Mission 83-10, National Undersea Research Program, Rockville, MD.

Thompson, M.F., R. Sarojini, and R. Nagabhushanam. 1987. *Biology of Benthic Marine Organisms.* A.A. Balkema, Rotterdam.

Tribble, G.W., F.J. Sansone, Y. Li, S.V. Smith, and R.W. Buddemeier. 1988. Material fluxes from a reef framework. Vol. 2, Pages 577–582. In: J.H. Choat, D. Barnes, M.A. Borowitzka, J.C. Coll, P.J. Davies, P. Flood, B.G. Hatcher, D. Hopley, P.A. Hutchings, D. Kinsey, G.R. Orme, M. Pinchon, P.F. Sale, P. Sammarco, C.C. Wallace, C. Wilkinson, E. Wolanski, and O. Bellwood, eds. *Proceedings, 6th International Coral Reef Symposium.* Symposium Executive Committee, Townsville, Australia.

USEPA. 1983. *Methods for Chemical Analysis of Water and Wastes.* Environmental Monitoring and Support Laboratory, United States Environmental Protection Agency, Cincinnati, OH.

Valiela, I. 1995. *Marine Ecological Processes.* 2nd Ed. Springer-Verlag, New York.

Zedler, J.B. 1996. Ecological issues in wetland mitigation. *Ecological Applications* 6:33–37.

Zieman, J.C. and R.T. Zieman. 1989. The ecology of the seagrass meadows of the West Coast of Florida: a community profile. United States Fish and Wildlife Service Biological Report 85 (7.25).

CHAPTER 5

Fish and Macroinvertebrate Evaluation Methods

**Stephen A. Bortone, Melita A. Samoilys,
and Patrice Francour**

CONTENTS

0-8493-9061-3/00/$0.00+$.50

5.1 SUMMARY

 This chapter discusses the importance and need for evaluating fish and invertebrate faunas associated with artificial reefs, reviews the biotic and abiotic factors impacting the faunas, examines the criteria involved in designing relevant studies, and describes the methods (both destructive and nondestructive) used in assessing faunal assemblages. Terms employed in these studies are defined. This is followed by some actual and hypothetical examples relating to assessment studies. The chapter concludes with a discussion of future needs and directions in artificial reef research.

5.2 INTRODUCTION

A principal reason for artificial reef deployment is to improve, increase, or at least maintain the fishery resources in a local area. Polovina (1991: 164) depicted how artificial reefs theoretically affect fisheries. Figure 5.1 illustrates how artificial reefs may impact fisheries resources through the added surface area they provide for the attachment of grazers and filter feeders, as a basis for transfer of energy derived from the water column to reef-associated fish and macroinvertebrate predators. Thus, evaluation of a reef's effect on fisheries resources can be based on the biological attributes of those resources, such as their abundance, size, and biomass as well as species richness and relative species diversity.

Assessment methods for artificial reef fishes and macroinvertebrates have been examined in publications by Bortone and Bohnsack (1991); Bortone and Kimmel (1991); and Seaman et al. (1992). To date, most assessment methods have been developed and modified from studies on natural tropical coral reefs (e.g., Sale 1991a), temperate rocky reefs (e.g., Kingsford and Battershill 1998), or other irregular nearshore biotopes. Thus, many of the methods, protocols, and studies cited herein are from the nonartificial reef literature. However, many of the attributes (e.g., spatial heterogeneity, and species diversity) of natural reefs have features directly related to difficulties and problems associated with sampling faunas associated with artificial reefs.

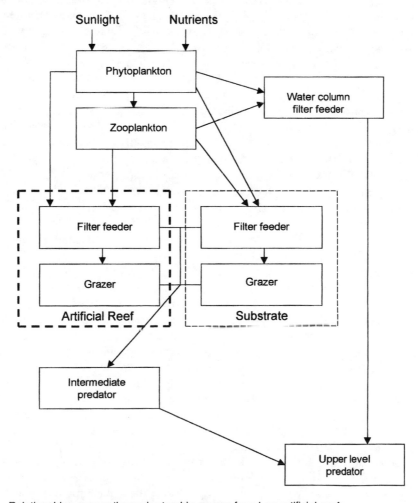

Figure 5.1 Relationships among the major trophic groups found on artificial reefs.

5.2.1 Objectives

This chapter has six objectives: (1) explain the need to evaluate the fishes and larger invertebrates that are potentially impacted by the presence of an artificial reef; (2) define terms frequently used in artificial reef faunal assessments to clarify terminology used in this chapter and the artificial reef literature in general; (3) describe the abiotic and biotic factors or variables that have an impact on the faunal assemblage associated with artificial reefs; (4) within the context of these factors, discuss design criteria for studies on the fishes and macroinvertebrates that associate with artificial reefs; (5) present various methods of destructive and nondestructive assessment; and (6) discuss the application of such assessments by providing case studies and examples.

5.2.2 Definitions

The great majority of fish species found on artificial reefs are in the taxonomic Division Teleostei of the Class Actinopterygii (formerly the Class Osteichthyes; Nelson 1994), and most of these are from the advanced Order Perciformes (Choat and Bellwood 1991). In warm and cool temperate marine and estuarine waters worldwide, these often include species in families such as the Serranidae (groupers, seabasses, and rockcods), Sparidae (porgies and seabream), Carangidae (jacks and amberjacks), Lutjanidae (snappers), Haemulidae (grunts), Chaetodontidae (butterflyfishes), Pomacentridae (damselfishes), Labridae (wrasses), and Acanthuridae (surgeonfishes). Species from less advanced orders also occur in marine biotopes, including the Anguilliformes (e.g., Muraenidae [moray eels]), Clupeiformes (e.g., Clupeidae [herrings, sardines, and pilchards]), Beryciformes (e.g., Holocentridae [squirrelfishes and soldierfishes]), and Syngnathiformes (e.g., Aulostomidae [trumpetfishes]). The family composition in freshwater reefs is determined, to a greater degree than in marine reefs, by the families that are native to a specific geographic area. For example, certain sunfishes and blackbasses (family Centrarchidae) are a dominant component of reefs in North America, along with families such as the Esocidae (pikes), Cyprinidae (minnows), and Percidae (perches). The last three families are also found associated with freshwater European reefs, whereas centrarchids are not native to that continent. None of these four families is native to South America, and freshwater reefs on that continent would be dominated to a great degree by cichlids, which in turn are absent from Europe and all but extreme southern North America. Taxonomic references should be consulted for a comprehensive classification of the marine, estuarine, and freshwater fishes specific to local areas.

Macroinvertebrates (i.e., larger invertebrates) are more difficult to define because their chief delimiter is size (i.e., any invertebrate that can be seen easily with the human eye, usually no smaller than 1 cm in total length). In most marine and estuarine areas, these groups often include representatives of the Superclass Crustacea (i.e., crabs, shrimps, and lobsters); the Phylum Mollusca (i.e., octopus, squid, cuttlefish, clams, oysters, snails, whelks, and conchs); and the less motile representatives of the Phylum Echinodermata (i.e., starfish, urchins, crinoids, and holothurians). In fresh waters, macroinvertebrates most often include crustaceans and molluscs.

The reef fauna can also be categorized according to intra- and interspecific associations. Thus, a *population* consists of a number of individuals of the same species, which is usually restricted to a defined area. For example, we may refer to the population of a particular grouper species on a reef or on an archipelago of reefs. *Demes* are considered to be local units, or subpopulations, of a population that normally interbreed. A *genetic stock* consists of several demes over a broad area, including multiple generations (e.g., larval forms and adults), and is termed a *metapopulation* (Roughgarden and Iwasa 1986; Roughgarden et al. 1988; Doherty 1991). A metapopulation is typically linked through a dispersive larval phase (e.g., Doherty 1991). Thus, metapopulations are comprised of several populations that are linked genetically, sometimes over several generations.

When two or more species occur together at a location the association of species is termed a *community*. The term assemblage is frequently used synonymously with community (Sale 1991b).

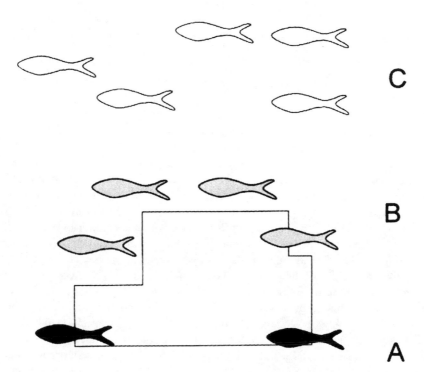

Figure 5.2 Fishes classified according to their typical position relative to the reef (modified from Nakamura 1985).

Here, however, we use the term *assemblage* (*sensu* Bohnsack et al. 1991) when referring to a multispecies group of organisms found on an artificial reef. This is because the organisms are found together, partly due to artificial circumstances and not necessarily because of long-term evolutionary adaptations, strategies, or coevolution.

For some analyses it may be more important to consider organisms based on their physical position relative to the reef. Nakamura (1985) used three categories to classify organisms with regard to their relative dependencies on the artificial reef structure itself (Figure 5.2). The first of these (Type A) comprises benthic fishes that tend to have direct contact with the reef itself and often occupy crevices, holes, or internal spaces within the reef. Examples include toadfishes (Batrachoididae), wolffishes (Anarhichadidae), and scorpionfishes (Scorpaenidae), as well as many invertebrates such as stone crabs (*Menippe*), lobsters (both *Panulirus* and *Homarus*), and many sea urchins and holothurians. The second group (Type B) includes organisms found closely around the reef but not coming into direct contact with it. Associated fishes include various seabasses and groupers (Serranidae), grunts (Haemulidae), and flounders (Bothidae). Type C species include organisms found above the reef in the midwater and pelagic zones. Examples are jacks (Carangidae [especially amberjacks of the genus *Seriola*]), certain herrings (Clupeidae [genera *Jenkinsia* and *Sardinella*]), certain silversides (Atherinidae [genera *Hypoatherina* and *Atherina*]), tunas and mackerels (Scombridae), as well as invertebrates such as squids.

5.3 ASSEMBLAGE DYNAMICS

The basic dynamics of artificial reef assemblages are discussed in Chapter 4, with an introduction to primary and secondary transfers of energy through their trophic relationships. Here we review some of the resulting generalizations, which will help provide an overall appreciation of fish and macroinvertebrate assemblage dynamics. Understanding these (or at least being aware of their existence) should help the reader in designing assemblage assessment studies and interpreting the results.

5.3.1   Fish and Macroinvertebrate Assessment Components

Artificial reefs are significantly different from coral reefs, but strongly resemble rocky reefs in that their structural components are nonliving and therefore do not provide direct energy to the associated assemblage of organisms. Artificial reefs, coral reefs, and rocky reefs all provide increased surface area for the attachment of primary and secondary producers. However, even the most intricately complex artificial reef surfaces provide an infinitesimally small addition to the overall surface area available in the natural surrounding environment. Thus, the majority of energy resources that are available as consumable items to the fishes and macroinvertebrate assemblages associated with artificial reefs come from the surrounding water column by way of filter-feeding organisms (Figure 5.1). Some of the zoo- and phytoplanktivores are substrate dwelling, benthic organisms that are attached to the reef surface or dwell on or in the nearby substrate.

Although the importance of energy flow may seem obvious, only a few studies have examined the trophic or guild structure around artificial reefs. Among these, Lindquist et al. (1994) noted that fishes associated with temperate-zone artificial reefs chiefly consumed prey items that were normally found in the sandy forage areas surrounding the reefs. Frazer and Lindberg (1994) found evidence that the amount of area available for foraging was directly related to populations of fishes (mobile, off-reef foragers such as seabass and triggerfish) and macroinvertebrates (i.e., stone crabs and octopuses). Bortone et al. (1998) and Bortone (1999) found evidence that foraging away from reefs could alter the surrounding community of organisms that dwell in the sand substrate near reefs. Additionally, they speculated that the amount of fish biomass supported on a reef might be directly dependent upon the size and productivity of the potential feeding zones surrounding a reef. Nelson and Bortone (1996) found that a guild structure among larger fish associated with artificial reefs included upper and lower structure pickers, upper and lower structure predators, water-column pickers and predators, and ambush predators. Their study indicated that a large amount of material consumed by these fish also came from sources away from the reef. Thus, the forage areas above and away from the reef become feeding zones or "halos" whose size and productivity may be critical to sustaining the biomass of secondary and tertiary trophic organisms associated with artificial reefs, either as permanent residents or as opportunistic invaders.

5.3.2   Patterns in Assemblage Structure

This section identifies some general features or patterns of species attributes and community structure that should be understood when contemplating the design and execution of a study on the fish and macroinvertebrates associated with artificial reefs. Specific local conditions (both spatially and temporally) may contradict some of these statements. However, general conditions may prove heuristic in that they permit a basis for forming alternative hypotheses regarding our current understanding of these assemblages. More importantly, they help organize and consolidate research questions so that assessment efforts become directed and purposeful.

5.3.2.1   Salinity

Salinity plays an exceptionally important role in determining the ecosystem affinities to which nearly all organisms must adhere, chiefly due to short-term osmoregulation and long-term evolutionary adaptation. The outstanding difference between fish and macroinvertebrate artificial reef assemblages in the marine environment and those from either estuarine or fresh waters is higher species richness (i.e., number of species) in the former. Marine artificial reef assemblages may have well over 50 different species (e.g., Ardizzone et al. 1989), not including the different sex or life-stage forms of many tropical faunas (e.g., Böhlke and Chaplin 1993). High species richness can add a significant complication to the assessment study design, especially relative to training procedures and data gathering. Choices may have to be made to restrict assessments to a few "target" species or families, especially when taxonomic expertise for quality identifications is not readily available. For example, Bohnsack (1982) used a modified point count method to survey for only piscivorous predatory fishes on a coral reef. In another example, Alevizon et al. (1985) surveyed artificial reefs in the Bahamas and targeted only selected groups or families of fishes, not the entire assemblage.

Although the species richness of fish and macroinvertebrates associated with artificial reefs in estuaries is generally lower than in adjacent marine environments, it can fluctuate dramatically. Especially in temperate areas, estuaries are noted for dramatic seasonal shifts in faunal composition, as species often migrate because of seasonally driven environmental preferences and reproductive behavior (Dovel 1971). This means that comparisons between assemblages must be made with the full knowledge that they are predicated upon both the spatially and temporally varying conditions offered by the estuary. Further, each estuary may possess these spatially and temporally varying conditions in combinations that are unique to each system, thus often making comparisons between assemblages in different estuaries problematic.

Faunal-assemblage diversity in freshwater reefs is relatively lower than for brackish or marine reefs and is determined by factors such as age and diversity of the continental areas from which the fauna was derived. Based on this reasoning, the fauna of an artificial reef in southeastern Asia should be more diverse than one in northern Europe. Also, older and larger bodies of fresh water usually have more diverse communities of organisms available for potential colonization of a local artificial reef (i.e., species/area relationship [Barbour and Brown 1974]).

5.3.2.2 Latitude

A general paradigm of both terrestrial and aquatic ecology is that species richness and diversity tend to be higher closer to the equator, the principal reason being that tropical regions have on average remained more ecologically stable than temperate regions (Emlen 1973). This is most probably due to the effect of Pleistocene glaciation in the Northern Hemisphere during the past 2 million years, which not only eliminated or displaced a high percentage of the floras and faunas in northern regions, but also had profound effects (mainly through changes in temperatures and weather patterns) in more distant areas (e.g., the Sahara region has undergone desiccation within the past million years). Study designers should recognize that comparisons made along extensive latitudes might have to allow for the natural faunal differences likely to be present.

5.3.2.3 Depth

Natural fish and macroinvertebrate communities tend to form fairly strict zonations with regard to depth regimes in various biotopes (Rosenblatt 1967). Variation in environmental features may tend to be greatest near shore and in surface waters. Variation in environmental attributes may become dampened with depth. It is in the tidal zone that the impact of short-term environmental changes induced by weather may be the greatest due to dramatic changes in salinity, temperature, and wave action. Under these more rugged conditions, species diversity is often reduced. Fish and macrobenthic diversity tends to be higher from the littoral zone out to the continental shelf edge. With increasing depth (off the shelf edge), light penetration progressively decreases and the bottom tends to become more homogeneous, which results in a lower biodiversity of species available to form artificial reef assemblages.

5.3.2.4 Day/Night

The time of day during which a survey is conducted can greatly influence perceived species diversity, due to the varied natural activity patterns among fishes (Hobson 1991). For example, Sanders et al. (1985) noted that the time of day a survey was conducted influenced the measured fish species diversity on an artificial reef in the northern Gulf of Mexico. In the same general region, Martin and Bortone (1997) found that the observed abundance and number of taxa of epifaunal organisms associated with artificial reefs were also associated with time of day. This implies that potential food resources also show activity patterns that affect or are related to fish and macroin-vertebrate assemblages on artificial reefs.

5.3.2.5 Season

Bortone et al. (1994a) found species diversity and individual fish abundance varied with the seasons on warm-temperate artificial reefs. Tropical areas of the world display seasonal variations in species and individual abundance in inland fish faunas associated with wet and dry seasons (Lowe-McConnell 1979). Cold-temperate areas may also display dramatic differences associated with seasonal variations in plankton abundance, which in turn impacts the distribution and abundance of fish and macroinvertebrate faunas in those areas.

5.4 PURPOSES AND STRATEGIES OF ASSESSMENT AND ANALYSIS

The overall objective of a study should be directed toward evaluating the rationale for constructing the reef. It is presupposed that no artificial reef should be built without some intended purpose. However, not all reefs have been constructed with a specific purpose in mind, and many *a posteriori* studies have provided insight into the nature of artificial reefs that have been previously deployed without any specific goal, objective, or rationale other than to "make fishing better." Nevertheless, we will examine some important aspects that should be considered when designing a study to evaluate or assess fishes and macroinvertebrates on artificial reefs.

Most laypersons perceive that the overall objective for assessing the fish and macroinvertebrates is to determine the reef's effectiveness in increasing the biomass, number, or size of all species, some species, or a preferred species. While laudable and perhaps obvious, goals such as these are far too simplistic and lack any real chance of being scientifically evaluated because of their over-generalized nature. A far better series of objectives for a reef with regard to fish and macroinver-tebrates is to provide some aspect or feature that is required of a species or assemblage that would enhance the overall goal of increasing biomass, number, size, and/or diversity. Thus, the purpose of evaluation is to determine the effectiveness of a reef in providing features that enhance survival, increase colonization, facilitate energy transfer, or maximize habitat diversity. Once broken down into smaller and more understandable components, the overall framework for the evaluation becomes directed and focused.

5.4.1 Types of Information

Another way of giving more concrete direction to an assessment and evaluation of artificial reefs is to organize the study design into levels of information needed to meet the evaluation needs. These levels increase in complexity as the information and degree of sophistication increase between successive levels. (See Section 5.6.1 for examples.)

The first level of information allows a *description of the assemblage.* This can include descriptive statistical attributes of the variables — the maxima/minima, number of observations, measures of central tendency (i.e., mean, median, and mode), and measures of dispersion (e.g., variance, standard deviation, standard error of the mean, and confidence limits) for each of the variables associated with the individual characters (e.g., abundance, size, biomass, and reproductive condition). It also includes statistically descriptive parameter values for the overall assemblage (e.g., number of species, number of individuals, and diversity indices). Thus, the first level of information should permit a clear picture of the variables and some idea of how they vary within a site at a particular time.

The second level of information is to make *temporal/spatial comparisons* of faunal assemblages between artificial reefs or artificial and natural reefs. The basis for making these comparisons is to test for statistically significant differences in the descriptive measures recorded at alternative times and places. Thus, questions can be answered relative to changes in the fauna between places or due to time. This information can form the basis for more detailed and expansive comparisons between habitats, conditions at several sites, or in a time series (days, months, years, and decades).

The third level of information involves taking the faunal parameter estimates and *identifying significant and interactive environmental factors*. Thus, more advanced levels of information are needed in an attempt to relate the variation found in faunal descriptions to the variation in environmental factors and attempt to recognize associations among them. The ultimate goal is to determine, with some degree of certainty, if there is a relationship between a factor or factors and the dependent faunal variables. Ultimately, if a clear relationship can be shown, then future artificial reef design, construction, and deployment can make use of such information to optimize the desired objective for constructing a reef.

5.4.2 Strategies of Assessment

When designing and conducting studies on artificial reefs, one must first define the question(s) being asked. This is the first step necessary to ensure that the study methods and statistical procedures can be aligned to specifically answer the question(s). Defining a question involves a logical process in which observations or knowledge are converted into a model that can then serve as the foundation for a hypothesis, a null hypothesis, and eventually an experiment or test. This process is presented in Chapter 2 and elsewhere (e.g., Lincoln-Smith and Samoilys 1997: Figure 2.1) and describes the falsificationist or refutationist test introduced by Underwood (1990).

5.5 ASSESSMENT METHODS

Properly choosing, developing, or designing methods appropriate to the evaluation of fishes and macroinvertebrates on artificial reefs requires an understanding of some general features or aspects specific to artificial reefs. This refers to both the reef and its associated fauna and the general classes and features of the methods themselves. This is important because each of these features (i.e., all aspects of the reef environment and methods used to assess them) are interdependent and can have a significant impact on each other. Understanding the nature and degree of this interaction is key to developing an effective study design.

5.5.1 General Considerations

Included here are those aspects of the artificial reef that theoretically are independent of the measuring technique. While many factors may never be completely independent, for practical purposes it is necessary to identify those environmental attributes that are apparently unrelated to sampling method.

The factor overshadowing all others in determining the most effective sampling technique is the physical nature of the reef itself. Reef profile irregularity affects the amount of effective reef-surface area through increased habitat heterogeneity. Unfortunately, this irregularity significantly impacts the effectiveness of sampling techniques (trawls, dredges, etc.) and interferes with visual censusing techniques as well.

5.5.1.1 Factors Affecting Fishes and Macroinvertebrates

Before an evaluation of the assemblage of fish and macroinvertebrates inhabiting artificial reefs can be contemplated, there must be some appreciation of the factors that may impact species composition of an assemblage and the condition or fitness of the individuals. These factors can be classified in terms of their presumed cause and effect relationship (i.e., as independent or dependent factors). Dependent factors are generally those we are trying to measure as having been influenced by some other, usually independent, factor or factors (Figure 5.3). Independent factors are those thought to be responsible for having an impact on the dependent factors. For example, fish growth is thought to be partially affected by temperature (i.e., growth is slower at colder temperatures). Growth (as measured by length increase over a certain time span) is the dependent variable and temperature is the independent variable. Our hypothesis is that fish growth (the dependent variable) is likely to be impacted by water temperature (the independent variable), whereas the converse is not true — water temperature will not be altered by fish growth. In other words, the independent variable, water temperature (the cause), has an effect on the dependent variable, fish growth (the effect).

Such cause and effect relationships can be complex. Both independent and dependent factors can interact simultaneously (Figure 5.3). Here we categorize these factors into two groups: abiotic (nonliving) and biotic (living) factors. Generally, in artificial reef studies we are interested in how abiotic factors affect biotic factors. There may also be a need to determine how some biotic factors impact each other. In contrast, an engineer might want to know what impact the living resources on a reef may have on the substrate composition surrounding a reef. In this case the substrate composition (an abiotic factor) is the dependent variable and the biotic assemblage components

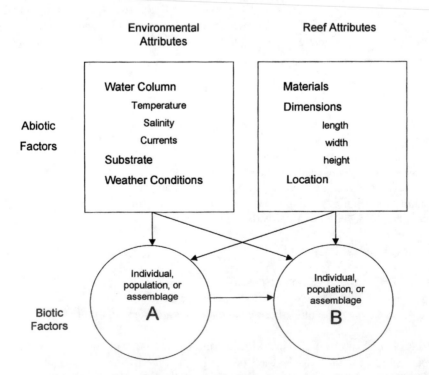

Figure 5.3 Schematic diagram of the relationships between biotic and abiotic factors on artificial reefs.

are the independent factors. Some factors may seem apparent, whereas others are merely suspected of having an impact and await further study.

5.5.1.1.1 Abiotic Factors — Abiotic factors are variables that may be understood more readily if we organize them into two distinct groups: (1) environmental factors (i.e., having features related to the natural environment); and (2) reef-attribute factors (i.e., related to the artificial reef itself; Bortone et al. 1997). This distinction is important if we are to determine those factors on a reef that can be controlled and how they interact with those factors that cannot be controlled. This perspective is essential when trying to improve artificial reef design. Environmental factors generally cannot be easily manipulated in the reef design, construction, or deployment process. This is in contrast to reef–attribute factors where manipulation is possible. An important consideration, however, is that it may be possible to make choices with regard to deployment conditions to maximize the potential of an artificial reef with regard to environmental factors.

5.5.1.1.1.1 Environmental Factors. There are potentially a great number of environmental factors that may be responsible for affecting fish and macroinvertebrate assemblages and their associations on artificial reefs. These environmental factors may have immediate impacts on the dependent variables, and consequently, may severely limit or impact the effectiveness of some assessment techniques. Such factors range from local weather and aquatic conditions, particularly water temperature, to more broad-scale factors such as season, lunar phase, and time of day (Table 5.1). There are numerous, generally localized, studies that have examined the effects of one or more of these environmental variables on artificial reef fauna assemblages, although to our knowledge this information has not been synthesized and reviewed. Selected works for coral reefs include Ebeling and Hixon (1991); Sale (1991b); Bouchon-Navaro et al. (1997); and Halford (1997). Special attributes of the effects of environmental variables are discussed below.

Fishes and macroinvertebrates display varying degrees of mobility. Many species are found on the reef, whereas others are located above the reef structure itself (i.e., are "reef-associated" *sensu* Choat and Bellwood 1991; Nakamura 1985). Therefore, it is important that environmental measures include information on attributes of the water column. Water-column changes in temperature, salinity, and turbidity can alter the distribution of many organisms. The depth at which other specific attributes or parameters change can be significant as well. A thermocline (i.e., temperature stratification related to depth) is known to occur seasonally in many areas. Although most often described as a temperature-related phenomenon, it may more properly be described as a measure of water masses of different density. Different water masses characteristically have different salinity and water clarity (among other factors such as productivity, plankton components, etc.).

Visibility is most often measured vertically in the water column by turbidity (using a Secchi disk) or by the amount of light. The latter is usually measured with a photometer, but may also be measured using the f-stop reading from the light meter of an underwater camera if done at fixed film and shutter speeds, always aimed in the same direction and plane. While measuring the vertical penetration of light has some value, horizontal visibility when recorded at depth may be more meaningful in assessing fish and macroinvertebrates. In some study areas, horizontal visibility (as measured *in situ* by divers using a Secchi disk or remotely by a horizontally directed light meter shielded from direct surface light) can give an important measure of the amount of light available for horizontal visual acuity. The amount of horizontal visibility can affect the effectiveness of a sampling device such as a trawl, and can have an obvious impact on an *in situ* observer's ability to perceive objects underwater (Nelson and Bortone 1996; Bortone and Mille 1999).

Many species feed on the infauna associated with the substrate surrounding a reef. Conversely, the type of substrate will determine the infaunal composition and relative abundance. A measure of the substrate can become an essential element in being able to associate fish and macroinvertebrate assemblage features with environmental variables. Composition of the substrate should therefore be recorded and should include information such as percent composition of inorganic vs.

Table 5.1   Abiotic and Biotic Variables Often Measured during Surveys on Artificial Reefs When Sampling for Fishes and Macroinvertebrates

Abiotic
    Environmental
        Water column
            Temperature
            Salinity
            Oxygen
            Visibility
                Horizontal
                Vertical
            Water Chemistry
            Currents
                Direction
                Intensity
            Surface conditions
                Wave height
                Wave length
            Depth
                Water depth
                Thermocline depth
        Substrate
            Composition
            Grain size
            Sand waves
                Crest to crest distance
                Crest to trough height
                Orientation
            Undermining of modules
            Vicinity substrate
        Weather conditions
            Air temperature
            Wind
                Direction
                Intensity
            Sky conditions
    Artificial reef attributes
        Materials of composition
            Type (metal, plastic, rubber, wood, etc.)
            Percent of each
        Module
            Dimensions
                Area
                    Length
                    Width
                Volume
                    Length
                    Width
                    Height
                Habitat
                    Hole size
                    Rugosity
                    Complexity
                    Void space
            Surface
                Color
                Texture
       Artificial Reef — Set and Group
            Module
                Types

Table 5.1   Abiotic and Biotic Variables Often Measured during Surveys on Artificial Reefs When Sampling for Fishes and Macroinvertebrates (Continued)

<pre>
                      Number
                  Total area
                  Total volume
                  Distance between modules, sets, and groups
              Location
                  Latitude/longitude
                  Distances
                      Shore
                      Other reefs
                          Natural
                          Artificial sets, groups, and complexes
                  Geological features
                      Shelf edge
                      Estuary
                      Anthropogenic structures
      Other
              Date of Deployment
                  Day
                  Month
                  Year
              Date of survey
              Time of survey
                  Local time
                  Absolute time from noon
                  Sunrise/sunset
              Lunar phase
                  Equatorial moon position
                  Tide
  Biotic
      Individual
              Life stage
              Size
                  Length
                  Weight
              Age and growth
              Condition factor
              Feeding
              Behavior
              Microhabitat preference
              Reproduction
                  Maturity
                      Season
                      Age/size
                  Fecundity
              Migration
      Population
              Abundance
              Density
              Biomass
              Structure
                  Size
                  Life stage
              Sex ratio
      Assemblage/community
              Diversity measures
                  Number of taxa
                  Assemblage indices
                      Shannon Diversity Index (H')
</pre>

Table 5.1   Abiotic and Biotic Variables Often Measured during Surveys on Artificial Reefs When Sampling for Fishes and Macroinvertebrates *(Continued)*

Evenness (J)
Similarity/association measures
Presence/absence
Relative abundance
Index of Biotic Integrity (IBI)
Colonization
Succession
Associated biotic variables
Infauna and flora/fauna measures
Diversity
Abundance
Biomass
Cover
Density
Fishing measures
Catch/landings
Number
Size/age composition
Biomass
Effort
Catch-per-unit effort (CPUE)

organic material, as well as a more detailed examination of the organic component (e.g., percent mollusk shell and percent grasses). Moreover, measures of substrate particle size, along with measures of sand-wave dimensions, can add a great deal of information about bottom currents with regard to their direction, strength, and duration. Likewise, examination of bottom sand waves near reefs can provide evidence that the reef structure is possibly being undermined by strong currents, with eventual destabilization of the artificial reef modules.

Surface weather characteristics should also be recorded as important environmental factors. These may include measures of meteorological conditions (e.g., wind direction and strength; clear weather, partly cloudy, or overcast; and rain), as well as surface water and wave conditions. Local weather may have a negligible effect on deep-dwelling benthic organisms, but it may impact the effectiveness of assessment methods. Moreover, weather conditions may affect the movement (direction and distance) of motile organisms. Therefore, the inclusion of weather conditions in the analysis of variables may help qualify or give meaning to the study results as well as be of interpretive value for the study results.

5.5.1.1.1.2   Reef Factors. Reef-associated factors are those variables that are controlled by the reef designer, builder, and to some extent by the deployer (Table 5.1). Since reefs often are deployed as modules in groups or sets (Figure 5.4; *sensu* Grove and Sonu 1985), it is more accurate and easier to measure each individual module before deployment. Additionally, because reef deployment may not always proceed as planned, actual postdeployment dimensions also must be measured, at least with regard to the set and group dimensions. Postdeployment measurements can be made *in situ* by a diver using a fiberglass or waterproof tape measure or even a length of line (preferably weighted), marked off in meter intervals. With the recent advent of GPS (Global Positioning System) and its more sophisticated counterpart enhanced by radio signals, DGPS (Differential Global Positioning System), when used in conjunction with Sidescan Sonar (Lukens et al. 1989), or with aerial detection using LIDAR (Light Detection and Ranging, i.e., laser radar; Mullen et al. 1996), it is possible to accurately determine reef-set or group dimensions and relative positions remotely.

Linear dimensions of the reef modules, sets, and groups are important, but equally important are the area and volume aspects of the reef. Again, predeployment determinations of these parameters are much easier and more accurate than those conducted by postdeployment measures. Most

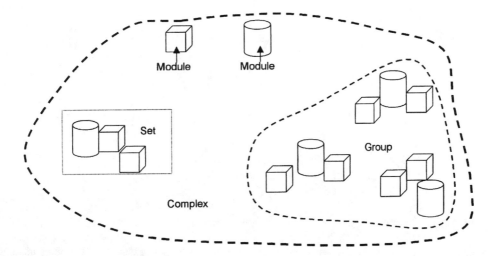

Figure 5.4 Groupings of reef to indicate relationships between modules (units), sets, groups, and complexes (modified from Grove and Sonu 1985).

postdeployment measures of area and volume are mere calculations based on the linear dimensions. For some objects these determinations can be simple and straightforward, but for other modules and configurations these measures are extremely problematic and potentially inaccurate.

Module composition may also affect the fauna associated with a reef. Actual measures of the compositional components of artificial reefs (or even estimates of their percent composition), either by volume or weight, are certainly important items of information. Usually estimates of reef composition are straightforward, as most reef modules are composed of the same materials. Occasionally, however, it is important to determine the percent composition of the modules themselves (either by volume or weight), such as occurs when tires are imbedded in concrete. Also, the reef set may be composed of several types of modules, each of a different material. Predeployment determinations of this are obviously the most accurate measure.

Just as significant in artificial reef assessment may be reef surface texture. Chemical composition of the surface as well as its porosity, roughness, and rugosity are all factors that can affect bacterial and subsequent sessile invertebrate colonization. Few studies have examined the importance of reef material color in the colonization and succession of reefs, but it may prove to be a significant factor, especially during the initial stages of colonization before complete encrustation occurs.

Some measures of reef dimension that can prove useful in assessing a reef fish and macroinvertebrate assemblage are the architectural complexity (i.e., rugosity) of the reef itself (Luckhurst and Luckhurst 1978). While complexity is usually an approximation, it can be loosely defined as the amount of irregularity to the structure. It can be more accurately defined as the ratio between surface perimeter and linear distances between two points (Figure 5.5). Bortone et al. (1997) used an arbitrary qualitative scale of 1 to 20 to indicate complexity, especially as it relates to cryptic habitat. For example, a box with no holes would have no cryptic places and therefore a complexity of 1. Conversely, a bridge trestle with numerous openings and passageways would have a complexity of 20. In a nonartificial reef environment, a similar arbitrary scale was used by Francour et al. (1995) to classify rocky biotopes. Users may be able to devise and define their own definitions and scaled scores of complexity to meet the specific objectives of a particular study.

Another useful measure relative to reef structure is referred to as void space. Void space can be an indicator of the relative amount of habitat preferred by some species. Some grouper species, for example, may prefer to "hide" within crevices and the size of the crevices may limit the size of the grouper that can reside in association with the reef. Void space is usually a qualitative estimate of the relative percent of empty reef. For example, an empty barrel would have 100% void space,

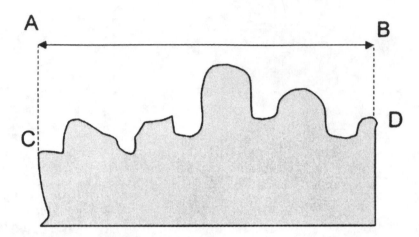

Figure 5.5   Rugosity or surface irregularity, measured as a ratio between the distance along the surface (C to D) and the straight-line shortest distance between two perpendicular points (A to B).

while a barrel filled with concrete would have 0% void space. The scale of estimation becomes somewhat subjective if done *in situ,* although if done during predeployment it can be an objective measure. On large heterogeneous substrata such as harbor dikes, Ruitton et al. (in press) estimated the degree of anfractuosity (Anf.) as the ratio between the cavity surface and the total surface expressed in percent: $Anf = 100 * b^2/a^2 + b^2)$. The cavity parameter is measured *in situ* by 50 separate measures of the distance between modules and the total surface area estimate of the module (Figure 5.6).

Location variables are essential to reef assessment studies. Longitude and latitude should be recorded, preferably with a differentially corrected GPS (i.e., DGPS), and the units, whether recorded as degrees, minutes, and seconds, or decimal degrees, should be clearly stated. Loran estimates of location are still used and may be found in the older literature; however, the continued use of Loran (or more accurately, Loran C) should be discouraged. Moreover, if Loran is used, the

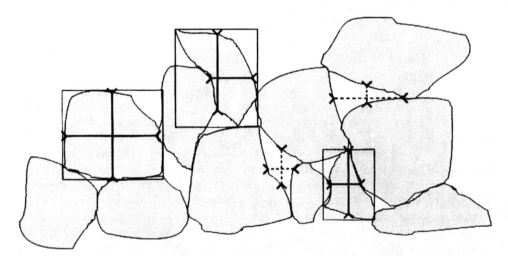

Figure 5.6   Anfractuosity is a cavity parameter, measured as a ratio between the distance between the modules and the estimated total surface area of the modules as an average from 50 separate measures. The solid lines indicate the measurements on the modules for the length and width dimensions to estimate area. The dashed lines are linear distances between modules.

delay in microsecond data should be converted into latitude/longitude. Users should be aware that conversion algorithms often do not produce a result that is an exact conversion of Loran to GPS latitude/longitude.

Geographic/geological information can be especially important since there are factors associated with formations, structures, and features that can significantly impact artificial reef faunas. The database can include geographic/geological reference information such as distances to natural reefs, estuaries, river mouths, the shelf edge, or other submarine structures such as canyons and outcroppings. This can include distances to biogeographic boundaries. Distances to unique anthropogenic features such as cities, effluent outfalls, harbors, etc. can also be important when trying to determine the factors that affect the fish and macroinvertebrate assemblage.

Additional variables that should be considered as part of the environmental database are date and time of the survey and date the reef was deployed. The age of a reef is important inasmuch as succession following initial colonization is as likely to dictate assemblage structure as month or season of deployment. The time at which a survey is conducted is also critical. Activity rhythms of organisms often fit into patterns of social organization that may positively or negatively affect observation or capture. Local time should be recorded. More importantly, it is necessary to calculate the time relative to natural timing events (i.e., zeitgebers) such as time from sunrise/sunset or time from absolute noon. Sanders et al. (1985) and Martin and Bortone (1997) used the absolute-time-from-noon variable to standardize the natural timing event of noon. Likewise, lunar phases should also be recorded. Not only do these affect tides (and therefore currents), but activity patterns (e.g., reproductive aggregations) are strongly influenced by lunar phases (e.g., Hastings and Bortone 1980). To facilitate numerical analyses, equatorial moon position (EMP) can be calculated to assist in providing magnitude to the lunar phase variable (Sanders et al. 1985; Martin and Bortone 1997). If accurate calculation of the lunar phase proves too cumbersome, a simpler approach would be to record the lunar stage as reported in the local newspaper or almanac.

5.5.1.1.2 Biotic Factors

Biotic factors may be defined as the dependent variables to be measured in assessing a reef's macrofauna. These factors may be categorized logically into three main types: individual, population, and assemblage/community. Although usually considered a dependent (i.e., response) variable, in some cases biotic factors can also be treated as independent variables. For example, a variable such as amount of algal cover, coupled with number of sea urchins, also may be used as a predictive variable to explain the presence, abundance, or condition of the fish and macroinvertebrates (especially herbivores) in which we are interested. Here we term these *associated biotic variables*.

Associated biotic variables such as measures of the infauna and flora (Table 5.1) are important factors since they may impact the fish and macroinvertebrates associated with artificial reefs. These organisms can be measured on individual, population, or assemblage levels. Other associated biotic factors include measures of fishing pressure or effort. Although often difficult to obtain, catch-per-unit effort (CPUE) can be significant in explaining variation in dependent response variables. Other important factors include catch/landings from the local area, as well as individual, population, and assemblage characteristics of the catch.

5.5.1.1.2.1 Individual Factors. Organism variables or factors associated with the individual are generally considered to be life-history features that characterize the status or condition of an organism. These include life stage (i.e., larva, juvenile, or adult), size (length or weight), growth rate, condition (ratio of length to body weight; Prince et al. 1985), reproductive condition, reproductive season, and fecundity (number of eggs produced). Other variables include local and long-term movements as well as species-specific behavior. Of the variables listed above, life stage and size estimates seem to be those most commonly recorded in assessing macrofaunas. Above all, the parameters chosen should reflect the questions asked in a reef-assessment study.

In any assessment involving living organisms, inherent features of the species must be considered when designing and conducting an assessment. Behavioral features are among the most important factors affecting survey results (e.g., Lindquist and Pietrafesa 1989). They represent an immediate response that an organism makes to a circumstance and can be of value in assessing the effects of environmental change. Although these behaviors can be difficult to control or manipulate, an understanding of a species' behavior is essential to effective project design. Some species are more likely than others to appear in collection samples (or observed during visual surveys) during an artificial reef assessment. By contrast, the secretive behavior of other species may result in underestimation of actual densities. For example, some of the most important commercial species in the northwestern Mediterranean region belong to the families Sparidae, Moronidae, and Carangidae. These fish are strong swimmers and are readily observed, even though they may be less numerous than other cryptic reef-dwelling species. Any assessment of an artificial reef's effect on small-scale fisheries should include a sampling design that takes into account the behavior of individual species. Otherwise, the data may be biased. On the other hand, a species' behavior can be used advantageously, thus increasing its chances of detection. This is especially useful when the objective of a survey is to obtain information on a single species or group of species with similar behavioral attributes.

Other sampling biases may be related to physical attributes of the species, including body morphology and color, coupled with the habitat in which they live. Visual observation can be affected by camouflaged species whose body shape and/or color merge with the background, such as frogfishes (Antennariidae), pipefishes (Syngnathidae), and several families of flatfishes (Bothidae, Soleidae, etc.), as opposed to the brightly colored species of butterflyfishes (Chaetodontidae) and tropical damselfishes (Pomacentridae).

5.5.1.1.2.2   Population Factors. Population variables include estimates that pertain to the overall status of a single species on a reef. Included among these variables are estimates of abundance, density (number of individuals or weight per unit area or time), biomass (estimated weight of entire population), population structure (i.e., the proportion that each life stage or size group makes toward the entire population), and sex ratio. Here we use "abundance" as a general term for overall number of individuals, whereas density refers to numbers per unit of measurement, usually area. "Biomass" refers to overall weight of the population, which more specifically may be defined as weight per unit area (e.g., kg/ha).

5.5.1.1.2.3   Assemblage Factors. Assemblage or community variables, such as species richness and abundance, can be used to calculate coefficients or indices that indicate some scale of relationship among the species or assemblages under consideration. The most commonly used indices involve species diversity (Shannon Diversity Index [H']) or other assemblage diversity indices such as evenness (J). Similarity (or association) indices are often used to gauge the similarity (or dissimilarity) of one assemblage to another (e.g., cluster analyses based on a variety of indices [Ludwig and Reynolds 1988]). These comparisons can be done over time in the same place or between places, and can include estimates of abundance or simple measures of presence/absence of species within an assemblage. Indices of biotic integrity (IBIs) have been developed in freshwater systems in order to establish some level of comparison with other assemblages in time and space (e.g., Karr 1981). It can be anticipated that IBIs (which are multidimensional and involve both biotic and abiotic parameters) will be used in estuarine or nearshore and coastal areas (where artificial reefs predominate). Multivariate analyses, which can statistically compare communities, are sophisticated (Manly 1986; Clarke 1993) and have been applied successfully to a number of species assemblages in tropical fisheries (Fennessy et al. 1994; Anderson and Gribble 1998; Connell et al. 1998).

The multidimensional nature of artificial reef surveys implies that water-column height above the reef adds a component of complexity to the survey techniques to be adopted. Sampling of organisms throughout the entire water column is important, but at the same time poses special problems.

Artificial reefs often contain a unique mixture of species different from those found in natural reef communities. Consequently, extra care should be taken to evaluate habitat associations, as well as microhabitats, that may be somewhat altered from their presumed natural state. Although schooling and loose aggregation behaviors of fish species are not unique to artificial reefs, the assemblages associated with artificial reefs often include schools comprising several species, in which individuals of different species are of similar appearance and behavior. Considering this, care should be taken in identifying the correct proportions of the component species.

5.5.2 Sampling Scale

Recognition of the scale at which one is conducting a study is critical toward designing and conducting an artificial reef assessment study (Schneider 1994). Scale relates to time and space and must adequately address the problem at hand. Adequacy of scale will also ensure that the sampling design is sufficiently robust to permit effective evaluation of unanticipated responses or conditions. For example, it is unlikely that one could conduct a study at one time on a 1 ha reef and then extrapolate the results to all reefs and all years. Underscaling or overscaling the sampling and assessment design can be counterproductive and lead to misdirection of effort. The two most critical aspects of sampling scale (i.e., time and space) are discussed below.

5.5.2.1 Time

Since faunas are comprised of active and interactive organisms, their behavior often varies considerably on several time scales. Thus, time of day, season, and year (because of possible long-term cyclical changes) are critical to sampling results and the survey as a whole. Time scales can be closely related to two environmental factors, lunar phase and tidal state, both of which can have significant impacts on a reef fauna. If season, lunar phase, and tidal state are relevant to the question(s) being asked, sampling design must accommodate these time intervals. For example, to effectively gauge the impact of lunar phase on an assemblage, sampling must not only occur within every 28-day period, but also be done during several lunar phases. Observations and/or sampling should ideally be conducted during the same general time of day. This will help account for habits of individual species, some of which are diurnal (i.e., active during the day), others are nocturnal (active at night), and others are crepuscular (periods of change from day to night or vice versa, when light is dim). Visual observations, for example, are best conducted during the day under optimal lighting conditions. However, many species are more active and the visible faunas can be greater during crepuscular periods (Hobson 1991).

Tidal cycles can be an important factor, particularly in shallow inshore areas and in geographical areas where tidal amplitude (and current) are likely to be high. Even if these factors are not obviously relevant to the study objectives, it is nevertheless desirable to standardize sampling or observational procedures to minimize any potential unknown bias. If the surveys are to have long-term general management applications, it is desirable to repeat the surveys over a number of years.

Artificial reefs are often placed in inshore areas and are thus more subject to the vagaries and influence of seasonal changes in water temperature, salinity, and water clarity. When comparing reefs from different locations, it is important to conduct the surveys during the same seasons. Oppositely, it is preferable to conduct faunal evaluations over all seasons in order to assess the impact that seasonal differences may have on the assemblage composition and faunal structure.

5.5.2.2 Space

There are a number of spatial scales possible in reef assessments that relate both to the study's question(s) and methodological bias. A large-scale study may involve several artificial reefs, whereas, at the opposite end of the scale, a study may consider faunal assemblages in microhabitats within a reef. More typically, an artificial reef may be partitioned into spatial strata that relate to its orientation and shape. For example, one might divide a reef into leeward and windward faces, or reef tops and reef sides. Certain questions may require that surveys conducted over extended areas include many microhabitats, while some survey objectives may involve only one or a few microhabitats. Greater microhabitat diversity in a survey will result in increased species richness and diversity. Clearly, the choice of spatial levels relates to the question(s) being asked and, ultimately, deployment of the reef.

Once the appropriate level of spatial scale is determined, the appropriate assessment design can be established. Most survey method protocols have a spatial component. Typically, surveys are conducted over a predetermined area that is measured as accurately as possible. Thus, survey data are collected and expressed per survey area (e.g., density or number of individuals per unit area). The survey area is the smallest spatial sampling unit (by area) in a study and defines the dimensions of the replicate sampling unit. Thus, if a visual census encompasses an area of 250 m^2 (a 50-m long × 5-m wide transect), then all subsequent visual censuses should be identical in survey area.

5.5.3 Assessment Techniques

There are numerous problems and circumstances that may dictate the methods that can be used; the specific methods employed often define the limitations of the data. For example, if a visual census is used, generally only diurnal species will be assessed. It is obviously important for researchers to be aware of features of the subjects, environment, and other related factors that may *a priori* put limitations on a particular method. Moreover, understanding these limits will ensure meaningful interpretation of the results. An evaluation of the relevant abiotic and biotic factors (see Section 5.2 above) is critical to this understanding.

The assessment methods described below have been applied to artificial or natural reef assessments of fishes and macroinvertebrates. These are reviewed with the positive and negative attributes of each noted. It is suggested that assessors use methods that have been previously employed, especially if comparisons are to be made with data previously collected in the same area. In general, there should be strict adherence to protocols when using a particular method (e.g., amount of survey time, area survey, time of day, or associated environmental measures). Failure to do so may invalidate even the most basic comparative analysis.

Having stated this, it is appropriate to make the seemingly contradictory observation that experienced investigators who are familiar with a particular method's strengths and weaknesses sometimes are able, in special situations, to introduce changes or modifications that may enhance the utility of a particular assessment technique. This is especially true when certain questions require innovative data collecting methods. Thus, while it is generally important to follow established procedures, flexibility and innovation can lead to the development of new or improved assessment methods.

In this section, we have divided the methods into two groups — capture methods that generally involve destructive sampling, and observations that are nondestructive and range from underwater visual censusing techniques to remote sensing methods such as hydroacoustic technology. Table 5.2 summarizes the most often used techniques.

5.5.3.1 Destructive Assessment

This general class of methods has been historically used by commercial, recreational, and subsistence fisheries, and includes trapping, hook-and-line fishing, trawling, netting, spearfishing, poisoning, and explosives (Cappo and Brown 1996). It is important to note that capture methods are not always destructive because fish can be removed unharmed from hooks, traps, and some nets, and these methods are regularly used for obtaining certain kinds of life-history data (e.g., movements using tag-and-release techniques). Although these have proved effective, one should be aware that significant alterations in behavior might result from capture, handling, and release

**Table 5.2   Sampling Methods (by General Classes) Used to Survey and Assess the Fishes and
Macroinvertebrates Assemblages on Artificial Reefs**

Destructive
    Fishery Dependent
        Hook-and-line
            Rod-and-reel
            Long line
            Trotline
        Netting
            Active
                Trawl
                Towed nets
            Passive
            Gill nets
            Trammel nets
            Pound nets
            Fyke nets, etc.
    Fishery independent
        (Same gear as above but under more controlled conditions as to time, area, gear size, etc.)
        Ichthyocides
        Electrofishing
Nondestructive
    Visual
        Transect
        Quadrat/point-count
        Random swim
        Interval
        Total counts
        Combinations
    Nonvisual
        Hydroacoustic

of individuals. Most capture methods (e.g., trawls, gill nets, and ichthyocides) will result in permanent removal of individuals, thus altering the assemblage and perhaps affecting faunal surveys in the same area in the immediate future; however, over an extended period (e.g., more than a year) faunal surveys based on capture techniques can be effective.

The great advantage of capture techniques relates to the collection of specimens that, after being sacrificed, can yield certain life-history data not obtainable from living individuals, such as reproductive condition (including fecundity), age (based on examination of otoliths or bones), and feeding habits (through analysis of stomach contents). This also results in the acquisition of voucher specimens that serve as permanent records (when properly preserved) and from which additional data can be gathered over the indefinite future. Another advantage of these destructive techniques is that they are effective at night. The researcher can use these techniques to sample nocturnal assemblages. When coupled with diurnal sampling, this permits an assessment of the entire artificial reef assemblage.

Overall, probably the most common fishing method used in marine sampling is trawling. However, because the effectiveness of trawls is severely compromised by uneven bottoms that characterize reefs, traps and other passive gears are more effectively used in, or contiguous to, the reef itself.

Destructive or capture-sampling techniques may be of two types: those that are fishery-dependent and those that are fishery-independent (Samoilys and Gribble 1997). Fishery-dependent techniques rely on commercial and/or recreational fishing activities, which employ gear such as hook-and-line, traps, and nets to obtain both catch and effort data (including area, time, type of gear used, etc.), as well as organisms for life-history assessment and voucher specimens. Researchers often accompany fishers and record faunal information directly from the catch, sometimes supplemented by information from the fisher's logbooks. Fishery-dependent techniques can also provide a very realistic measure of the actual fishing mortality to which faunas are subjected. As discussed above, however, these fishery-related techniques normally target fishes around the outer areas of artificial reefs, rather than inside or between artificial modules, sets, or groups. Thus, only one part of a fish assemblage can be sampled using these methods. Although these techniques may provide only rough estimates of density and biomass, if the main objective of a study is to assess the value of an artificial reef to the local fisheries, sampling proximate to the reef is a relevant assessment method protocol.

Fishery-independent methods are employed by researchers to sample the fishery using structured, controlled, and special sampling designs. Capture methods are often the same as those employed by both commercial and recreational fishers, but are under the control of researchers and with specific objectives in mind. The structured sampling may include designs based on areas and times of day or at depths not normally sampled or fished by a fishery. Special sampling designs may include randomized sampling strategies based on time and/or area, as well as use of special sampling gears. Some of these gears may target limited portions of the fauna. For example, a fine-meshed net (which could not be legally used by recreational or commercial fishers) may facilitate the capture of small specimens or species not normally obtained by legal gear. The advantage of using fishery-independent sampling techniques is that these methods can be exploratory in nature and thus assess broader spatial, temporal, or age-size specific components of the fauna. They also reduce the bias associated with fishery-dependent data. Additionally, their controlled deployment permits investigations of the catch/effort potential of particular gear. Conversely, these techniques do not permit accurate comparisons with sampling results based on fishery-dependent data. These techniques also may prove to be more destructive to the habitat and faunal components than fishery-dependent techniques. As for fishery-dependent methods, they are usually directed toward certain types of organisms (e.g., purse seines for schooling species; hook-and-line for larger predators), and generally do not sample for the broader components of the assemblage. Ichthyocides and explosives (especially the former) are generally considered to be the least biased sampling methods (Smith and Tyler 1972; Goldman and Talbot 1976; Russell et al. 1978; Thresher and Gunn 1986; Samoilys and Carlos in press). Ichthyocides will result in assessment of the broadest components of an assemblage, but are less effective on air-breathing fishes or species living in low-oxygen environments. However, the consequences of complete faunal removal may be too environmentally intrusive. Explosives have sometimes been used, and may result in a complete kill for certain species (particularly free-swimming species having an air bladder; Samoilys and Carlos in press). They are much less effective for benthic species in which the air bladder is absent. Also, explosives can result in extensive physical damage to the environment.

Readers should refer to fisheries biology methods references (e.g., Schreck and Moyle 1990; Cappo and Brown 1996; Murphy and Willis 1996) on how to plan and execute fishery-independent and dependent sampling techniques. Most references to the standard fishery literature can be applied (with some modification) to assessments of artificial reef-fish faunas.

5.5.3.2 *Nondestructive Assessment*

Nondestructive assessment methods (which include observational or visual methods) are generally preferred for sampling artificial reefs because they do not result in destruction of fauna or remove them from the environment. Their major drawback is that they do not permit capture of specimens to obtain life-history data and often do not permit accurate taxonomic identification. Nondestructive techniques are often more easily adapted to survey heterogeneous habitats and can relate more readily to species with specific microhabitat affinities. In addition, they can be targeted to certain portions of the assemblage without impacting the other affiliated organisms. Clearly, the major advantage of using nondestructive techniques is that repeated sampling can occur in the same place with brief intervals between samplings. The assessment method itself should have little or no impact on the faunal structure and life-history aspects of the organisms. Observational techniques can currently be categorized into two groups: (1) visual census techniques; and (2) hydro-acoustics. The survey method must be chosen with care. Some of the factors that should be taken into consideration when making this choice include: past experience, costs, training requirements, circumstances, objectives of the study, and the need to be consistent with other studies.

5.5.3.2.1 Visual Census — Visual assessments of fish and macroinvertebrates are the techniques most frequently used for obtaining population, assemblage, and (more recently) life-history data on these organisms (Barans and Bortone 1983; Harmelin-Vivien et al. 1985; Thresher and Gunn 1986; Bortone and Kimmel 1991; Samoilys 1997; Samoilys and Carlos in press). There are some obvious advantages to using a visual assessment technique when assessing fish and macro-invertebrates on artificial reefs, including adaptability of the techniques to a wide range of conditions. Since reefs are of varying sizes, dimensions, compositions, and arrangements, visual techniques quickly can be modified or adjusted to accommodate most circumstances. Visual assessments permit the survey of high-relief structures, which would otherwise ensnarl or render ineffective most surface-tended sampling gears. Another advantage is that visual assessments can follow very specific protocols, with strict limitations on time and area, in order to conform to specific community parameters. Other advantages include putting the observer in intimate contact with the organisms and habitats being surveyed, thus permitting the collection of considerable ancillary information.

The disadvantages of visual sampling are depth constraints and dependence on visual cues. Poor visual conditions result from lack of light and/or poor water clarity (Bortone and Mille 1999). Light reduction may occur at night, on overcast days, or in deeper waters. Turbid water may result from the presence of suspended sediments, mixing of water masses of different salinities, or high plankton levels. Most visual assessments to date have been done by an *in situ* observer using scuba gear, an oxygen rebreather, or snorkel. Each of these types of gear has its own limitations and associated costs and risks, and training is necessary (particularly for the first two) for implementation.

There are five broad categories of visual census methods: (1) strip transects; (2) stationary point counts; (3) line transects; (4) interval counts, and (5) total counts. Strip transects and stationary point counts are the most widely used for fishes (Sale 1997). There are many variations on these, and individual aspects of one method are often borrowed to form new or modified assessment methods (e.g., Kimmel 1985). Here we present a general description of each, with an indication of their use and circumstances pertaining to their application.

5.5.3.2.1.1 Strip Transects. Strip transects, sometimes called belt transects, were first adopted by Brock (1954) for underwater assessment of fish. They have since been used extensively in studies on coral reef fish populations (Sale and Douglas 1981; Sale and Sharp 1983; Thresher and Gunn 1986; Cappo and Brown 1996; Samoilys and Carlos in press). The general protocol for data gathering calls for two observers (paired for safety purposes) to move along a transect of specified length and width. Specific aspects of the protocol call for standardizing the observation zone (i.e., width, length, and height), its duration, and direction of the transect (Figure 5.7). Methods using transects often include protocols in which only the width (typically from 1 to 5 m wide) and length (which varies considerably between studies, from 5 to hundreds of meters) dimensions are specified. Recently, users have defined the upper visual zone of the transect as well. It is recommended that time of observation should be standardized (Samoilys 1997) because the amount of time one spends within a survey zone can dictate both the species observed (Lincoln-Smith 1988) and accuracy of density estimates (Watson et al. 1995). Some studies indicate duration of a survey to be one of the most significant variables affecting survey results (Bortone et al. 1986; Bortone et al. 1989). It stands to reason that the more time devoted to a survey the greater the opportunity to "catch" an individual. A pilot study should be conducted to determine optimal survey times for individual species (Samoilys 1997).

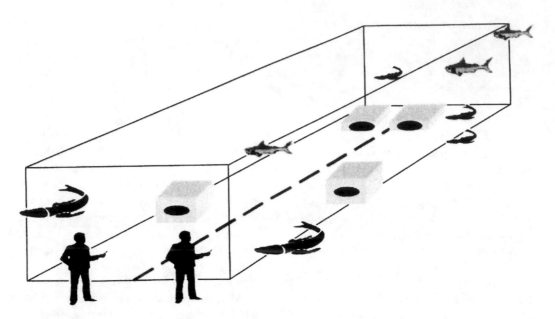

Figure 5.7   Illustration of transect dimensions.

Measuring tapes, lines, or fixed stakes are used to mark transect boundaries. Although pre-deployed transect lines often have been used in underwater surveys (Bortone et al. 1986), this procedure is not recommended because the amount of time involved in setting out a line can severely limit one's observation time and disturb the fauna (Samoilys 1997). The advantage of strip transects in visual assessments is that observers can study some individuals "close-up" and conduct replicate surveys in specific areas and habitats (e.g., natural reefs and grassbeds).

Although widely used in artificial reef surveys, strip transects are not always easy to deploy because most artificial reefs are not sufficiently long to accommodate a transect of more than 10 m in length. Surveys that take place on large artificial reefs or in areas with linearly deployed modules are best surveyed using this technique. Transect techniques also can be used to sample areas around artificial reefs for assessing the reefs' impact on the local fishery.

5.5.3.2.1.2   Stationary Points. The point-count method (Bohnsack and Bannerot 1986; Samoilys 1997; Samoilys and Carlos in press) is basically a quadrat type of survey in which a prescribed area is surveyed by a relatively stationary observer (Figure 5.8). The area is usually of geometric shape (usually circular) with specific dimensions surveyed over a stipulated amount of time (Bortone et al. 1994b; Falcón et al. 1996; Bortone et al. 1997). Dimensional lines may or may not be deployed at the time of the survey, although some investigators prefer to use some type of guide line to estimate dimensions (Samoilys and Carlos in press). In some cases the protocol will call for some limitations on the upward boundaries of the survey area, and in other cases the time of the survey may not be precisely stipulated. One of the most frequently cited examples of a point-count survey (Bohnsack and Bannerot 1986) is limited to 5 min, but is indefinite with regard to time required for identification of cryptic species. This can add additional variance to survey methodology, especially when comparisons are made between areas with unequal hiding spaces

To conduct a point-count survey, a diver/observer occupies the center of the circle or quadrat, and slowly turns for a prescribed period of time (e.g., 5 min), while simultaneously recording the fauna observed within the area and height dimensions dictated by the protocol. Survey duration should be dictated by the size of the census area and the number of species being censused. For example, few species and a small area can be censused more quickly. The fish census in point-count surveys should approximate an instantaneous count (i.e., the density reflects the number of

Figure 5.8 Illustrations of dimension and survey area of a typical point-count survey.

individuals counted in the survey area at one moment). This is also true of strip transect surveys. However, with point-counts it is readily apparent that for the observer to complete a rotation takes time, during which new fish can swim inside the circle and fish can swim outside the circle. To avoid the former, the observer should not count individuals that enter the survey area after the count has started (i.e., "incoming fish"; *sensu* Samoilys and Carlos 1992; Samoilys 1997). To avoid missing fish that leave the area (i.e., underestimating density), the census should be as rapid as possible, and the number of species being censused simultaneously should be minimized (Samoilys 1997). Alternatively, Brock (1954) established a widely followed protocol in which once a visual survey is initiated, all fishes are counted that enter the visual zone (even after the enumeration process has begun), but care is taken to count fish only once. Thus fish that leave and reenter the survey zone are not counted again.

The point-count method has many of the advantages of the transect method (Samoilys and Carlos in press), with the added advantage that the survey often does not extend beyond the limits of the reef. Most radii are between 3 and 15 m, depending upon visibility and study objectives, but Bortone et al. (1994a) established the 5.64 m radius as a standard, inasmuch as it encompasses exactly 100 m² and is still within the distance of visual limitation one is likely to encounter. Also, the limited dimensions of the survey area permit assessment of a reduced number of character states within the habitat. Thus, microhabitats within the reef or module within a set can be sampled during a point-count survey. Habitat attributes are thus easily ascribed to an assemblage.

When the dimensional area is large (i.e., at the limits of visibility), it is often difficult to accurately identify or even observe some individuals. Although this is true of the transect method as well, that method requires the observer to move along the transect and thus decreases the distance between the observer and the object, whereas most point-count protocols call for a stationary observer. Care must therefore be taken in selecting appropriate dimensions for the species and the conditions being surveyed. Moreover, those using the point-count method can cause damage to the substrate if divers are not appropriately buoyant to avoid contact with the delicate substrate. Transect observers are swimming and thus off the bottom.

5.5.3.2.1.3 Line Transects. Line transects have been used extensively in terrestrial studies, and have potential in aquatic surveys (Thresher and Gunn 1986; Kulbicki 1988; Cappo and Brown 1996). Briefly, the observer records each individual observed and its distance and angle from the line of the transect. The advantage of this method is that differences in subject detectability can be readily determined, whereas this is an unknown aspect of the strip transect. Conversely, this method is particularly dependent upon the observer's ability to make accurate estimates of distances

Figure 5.9 Illustration of "random" survey path by a diver during an interval count.

underwater. Bortone and Mille (1999) have outlined the numerous potential causes for errors in making distance and size estimations underwater. Thresher and Gunn (1986) have indicated that differences in distance estimations between observers can be substantial, especially when estimating distances for rapidly moving species such as those characteristic of the family Carangidae.

5.5.3.2.1.4   Interval Counts. The interval count or timed swim (Species-Time Random Count) involves a modification of the method first used by Beals (1960). In this case, a person "haphazardly" swims over the general survey area (Figure 5.9), listing species observed either within a set time period (e.g., Williams 1982; Russ 1984) or in the order in which they were initially seen (Jones and Chase 1975; Thompson and Schmidt 1977; Jones and Thompson 1978). In the original literature (Beals 1960) this method is referred to as "random," but as used it lacks the rigor of true random movement. Through repeated encounters, species are given scores of relative abundance based on the time interval in which they were initially observed in the survey. Theoretically, species that are very abundant and obvious would always be seen earlier during every survey, and rarer and less obvious species would be seen only once, usually at some later time during a survey. Although there could be a number of variations related to scoring of relative species abundance, many authors studying areas of high diversity choose to use 5, 10-min intervals, each of which is repeated eight times. Thus, a species that is very "abundant" would receive an abundance score of 40, whereas the rarest species would receive a score of 1. Those using this method in areas of lower diversity, or in a more limited area, would probably choose to reduce the time interval (e.g., 5 min).

This method has gained some limited acceptance by fish-assemblage surveyors for several reasons. It permits observers to move freely about the reef area, thus resulting in discovery of more species and allowing for observation of additional reef features than would be possible during a more limited visual survey. The species richness parameters generated from such surveys are very high (Bortone et al. 1989). The primary goal of this survey technique is qualitative and not quantitative. It offers the opportunity for the diver to record as many species as possible while not being distracted with the activity of counting individuals. It does provide an indication of relative abundance (similar to the rank order of some nonparametric statistical treatments) as a rank of abundance can be gleaned from the species scores, weighted according to the time interval in which they were first observed. Because of the high potential for misrepresentation, especially because of rarefaction, at least eight replicates are necessary. There are various factors that contribute to misrepresentation of relative abundance, of which two are cited here: (1) a rare species observed, by chance, at the beginning of a survey would be given a high score (minimally 5), whereas in fact it might well be missing from subsequent replicates; and (2) a numerically abundant schooling species could easily be missed if the school happens to be outside the viewing area, as might often be the case during any particular dive. This method is also time consuming. Five, 10-min replicates, each conducted eight times, means a total underwater observation time of 400 min. Usually a visual survey using a transect or point-count technique would last no more than 10 min for each replicate. Thus, with a minimum of five replicates each survey would have a duration of no more than 50 min. It is important to realize the extreme time investment necessary to effectively conduct the interval count technique.

5.5.3.2.1.5   Total Counts. Many researchers have chosen to use a "total-count" technique that literally has no bounds of time or area. Some workers have found this technique useful, especially when the reefs are small and relatively isolated (Hixon and Beets 1989). Under these circumstances, all species are identified and all individuals counted until one is satisfied that no other species remain and quantitative data cannot be improved upon. Under some special conditions and assumptions (e.g., that the fauna is relatively fixed and immobile), this method may have practical applications. Following this method, the observer positions himself/herself near a reef and counts all species and individuals until all resident species and individuals are believed to have been counted.

5.5.3.2.1.6   Combinations of Methods. Since no single method may yield a faunal assessment that is clearly more accurate or precise than others, many researchers have chosen to use two or more methods for obtaining the best results. For example, transect, point-count, and timed swim techniques can each be employed simultaneously, thus utilizing the positive attributes of each and also permitting a realistic assessment of the biases and shortcomings of each (Kimmel 1985; Bortone et al. 1986; Bortone et al. 1989).

5.5.3.2.1.7   Additional Aspects in Visual Census. Most fisheries scientists, to date, do not necessarily consider the visual gathering of data a valid exercise (Bortone 1998). While their bias is hardly justified, considerable effort has been made in recent years by the scientific community in "legitimizing" this sampling venue. Below are some special considerations that should be of interest to those pondering study designs for future, *in situ*, data gathering activities.

5.5.3.2.1.7.1   Recording Data Underwater. Underwater data recording may be compromised by the methods used, which in turn may interfere with the sampling protocol. For example, divers normally record data on a "slate" (a slate is usually composed of a roughened sheet of plastic or plasticized paper attached to a clipboard, with the writing done using a pencil). Utilizing this method, a 5-min observation does not result in 5 min of actual data taking, as the observer must look down to record his/her findings, thus reducing the true observation time. The advent of underwater video, with audio attachments, now makes it possible for a diver to visually record a survey while simultaneously making on audible record of his/her observations. These data are later transcribed into standard data-recording formats (computer spreadsheets, databases, and field sheets). Bortone et al. (1991) conducted a limited comparative study of slate, audio, and video recording methods conducted *in situ*. These results indicated that the combination of audio/video yielded the most information, whereas audio alone provided the most data using a single recording device. The slate method, while yielding less information during the same unit period of time, nevertheless provided faunal data that were statistically similar in certain respects.

5.5.3.2.1.7.2   Submersibles and ROVs. Manned submersibles have been used to obtain both qualitative and quantitative data on reef assemblages (e.g., Shipp et al. 1986). Their use permits *in situ*
observations under a wide range of environmental conditions not normally available to divers, for
example, areas characterized by extremely strong currents. They also can remain at deeper depths
for extended periods, so that repeated or replicate measures are facilitated. Although significantly
safer under extreme environmental conditions, they are also very expensive to use, and their very
nature may limit flexibility through decreased maneuverability and a more limited field of vision
than would be afforded a free-swimming diver.

Remotely operated vehicles (ROVs) have also been employed in surveys of artificial reefs (Van
Dolah 1983). ROVs may combine the advantages of human divers and manned submersible
vehicles. ROVs are usually surface tended, which means they have a prolonged observation time,
but they may be limited by sea-surface conditions (as is also true of all other visual techniques
used to date). ROVs can be used at great depths for prolonged periods, thus removing potential
risks to the observer. With practice, trained individuals can maneuver the vehicle and virtually
duplicate the movements of an *in situ* observer. Video images can be transmitted directly to the
surface vessel, and/or the video tapes can be viewed following retrieval of the vessel. Costs and
training times for an ROV are high, but are considerably less than for a submersible. One of us
(P.F.) estimated that it takes at least three times the actual recording time to effectively retrieve
data from video tape recordings.

In the future, we anticipate that very sensitive video tapes will permit sampling of artificial reef
assemblages under conditions of poor visibility, even at night using a vision-enhancing "night-
scope."

5.5.3.2.2   Hydroacoustics — The most recent advancement in nondestructive techniques for
fish and macroinvertebrates involves the use of hydroacoustic assessment technology (Thorne et
al. 1989; Thorne 1994). Hydroacoustics show particular promise in surveying fish assemblages
attracted to oil drilling platforms (Stanley et al. 1994). The advantages of using hydroacoustic
equipment around larger, semipermanent structures are that when used in "arrays" (multiple receiv-
ers strategically placed), they can literally "see" in three dimensions at specified depths. When
combined with computerized data recorders, it is possible to obtain long-term, time-series data on
assemblage abundance. Limitations on their use include (1) the difficulty in identifying some species
from their sonar signal; and (2) the fact that verification is often required. Future studies may see
the application of other technological advances, such as the use of underwater lasers.

5.5.3.2.3   Types of Data — Several types of data can be recorded using visual assessment
methods. We include here those that one is likely to record with a nondestructive method such as
a visual census. There are obviously many other types of data, but most of these are derived or
calculated from the basic information discussed below.

5.5.3.2.3.1   Species Sight Identifications. This information can provide data for a species pres-
ence/absence matrix, as well as an overall indication of the species associated with a reef. Since
species identifications are critical for most other analyses (e.g., species-specific abundance and size
estimates), it is vitally important that species be accurately identified. Considerable experience is
usually necessary for accurate species sight identifications. Although there is no real substitute for
this, reasonably accurate identifications are usually possible using the large number of field guides
and manuals available. The number of species may exceed many dozens in some areas, and
combining this diversity with the numerous sex/maturity morphs characteristic of certain families
(most notably the parrotfishes [Scaridae] and wrasses [Labridae]) complicates species sight recognition even further. Workers are advised to enlist the expertise of fish taxonomists or others
intimately familiar with the local fauna, and also to routinely keep selected voucher specimens
during faunal surveys. The importance of accurate species identifications cannot be overstressed,

since these form the bases for all reef studies. Some studies may focus on a few or even a single species, which can reduce identification problems. Another approach is for several individuals to conduct the surveys, with each observer responsible for a limited number of taxa.

5.5.3.2.3.2 Abundance Estimates. Next to species presence/absence, the most commonly recorded type of data in a visual survey is number of individuals. Although counting moving and/or cryptic species (either singly or in mixed species schools) can be a formidable task, with practice, abundance estimates can be made with accuracy. When populations are small (usually less than 20), it is usually preferable to count each individual. For larger populations, it is often preferable to estimate populations in groups or sets of individuals. A diver can form a visual image of what constitutes a group of 20, 50, 100, or even 1000 individuals, and then visually "multiply" these counts in order to obtain an estimate of the entire population. For example, a diver might count 10 groups of 50 individuals, resulting in a total estimate of 500 individuals. Also, for mixed species schools, it is often easier to estimate the number of individuals in the entire school, followed by estimates of the relative percentages of each species in the school. A similar method involves estimation of abundance using logarithmic abundance categories (Williams 1982; Russ 1984). Many surveys have been conducted by assigning species abundance estimates to groups (Harmelin-Vivien et al. 1985). Many schemes have been devised, such as Rare (1 to 2 individuals), Few (2 to 5), Common (5 to 10), and Abundant (10 to 50).

Regardless of the scale used to estimate abundance, it is important to be consistent with regard to abundance-class definitions within a study. It is never acceptable to merely define relative abundance with such terms, without including explanatory information. For an example of how this should not be done see Bortone (1976).

5.5.3.2.3.3 Individual Size Estimates. Size information is essential for characterizing the fauna at a site. If taken with precision, it is even possible to obtain reasonable estimates of growth and biomass (Bortone et al. 1992). Estimating size of individuals underwater may seem like a formidable task, fraught with possibilities for error (especially optical distortion, which causes individuals to appear larger than they actually are). However, with constant practice, many observers are able to obtain reliable size estimates (Bell et al. 1985; Samoilys 1997; Bortone and Mille 1999). For most fishes it is preferable to base estimates using total length (tip of snout to end of caudal fin), although for certain fishes alternative measurements must be used (e.g., skates and rays, in which disk width is the normal basis of measurement). It is always important to define the basis of length and specify the units (preferably in cm or mm). In some studies, size estimates have been reduced to the nearest cm (estimates to the nearest mm are essentially meaningless), but some investigators prefer to base this on 5- or even 10-cm size groups. Estimates that do not define size intervals (e.g., small, medium, or large; or juveniles and adults) should be avoided.

Several techniques can be adopted for estimating size. For example, a ruler mounted on a stick or attached to the recording slate can be used (Bohnsack and Bannerot 1986). Some authors have used paired, mounted laser beams to serve as an underwater rule (McFall 1992), and in other studies workers have employed fish silhouettes (Mille and Van Tassell 1994). In some cases preserved fish have been mounted on an underwater glass in order to compare censused fish to a model of known length (Galzin 1985). Previously measured, surrounding inanimate objects can be used as a basis for comparison. Voucher specimens can be measured after capture. Constant practice is important for obtaining reliable size estimates.

5.5.3.2.3.4 Biomass. Biomass is actually a calculated variable, but it is included here since it is often considered an essential variable for describing the productivity of an area. Recent fish surveys on artificial reefs have estimated biomass (Bortone et al. 1994a; Bortone et al. 1997). It is impossible to estimate the weights of individuals or schools of fish from mere sight inspection (although experienced commercial fishers may be able to do this for groups with which they are especially familiar). A reasonable alternative is to estimate average size and abundance of a species and then calculate total biomass using previously published tables of length/weight relationships. It should be noted that these tables are not available for all species, although tables for similar species from the same geographic area could be used. It is also possible to use equations based on similarly shaped species.

5.6 EXAMPLES AND CASE STUDIES

In the following we present references of several basic studies, utilizing all three levels of information, which can serve as models for future study development for fish and macroinvertebrate assessment. While these examples may serve as guides for development of a study, it should be remembered that these are always subject to modification and improvement.

5.6.1 Previous Work on Artificial Reefs

Bortone and Kimmel (1991) documented much of the earlier work on methods of fish and macroinvertebrate assessments on artificial reefs. Since then, numerous other publications on the subject have appeared. Of particular note is the most recent compendium of papers from the Fifth and Sixth International Conferences on Aquatic Habitat Enhancement (Grove and Wilson 1994; Sako and Nakamura 1995). In addition, recent symposia held in Manzanillo, Mexico in 1992 (Secretaria de Pesca 1992) and in Loano, Italy in 1994 (Cenere and Relini 1995) include some of the most recent examples of surveys and assessments of artificial reefs.

In Section 5.4.1 we indicated that three types of information or levels of assessments can be performed during an artificial reef survey. Below, each of the three types or levels are addressed with examples from the recent literature. These can serve as examples of how to conduct assessments of fishes and macroinvertebrates. Also, they may serve as a starting point when beginning to plan an assessment study design. It should be noted that many studies may encompass more than one level. For example some studies that purport to compare two reefs over a period of time (level two — comparative study) may also present a detailed description of the reef (level one — description).

5.6.1.1 Level One — Description of the Assemblage

Moreno et al. (1994) offered a basic but useful description of the artificial reef program off the Balearic Islands. At the same time they presented information that allows a description of the fish fauna associated with these reefs in the area. A study such as this can be thought of as preliminary but essential in that it helps guide future studies that determine trends (level two) and relationships (level three).

Haroun et al. (1994) conducted a detailed description of a plethora of abitotic and biotic variables associated with artificial reef groups off the Canary Islands. Part of the description included information on the abundance of species associated with the reefs as well as with the natural environment proximate to the reefs. Data from both these habitats are essential to further develop studies that will examine the factors responsible for potential differences in the assemblages.

Bell and Hall (1994) studied the effects of a hurricane on the marine artificial reefs off South Carolina in the southeastern coastal U.S. While generally descriptive of the effects of the storm on the associated fish fauna, it necessarily provided the database essential to make time-series comparisons with reestablishment of the fauna in the successional stages that followed the storm.

DeMartini et al. (1994) presented an extremely detailed description of the productivity associated with an artificial reef off southern California. Although classified as a level one descriptive study, it offers a sophisticated and detailed presentation of the relative productivity of the reef.

5.6.1.2   Level Two — Temporal/Spatial Comparisons

The catch rates and species composition of the fishes associated with both natural and artificial reefs in Kerala, India, were compared by d'Cruz et al. (1994). While very basic, it represents the essence of a spatial comparison study on artificial reef assemblages.

Jensen et al. (1994) reported on the colonization of fishes on coal-ash artificial reefs in Poole Bay, U.K. Colonization and successional studies such as this represent a general study design for time-series (temporal) comparisons.

Bortone et al. (1994b) presented both temporal and spatial comparisons of artificial reef faunas off the Canary Islands. Faunal comparisons were made both between sites (reef and nonreef) and over time (before, shortly after, and long after deployment). Most comparative study designs follow this scenario.

5.6.1.3   Level Three — Identifying Significant and Interactive Environmental Factors

Friedlander et al. (1994) investigated the interactive aspect of reef type and location to determine their effects in aggregating fishes off the U.S. Virgin Islands. This study represents an elemental but essential step to dissecting features of reefs and environment that may lead to understanding the dynamics of their interactions.

On a larger scale, Bombace et al. (1994) compared multiple reef designs at a variety of sites to determine features important to the enhancement of artificial reef assemblages. This study can be considered a more elaborate version of the study cited above (Friedlander et al. 1994).

Bortone et al. (1994a) set out to determine the features of reefs that would help explain the associated fish fauna in an estuary in the northern Gulf of Mexico. Using a series of reefs of different sizes, orientations, and configurations, they analyzed the biological response variables (species number, abundance, size, and biomass) with respect to both reef and environmental variables over time. Thus, the study design was organized around the deterministic hypothesis that some set of conditions (both environmental and reef attribute) may interact to affect the associated fish community. Martin and Bortone (1997) conducted a similar analysis using the macroinvertebrate community attributes as the response variables.

5.6.2   Case Studies

Examining case studies can be extremely useful in planning a study to investigate artificial reef fish and macroinvertebrate fauna. Their utility comes from learning from the experience of previous investigators and how they applied their well-documented methods toward obtaining the data needed to answer specific questions. Below are presented three case studies that we believe demonstrate adequate but generalized procedures that should help guide future artificial reef assessments. The first case study below (Bohnsack et al. 1994) enables the reader to see the application of methods to a very specific set of objectives. The second case study (Bortone et al. 1997) takes a more general approach to determining faunal features associated with environmental and reef attributes. The third case study (Frazer and Lindberg 1994) is a model of how to investigate large-scale problems associated with spatial aspects of reef design over an extended time frame.

Bohnsack et al. (1994) set out to establish the effect of reef size on colonization and assemblage structure on artificial reefs off the coast of southeastern Florida. The study question was simple and straightforward with clearly stated, specific objectives: "… (1) to quantify the relationship between reef size and the biomass and numbers of reef fishes that settle and grow on artificial reefs versus those that arrive at older stages; (2) experimentally test whether multiple small reefs can support more fish than one large reef using equal quantities of materials; (3) compare the artificial reef assemblages of reef fishes with nearby natural and artificial habitats."

The methods used to measure the environmental and reef attributes were clearly stated as well as the methods for measuring the biological response (i.e., dependent) variables. The study results presented the survey data and analyses in enough detail to permit an assessment as to their efficacy in meeting the posed objectives. Lastly, but importantly, the study is mindful of the extreme variation inherent in artificial reef assessments by indicating the replicate responses as well as the long-term, time-series responses of the dependent variables. Although elaborate and intricate, this study should serve as a guide to study designers.

Bortone et al. (1997) took a slightly different approach in trying to determine factors associated with artificial reef fish assemblages in the northern Gulf of Mexico. Their overall goal to establish the factors associated with faunal assemblages was similar to that offered by Bohnsack et al. (1994). However, instead of constructing and deploying reefs in a well-designed array specifically directed toward the project objectives, they made use of previously deployed reefs to make detailed measurements of both environmental and reef-attribute factors in association with a host of biological response variables (number of species, number of individuals, size of individuals, biomass, species diversity, etc.). The study design included examining 64 reefs for 3 years with replicate biological response measures during each survey. Similar to Bohnsack et al. (1994), detailed references to the methods of measurement were presented. The results present both summaries of the descriptive data as well as evidence for the variation encountered on all variables. They used multivariate analyses to further elucidate and meet the study objectives. Moreover, while the study focused on helping to explain community assemblage features, it was also able to elucidate the populational aspects of several important target species in the assemblages.

Lastly, a study by Frazer and Lindberg (1994) warrants inspection as a case study as it painstakingly and meticulously sets out to answer well-crafted questions on specific artificial reef attributes. The hypothesis that widely spaced reef patches would have higher populations of three specific target species (i.e., octopus, black seabass, and triggerfish) than on some other reef deployment configurations is tested with data gathered with specifically defined protocols from an array of reefs that adequately offer alternative scenarios. Details of the time-series analysis of population data are presented in a clear and understandable manner. Most importantly, however, is that the paper concludes with a well-defined statement of finding: "We find that hard bottom-refuge and soft-bottom food resources are inextricably linked in this community …" This is the essence of "good science" in that the study results lead to a well-developed, testable hypothesis. Whether the statement is true or not remains to be seen. What has been established, however, is pattern and direction for future artificial reef research.

5.7 FUTURE NEEDS AND DIRECTIONS

Bortone (1998) recently editorialized about the positive and negative features of artificial reef research. Among the achievements have been collegial interaction and networks of communication that have developed among researchers in the scientific community. In addition, some methodological differences have been resolved, and more recently there have been attempts to analyze the data at the most sophisticated levels of data analysis. On the other hand, it was noted that the broader dilemma of resolving the attraction and production hypothesis still requires more work. With the application of ecological models and theory, along with the cooperative interaction of reef scientists and ecologists from other areas, the resolution of the dilemma may be within our grasp.

With regard to methodology, more testing of methods should lead to more effective surveys and reduced sampling error, while increasing information for the same unit of effort. It also is important to recognize that biases exist, and that it is important to correct these to ensure the future development of artificial fish and macroinvertebrate assessment technology.

Continued development of remote sensing seems to be an obvious and essential area for progress. Our present understanding of artificial reefs is mostly derived from studies conducted in shallow and warm areas of the world. The generalization paradigms of artificial reefs need to be expanded and broadened. This will not happen until we attain the ability to regularly sample the deeper, colder, and less hospitable parts of the aquasphere.

Not often acknowledged in the literature is the interesting and obvious premise that most artificial reefs constantly experience artificial faunal removal (i.e., fishing). We know of no other situations where scientific studies are attempted while the prime object of the evaluation is being uncontrollably manipulated, often without any documentation. This circumstance is unique in environmental science and should be addressed. At a minimum, some appreciation of the kinds, sizes, and rate of removal should be known so as to validate study results. What we are left with is the probably false assumption that all reefs are being fished equally. Future research efforts to assess artificial reef associated fishes and macroinvertebrates should be done in cooperation with economic and social evaluations of the fishing activity and success attributable to the reefs being studied: a monumental but necessary task.

5.8 ACKNOWLEDGMENTS

S.A.B. gratefully acknowledges the help of numerous graduate students at the University of West Florida, in particular Robert Turpin, Tony Martin, Richard Cody, and Ron Hill, who helped conduct field surveys on artificial reef fishes. He also thanks colleagues C.M. (Mike) Bundrick, Keith Mille, Robert W. Hastings, and especially Joseph J. Kimmel for their help and cooperation in much of the work cited herein. M.A.S. sincerely acknowledges the Fisheries group, Queensland's Department of Primary Industries, for its support.

REFERENCES

Alevizon, W.S., J.C. Gorham, R. Richardson, and S.A. McCarthy. 1985. Use of man-made reefs to concentrate snapper (Lutjanidae) and grunts (Haemulidae) in Bahamian waters. *Bulletin of Marine Science* 37:3–10.

Anderson, M.J. and N.A. Gribble. 1998. Partitioning the variation among spatial, temporal and environmental components in a multivariate data set. *Australian Journal of Ecology* 23:158–167.

Ardizzone, G.D., M.F. Gravina, and A. Belluscio. 1989. Temporal development of epibenthic communities on artificial reefs in the central Mediterranean Sea. *Bulletin of Marine Science* 44:592–608.

Barans, C.A. and S.A. Bortone. 1983. The visual assessment of fish populations in the southeastern United States: 1982 Workshop. Technical Report 1(SC-SG-TR-01-83), South Carolina Sea Grant Consortium, Charleston.

Barbour, C.D. and J.H. Brown. 1974. Fish species diversity in lakes. *American Naturalist* 108(962):473–489.

Beals, E. 1960. Forest bird communities in the Apostle Islands of Wisconsin. *Wilson Bulletin* 72:156–181.

Bell, M. and W.J. Hall. 1994. Effects of Hurricane Hugo on South Carolina's marine artificial reefs. *Bulletin of Marine Science* 55:836–847.

Bell, J.D., G.J.S. Craik, D.A. Pollard, and B.C. Russell. 1985. Estimating length frequency distributions of large reef fish underwater. *Coral Reefs* 4:41–44.

Böhlke, J.E. and C.C.G. Chaplin. 1993. *Fishes of the Bahamas and Adjacent Tropical Waters.* 2nd ed. University of Texas Press, Austin.

Bohnsack, J.A. 1982. Effects of piscivorous predator removal on coral reef community structure. Pages 258–267. In: G.M. Cailliet and C.A. Simnestad, eds. *Gutshop '81: Fish Food Habits Studies. Proceedings, Third Pacific Workshop.* Washington Sea Grant Publication, University of Washington, Seattle.

Bohnsack, J.A. and S.P. Bannerot. 1986. A stationary visual census technique for quantitatively assessing community structure of coral reef fishes. U.S. Dept. of Commerce, NOAA Technical Report NMFS 41:1–15.

Bohnsack, J.A., D.L. Johnson, and R.F. Ambrose. 1991. Ecology of artificial reef habitats and fishes. Pages 61–107. In: W. Seaman, Jr. and L.M. Sprague, eds. *Artificial Habitats for Marine and Freshwater Fisheries.* Academic Press, San Diego.

Bohnsack, J.A., D.E. Harper, D.B. McClellan, and M. Hulsbeck. 1994. Effects of reef size on colonization and assemblage structure of fishes at artificial reefs off southeastern Florida, U.S.A. *Bulletin of Marine Science* 55:796–823.

Bombace, G., G. Fabi, L. Fiorentini, and S. Speranza. 1994. Analysis of the efficacy of artificial reefs located in five different areas of the Adriatic Sea. *Bulletin of Marine Science* 55:559–580.

Bortone, S.A. 1976. The effects of a hurricane on the fish fauna at Destin, Florida. *Florida Scientist* 39(4):245–248.

Bortone, S.A. 1998. Resolving the attraction-production dilemma in artificial reef research: some yeas and nays. *Fisheries* 23:6–10.

Bortone, S.A. 1999. The impact of artificial reef fish assemblages on their potential forage area: lessons in artificial reef study design. Pages 82–85. In: W. Horn, ed. *Florida Artificial Reef Summit '98*. Florida Department of Environmental Protection, Tallahassee.

Bortone, S.A. and J.A. Bohnsack. 1991. Sampling and studying fish on artificial reefs. Chap. 7, pages 39–51. In: J.G. Halusky, ed. *Artificial Reef Research Diver's Handbook*. Technical Paper 63, Florida Sea Grant College, University of Florida, Gainesville.

Bortone, S.A. and J.J. Kimmel. 1991. Environmental assessment and monitoring of artificial reefs. Pages 177–236. In: W. Seaman, Jr. and L.M. Sprague, eds. *Artificial Habitats for Marine and Freshwater Fisheries*. Academic Press, San Diego.

Bortone, S.A. and K.J. Mille. 1999. Data needs for assessing marine reserves with an emphasis on estimating fish size *in situ*. *Naturalista Siciliana* (Suppl.):13–31.

Bortone, S.A., R.W. Hastings, and J.L. Oglesby. 1986. Quantification of reef fish assemblages: a comparison of several *in situ* methods. *Northeast Gulf Science* 8(1):1–22.

Bortone, S.A., J.J. Kimmel, and C.M. Bundrick. 1989. A comparison of three methods for visually assessing reef fish communities: time and area compensated. *Northeast Gulf Science* 10(2):85–96.

Bortone, S.A., T.R. Martin, and C.M. Bundrick. 1991. Visual census of reef fish assemblages: a comparison of slate, audio, and video recording devices. *Northeast Gulf Science* 12(1):17–23.

Bortone, S.A., J. Van Tassell, A. Brito, J.M. Falcón, and C.M. Bundrick. 1992. Visual census as a means to estimate standing biomass, length, and growth in fishes. *Proceedings of the American Association of Underwater Sciences, Diving for Science* 12:13–21.

Bortone, S.A., T.R. Martin, and C.M. Bundrick. 1994a. Factors affecting fish assemblage development on a modular artificial reef in the northern Gulf of Mexico estuary. *Bulletin of Marine Science* 55(2–3):319–332.

Bortone, S.A., J. Van Tassell, A. Brito, J.M. Falcón, J. Mena, and C.M. Bundrick. 1994b. Enhancement of the nearshore fish assemblage in the Canary Islands with artificial habitats. *Bulletin of Marine Science* 55:602–608.

Bortone, S.A., R.K. Turpin, R.P. Cody, C.M. Bundrick, and R.L. Hill. 1997. Factors associated with artificial-reef fish assemblages. *Gulf of Mexico Science* 1:17–34.

Bortone, S.A., R.P. Cody, R.K. Turpin, and C.M. Bundrick. 1998. The impact of artificial reef fish assemblages on their potential forage area. *Italian Journal of Zoology* 65 (Suppl.):265–267.

Bouchon-Navaro, Y., M. Louis, and C. Bouchon. 1997. Trends in fish distributions in the West Indies. Vol. 1, Pages 987–992. In: H.A. Lessios and I.G. Macintyre, eds. *Proceedings, 8th International Coral Reef Symposium*. Smithsonian Tropical Research Institute, Balboa, Panama.

Brock, V.E. 1954. A preliminary report on a method of estimating reef fish population. *Journal of Wildlife Management* 18:297–317.

Cappo, M. and I.W. Brown. 1996. Evaluation of sampling methods for reef fish populations of commercial and recreational interest. Technical Report No. 6, Townsville, CRC Reef Research Centre, Australia. 72 pp.

Cenere, F. and G. Relini. 1995. Convegno de Loano per la difesa del mare (8–9 Iuglio 1994). *Biologia Marina Mediterranea* Vol. II, Fasc. 1:iii.

Choat, J.H and D.R. Bellwood. 1991. Reef fishes: their history and evolution. Pages 117–143. In: P.F. Sale, ed. *The Ecology of Fishes on Coral Reefs*. Academic Press, San Diego.

Clarke, K.R. 1993. Non-parametric multivariate analysis of changes in community structure. *Australian Journal of Ecology* 18:117–143.

Connell, S.D., M.A. Samoilys, M.P. Lincoln-Smith, and J. Leqata 1998. Comparisons of abundance of coral reef fish: catch and effort surveys vs. visual census. *Australian Journal of Ecology* 23:579–586.

d'Cruz, T., S. Creech, and J. Fernandez. 1994. Comparison of catch rates and species composition from artificial and natural reefs in Kerala, India. *Bulletin of Marine Science* 55:1029–1037.

DeMartini, E.E., A.M. Barnett, T.D. Johnson, and R.F. Ambrose. 1994. Growth and production estimates for biomass-dominant fishes on a southern California artificial reef. *Bulletin of Marine Science* 55:484–500.

Doherty, P.J. 1991. Spatial and temporal patterns in recruitment. Pages 261–293. In: P.F. Sale, ed. *The Ecology of Fishes on Coral Reefs*. Academic Press, San Diego.

Dovel, W.L. 1971. *Fish Eggs and Larvae of the Upper Chesapeake Bay*. Natural Resources Institute, University of Maryland, Solomons. 71 pp.

Ebeling, A.W. and Hixon, M.A. 1991. Tropical and temperate reef fishes: comparison of community structures. Pages 509–563. In: P.F. Sale, ed. *The Ecology of Fishes on Coral Reefs*. Academic Press, San Diego.

Emlen, J.M. 1973. *Ecology: An Evolutionary Approach*. Addison-Wesley, Reading, MA.

Falcón, J.M., S.A. Bortone, A. Brito, and C.M. Bundrick. 1996. Structure of and relationships within and between the littoral, rock-substrate fish communities off four islands in the Canarian Archipelago. *Marine Biology* 125:215–231.

Fennessy, S.T., C. Villacastin, and J.F. Field. 1994. Distribution and seasonality of ichthyofauna associated with commercial prawn trawl catches on the Tugela Bank of the Natal, South Africa. *Fisheries Research* 20:263–282.

Frazer, T.K. and W.J. Lindberg. 1994. Refuge spacing similarly affects reef-associated species from three phyla. *Bulletin of Marine Science* 55:387–400.

Francour, P., M. Harmelin-Vivien, J.G. Harmelin, and J. Duclerc. 1995. Impact of *Caulerpa taxifola* colonization on the littoral ichthyofauna of North-Western Mediterranean Sea: preliminary results. *Hydrobiologia* 300/301:345–353.

Friedlander, A., J. Beets, and W. Tobias. 1994. Effects of fish aggregating device design and location on fishing success in the U.S. Virgin Islands. *Bulletin of Marine Science* 55:592–601.

Galzin, R. 1985. Ecologie des poissons récifaux de Polynésie française: Variations spatio-temporelles des peuplements. Dynamique de populations de trois espèces dominantes des lagons nord de Moores. Evaluations de la production ichtyologique d'un secteur récifolagonaire. Thèsis de Doctorat ès Sciences, Université de Sciences et Techniques du Languedoc.

Goldman, B. and F.H. Talbot. 1976. Aspects of the ecology of coral reefs. Pages 125–154. In: O.H. Jones and R. Endean, eds. *Biology and Ecology of Coral Reefs*. Vol. 3. Academic Press, New York.

Grove, R.S. and C.J. Sonu. 1985. Fishing reef planning in Japan. Pages 187–251. In: F.M. D'Itri, ed. *Artificial Reefs Marine and Freshwater Applications*. Lewis Publishers, Chelsea, MI.

Grove, R.S. and C.A. Wilson. 1994. Introduction. *Bulletin of Marine Science* 55:265–267.

Halford, A. 1997. Recovery of a fish community six years after a catastrophic mortality event. Vol. 1, Pages 1011–1016. In: H.A. Lessios and I.G. Macintyre, eds. *Proceedings, 8th International Coral Reef Symposium*. Smithsonian Tropical Research Institute, Balboa, Panama.

Harmelin-Vivien, M.L., J.G. Harmelin, C. Chauvet, C. Duval, R. Galzin, R. Lejeune, G. Barnabé, F. Blanc, R. Chevalier, J. Duclerc, and G. Lasserre. 1985. Evaluation visuelle des peuplements et populations de poissons: Méthodes et problèmes. *Revue d'Ecologie* (Terre Vie) 40:467–539.

Haroun, R.J., M. Gómez, J.J. Hernández, R. Herrera, D. Montero, T. Moreno, A. Portillo, M.E. Torres, and E. Soler. 1994. Environmental description of an artificial reef site in Gran Canaria (Canary Islands, Spain) prior to reef placement. *Bulletin of Marine Science* 5:932–938.

Hastings, P.A. and S.A. Bortone. 1980. Life history aspects of the belted sandfish *Serranus subligarius* (Pisces: Serranidae). *Environmental Biology of Fishes* 5(4):365–373.

Hixon, M.A. and J.P. Beets. 1989. Shelter characteristics and Caribbean fish assemblages: experiments with artificial reefs. *Bulletin of Marine Science* 44:666–680.

Hobson, E.S. 1991. Trophic relationships of fishes specialized to feed on zooplankters above coral reefs. Pages 69–95. In: P.F. Sale, ed. *The Ecology of Fishes on Coral Reefs*. Academic Press, San Diego.

Jensen, A.C., K.J. Collins, A.P.M. Lockwood, J.J. Mallinson, and W.H. Turnpenny. 1994. Colonization and fishery potential of a coal-ash artificial reef: Poole Bay, United Kingdom. *Bulletin of Marine Science* 55:1263–1276.

Jones, R.S. and J.A. Chase. 1975. Community structure and distribution of fishes in an enclosed high island lagoon in Guam. *Micronesica* 11:127–148.

Jones, R.S. and M.J. Thompson. 1978. Comparison of Florida reef fish assemblages using a rapid visual technique. *Bulletin of Marine Science* 28:159–172.

Karr, J. 1981. Assessment of biotic integrity using fish communities. *Fisheries* 6:21–27.

Kimmel, J.J. 1985. A new species-time method for visual assessment of fishes and its comparison with established methods. *Environmental Biology of Fishes* 12:23–32.

Kingsford, M. J. and C.N. Battershill. 1998. *Studying Temperate Marine Environments: A Handbook for Ecologists.* Canterbury University Press, New Zealand. 335 pp.

Kulbicki, M. 1988. Correlation between catch data from bottom longlines and fish censuses in the SW lagoon of New Caledonia. *Naga* 92(2–3):26–29.

Lincoln-Smith, M.P. 1988. Effects of observer swimming speed on sample counts of temperate rocky reef fish assemblages. *Marine Ecology Progress Series* 43:223–231.

Lincoln-Smith, M. and M.A. Samoilys, 1997. Sampling design and hypothesis testing. Pages 7–15. In: M. Samoilys, ed. *Manual for Assessing Fish Stocks on Pacific Coral Reefs.* Department of Primary Industries, Townsville, Australia.

Lindquist, D.G. and L.J. Pietrafesa. 1989. Current vortices and fish aggregations: the current field and associated fishes around a tugboat wreck in Onslow Bay, North Carolina. *Bulletin of Marine Science* 44:533–544.

Lindquist, D.G., L.B. Cahoon, I.E. Clavijo, M.H. Posey, S.K. Bolden, LA. Pike, S.W. Burk, and P.A. Cardullo. 1994. Reef fish stomach contents and prey abundance on reef and sand substrata associated with adjacent artificial and natural reefs in Onslow Bay, North Carolina. *Bulletin of Marine Science* 55:308–318.

Lowe-McConnell, R.M. 1979. Ecological aspects of seasonality in fishes of tropical waters. *Symposium of the Zoological Society of London* 44:219–241.

Luckhurst, B.E. and L. Luckhurst. 1978. Analysis of the influence of substrate variables on coral reef communities. *Marine Biology* 49:317–323.

Ludwig, J.A. and J.F. Reynolds. 1988. *Statistical Ecology.* John Wiley & Sons, New York.

Lukens, R.R., J.D. Cirino, J.A. Ballard, and G. Geddes. 1989. Two methods of monitoring and assessment of artificial reef materials. Special Report 2-WB, Gulf States Marine Fisheries Commission, Ocean Springs, MS.

Manly, B.F.J. 1986. *Multivariate Statistical Methods: A Primer.* Chapman & Hall, London.

Martin, T.R. and S.A. Bortone. 1997. Development of an epifaunal assemblage on an estuarine artificial reef. *Gulf of Mexico Science* 15(2):55–70.

McFall, G.B. 1992. Development and application of a low-cost faired-laser measuring device. *Proceedings of the American Academy of Underwater Sciences, Diving for Science* 12:109–113.

Mille, K.J. and J. Van Tassell. 1994. Diver accuracy in estimating lengths of models of the parrotfish, *Sparisoma cretense, in situ. Northeast Gulf Science* 13:149–155.

Moreno, I., K. Roca, O. Reñones, J. Coll, and M.Salamanca. 1994. Artificial reef program in Balearic waters (western Mediterranean). *Bulletin of Marine Science* 55:667–671.

Mullen, L.J., P.R. Herczfeld, and V.M. Contarino. 1996. Progress in hybrid Lidar-Radar for ocean exploration. *Sea Technology,* March 1996:45–52.

Murphy, B.R. and D.W. Willis, eds. 1996. *Fisheries Techniques.* 2nd ed. American Fisheries Society, Bethesda, MD.

Nakamura, M. 1985. Evaluation of artificial reef concepts in Japan. *Bulletin of Marine Science* 37:271–278.

Nelson, J.S. 1994. *Fishes of the World.* 3rd ed. John Wiley & Sons, New York.

Nelson, R.D. and S.A. Bortone. 1996. Feeding guilds among artificial-reef fishes in the northern Gulf of Mexico. *Gulf of Mexico Science* 14(2):66–80.

Polovina, J.J. 1991. Fisheries applications and biological impacts of artificial reefs. Pages 153–176. In: W. Seaman, Jr. and L.M. Sprague, eds. *Artificial Habitats for Marine and Freshwater Fisheries.* Academic Press, San Diego.

Prince, E.D., O.E. Maughan, and P. Brouha. 1985. Summary and update of the Smith Mountain Lake artificial reefs project. Pages 401–430. In: F.M. D'Itri, ed. *Artificial Reefs Marine and Freshwater Applications.* Lewis Publishers, Chelsea, MI.

Rosenblatt, R.H. 1967. The zoogeography of the marine shore fishes of tropical America. Pages 579–592. In: *Studies in Tropical Oceanography,* Miami.

Roughgarden, J. and Y. Iwasa. 1986. Dynamics of a metapopulation with space-limited subpopulations. *Theoretical Population Biology* 29:235–261.

Roughgarden, J., S.D. Gaines, and P. Possingham. 1988. Recruitment dynamics in complex life cycles. *Science* 241:1460–1466.

Ruitton, S., P. Francour, C.F. Boudouresque. In press. Relationship between algae, benthic herbivorous invertebrates and fishes in rocky sublittoral communities of a temperate sea (Mediterranean). *Estuarine and Coastal Shelf Science.*

Russ, G.R. 1984. Effects of protective management on coral reef fishes in the Philippines. ICLARM Newsletter, International Center for Living Aquatic Resources Management, Manila, October: 12–13.

Russell, B.C., F.H. Talbot, G.R.V. Anderson, and B. Goldman. 1978. Collection and sampling of reef fishes. Pages 329–345. In: D.R. Stoddard and R.E. Johannes, eds. *Coral Reefs: Research Methods*. Monographs in Oceanography Methods No. 5. UNESCO, Paris.

Sako, H. and M. Nakamura. 1995. Preface. In: *Proceedings, International Conference on Ecological System Enhancement Technology of Aquatic Environments*. Vol. I–III. Japan International Marine Science and Technology Federation, Tokyo.

Sale, P.F. 1991a. (ed.). *The Ecology of Fishes on Coral Reefs*. Academic Press, San Diego. 754 pp.

Sale P.F. 1991b. Reef fish communities: open nonequilibrium systems. Pages 564–598. In: P.F. Sale, ed. *The Ecology of Fishes on Coral Reefs*. Academic Press, San Diego.

Sale, P.F. 1997. Visual census of fishes: how well do we see what is there. Vol. 2, Pages 1435–1440. In: H.A. Lessios and I.G. Macintyre, eds. *Proceedings, 8th International Coral Reef Symposium*. Smithsonian Tropical Research Institute, Balboa, Panama.

Sale, P.F. and W.A. Douglas. 1981. Precision and accuracy of visual census techniques for fish assemblages on coral patch reefs. *Environmental Biology of Fishes* 6:333–339.

Sale, P.F. and B.J. Sharp. 1983. Correction for bias in visual transect censuses of coral reef fishes. *Coral Reefs* 2:37–42.

Samoilys, M.A. 1997. Underwater visual census surveys. Pages 16–29. In: M.A. Samoilys, ed. *Manual for Assessing Fish Stocks on Pacific Coral Reefs*. Department of Primary Industries, Townsville, Australia.

Samoilys, M.A. and G. Carlos. 1992. Development of an underwater visual census method for assessing shallow water reef fish stocks in the south west Pacific, Australian Centre for International Agricultural Research Project PN8545 Final Report, April 1992. 100 pp.

Samoilys, M.A. and G. Carlos. In press. Determining methods of underwater visual census for estimating the abundance of coral reef fishes. *Environmental Biology of Fishes*.

Samoilys, M.A. and N. Gribble. 1997. Introduction, pages 1–6. In: M.A. Samoilys, ed. *Manual for Assessing Fish Stocks on Pacific Coral Reefs*. Department of Primary Industries, Townsville, Australia.

Sanders, R.M., C.R. Chandler, and A.M. Landry, Jr. 1985. Hydrological, diel and lunar factors affecting fishes on artificial reefs off Panama City, Florida. *Bulletin of Marine Science* 37:318–328.

Schneider, D.C. 1994. *Quantitative Ecology: Spatial and Temporal Scaling*. Academic Press, San Diego.

Schreck, C.B. and P.B. Moyle, eds. 1990. *Methods for Fish Biology*. American Fisheries Society, Bethesda, MD.

Seaman, W., Jr., G.A. Antonini, S.A. Bortone, J.G. Halusky, S.M. Holland, W.J. Lindberg, J. Loftin, J.W. Milon, K.M. Portier, Y.P. Sheng, A. Szmant, and L. Zobler. 1992. Environmental and fishery performance of Florida artificial reef habitats: guidelines for technical evaluation of sites developed with state construction assistance. Project Report to the Florida Department of Natural Resources (Contract No. C-6989), Florida Sea Grant College Program, University of Florida, Gainesville.

Secretaria de Pesca. 1992. I reunion internationales sobre mejoramiento de habitats acuaticos para pesquerias (arrecifes artificiales). Instituto Nacional de la Pesca, Centro Regional de Investigacion Pesquera, Manzanillo, Colima, Mexico.

Shipp, R.L., W.A. Tyler, Jr., and R.S. Jones. 1986. Point count censusing from a submersible to estimate fish abundance over large areas. *Northeast Gulf Science* 8:83–89.

Smith, C.L. and J.C. Tyler. 1972. Space resource sharing in a coral reef community Pages 125–170. In: B.B. Collette and S.A. Earle, eds. *Results of the Tektite Program: Ecology of Coral Reef Fishes*. Science Bulletin. Natural History Museum of Los Angeles.

Stanley, D.R., C.A. Wilson, and C. Cain. 1994. Hydroacoustic assessment of abundance and behavior of fishes associated with an oil and gas platform off the Louisiana coast. *Bulletin of Marine Science* 55:1353.

Thompson, M.J. and T.W. Schmidt. 1977. Validation of the species/time random count technique for sampling fish assemblages. Proceedings, 3rd International Coral Reef Symposium 1:283–288.

Thorne, R.E. 1994. Hydroacoustic remote sensing for artificial habitats. *Bulletin of Marine Science* 55:897–901.

Thorne, R.E., J.B. Hedgepeth, and J. Campos. 1989. Hydroacoustic observations of fish abundance and behavior around an artificial reef in Costa Rica. *Bulletin of Marine Science* 44:1058–1064.

Thresher, R.E. and J.S. Gunn. 1986. Comparative analysis of visual census techniques for highly mobile, reef associated piscivores (Carangidae). *Environmental Biology of Fishes* 17:93–116.

Underwood, A.J. 1990. Experiments in ecology and management: their logics, functions, and interpretation. *Australian Journal of Ecology* 15:365–389.

Van Dolah, R.F. 1983. Remote assessment techniques for large benthic invertebrates. Pages 12–13. In: C.A. Barans and S.A. Bortone, eds. *The Visual Assessment of Fish Populations in the Southeastern United States: 1982 Workshop.* Technical Report 1(SC-SG-TR-01-83), South Carolina Sea Grant Consortium, Charleston.

Watson, R.A., G.M. Carlos, and M.A. Samoilys. 1995. Bias introduced by the non-random movement of fish in visual transect surveys. *Ecological Modeling* 77:205–214.

Williams, D. McB. 1982. Patterns in the distribution of fish communities across the central Great Barrier Reef. *Coral Reefs* 1:35–43.

CHAPTER **6**

Social and Economic
Evaluation Methods

**J. Walter Milon, Stephen M. Holland,
and David J. Whitmarsh**

CONTENTS

6.1 SUMMARY

This chapter describes the methods used to evaluate the social and economic performance of artificial reef development with a primary emphasis on data collection and measurement techniques. The second section presents the chapter objective, discusses the background and purposes of socioeconomic assessment, and introduces the possible social objectives and relevant types of information for artificial reef assessment of these objectives. In the third section, the social and economic dimensions of artificial reefs are defined, including the potential scope of reef-user interests and the importance of legal and institutional considerations. The fourth section discusses in greater detail the types of social and economic assessments and highlights the studies that have employed these concepts. The fifth section reviews the techniques available to collect and measure social and economic data for analysis. The chapter concludes with a case study.

6.2 INTRODUCTION

6.2.1 Chapter Objective

This chapter deals with people's "demand" for artificial reefs, whereas previous chapters focused on the "supply" of artificial reefs and their related biological resources. The primary beneficiaries of artificial reefs are pelagic, benthic, and human species. Depending on the focus, however, many would argue that the primary reason reefs are deployed is to serve human uses, such as recreational and commercial fishing and scuba diving. Central to maximizing the human benefits from reef utilization is a clear understanding of the:

- Direct or indirect uses of artificial reefs,
- Economic or social impacts that accrue from reef usage,
- Social preferences for certain reef characteristics and/or related marine species,
- Reasons that specific artificial reefs and reef characteristics are preferred.

There is a general lack of reports or studies about the demand for artificial reefs and the socioeconomic efficacy of these projects. Most studies that have been conducted focus on areas with the greatest reef-building activity. For example, the socioeconomic implications of artificial reefs have been analyzed in Japan (Simard 1997) and, in the U.S., in Florida (Milon 1988; Bell et al. 1998), South Carolina (Buchanan 1973; Liao and Cupka 19 (Ditton and Graefe 1978; Ditton et al. 1995). A few authors hav various social and economic aspects of artificial reef development (e.g., Bockstael et al. 1986; Key 1990; Willmann 1990). Frameworks have also been specified to evaluate social and economic impacts and efficiency of artificial habitats in general (Milon 1989, 1991).

This chapter extends previous literature on the social and economic evaluation of artificial habitats to provide guidance on socioeconomic data collection and measurement techniques for artificial reef assessment. Upon reading this chapter, reef analysts and managers can gain a better understanding of the potential socioeconomic impacts, and benefits and costs associated with artificial reefs and the appropriate techniques available for the evaluation of social and economic research objectives.

6.2.2 Why Perform Social and Economic Evaluation?

Unless the purpose of an artificial reef is for research or to mitigate environmental damages, most decision makers will judge the value or performance of a reef on its contribution to human satisfaction, i.e., a reef that is not useful to (used by) people is not a successful reef. Socioeconomic data can be collected and evaluated to measure the dynamics of artificial reef usage and gauge the extent to which the "public benefit" is served. This documentation of usage and benefits helps to justify previous or future (public) expenditures on the construction and maintenance of artificial reefs and assists coastal planners in distributing access and service points for reef use. Thus, social and economic evaluation can be useful to government entities in demonstrating constituency service from an artificial reef and in providing information that is important for efficient reef management.

Socioeconomic data collection and evaluation are also an integral part of an "adaptive management" strategy for resource use whereby "monitoring and evaluation of the physical, biological, and social systems responses must be conducted to assess the initial working hypotheses, to reduce scientific uncertainty, to inform the public, and if necessary to develop alternative hypotheses and action plans" (Milon et al. 1997). Figure 6.1 shows a general framework for adaptive artificial reef management. Social and/or policy goals for artificial reef development are expressed through various political and governmental agencies. These goals or objectives are often generally stated and require further refinement by scientists to become the basis for research objectives. For example, alternative policies available to address a social goal of improving nearshore fishing opportunities can be examined in the context of research objectives that study the productivity and the costs–benefits of potential artificial reef materials and locations. The design and interpretation of research objectives are shaped by scientists' understanding of the relationships among the social and ecological effects of artificial reef systems and policymakers' interests in achieving specific social goals. When policymakers and scientists agree on research objectives, a set of hypotheses is formed about social and ecological processes in artificial reef systems that results in the selection of a preferred location and design. After deployment, these hypotheses can be tested through monitoring and evaluation and reformulated if necessary. The continuous testing and reformulation of system hypotheses are crucial features of the adaptive resource management strategy (Holling 1978; Walters 1986). For artificial reef research, this requires a consistent agenda of physical, biological, and social system monitoring; open communication between social and natural scientists; and feedback from scientists to resource managers.

The adaptive management process also provides a useful forum for stakeholders to voice their interests in artificial reef development. For example, fishing, boating, and diving clubs can provide a public service, display member support, and pursue club goals through participation in reef monitoring and evaluation. In addition, well-documented socioeconomic information about artificial reefs can be used to educate nonusers about the services provided by coastal resources (Ditton and Burke 1985).

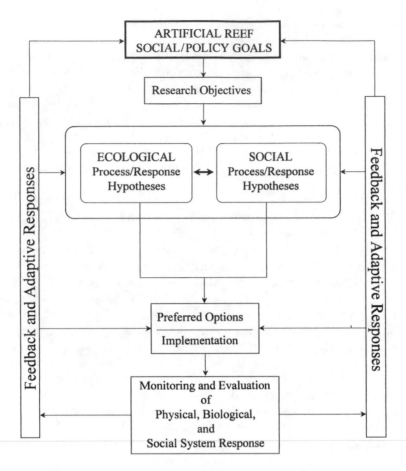

Figure 6.1   An adaptive management framework for artificial reef research.

6.2.3   Overview of Socioeconomic Goals, Objectives, and Assessment Concepts

Socioeconomic assessment of artificial reef projects is typically conducted by marine resource managers who have social science expertise, usually in conjunction with university economists or sociologists, or qualified private consultants. The stages of socioeconomic assessment involve: (1) objective and/or hypothesis identification; (2) survey instrument development; and (3) data collection and analysis. These stages are complex and easily can be biased in ways that can invalidate the results unless experienced and knowledgeable scientists guide or manage the research (see Chapter 2 on study design). This section describes how the potential socioeconomic objectives or hypotheses that are identified in the first stage of evaluation can be tested with specific research instruments and data collected in stages two and three. The primary purpose of the discussion is to introduce readers to the appropriate types of assessment strategies available to measure the attainment of socioeconomic objective(s) identified for artificial reef development.

6.2.3.1   Planning Goals, Objectives, and Socioeconomic Assessment

Socioeconomic goals for artificial reef development can be considered on several levels or scales. In general, socioeconomic goals differ in scale from ecological or environmental objectives. Socioeconomic goals tend to be broader, more abstract goals that include a series of more specific objectives, such as those related to ecological concerns. That is, ecological or environmental objectives typically are defined to guide data collection and evaluation efforts that also can be used to assess a broader socioeconomic objective. The same is true for lower level socioeconomic research objectives. Figure 6.2 depicts a framework for the socioeconomic assessment of an artificial reef project with four levels of goal and objective definitions shown across the top of the figure: social, policy, behavioral, and action. At the broad social scale, a reef may be proposed as a means to stimulate positive economic and/or social impacts in a local economy. However, decisionmakers must determine the policy objectives that can achieve this social goal. One policy objective could be to increase tourism to an area and/or to increase the number of recreational anglers and divers in an area. This could be achieved by pursuing a behavioral objective to enhance the local recreational fishing and/or diving satisfaction. Finally, an action objective could be proposed that seeks to enhance recreational fishing satisfaction by increasing the number of nearshore and shore-accessible recreational fishing sites. (See Chapter 7 for other examples.)

Once the development of an artificial reef is considered, the primary question is whether the project will be able to achieve the original social goal identified. The same question may also be posed for an existing reef project. In either case, the evaluation of the broad social goal requires prior or concurrent evaluations of the policy, behavioral, and action objectives. The lower portion of Figure 6.2 shows the general assessment techniques that are applicable for the evaluation of each goal/objective level. These assessment concepts are introduced in the next section and presented in greater detail in Sections 6.4 and 6.5.

6.2.3.2 Assessment Concepts

The type and quantity of information collected depends on the social goals, policies, and research objectives for artificial reef evaluation and the kinds of questions to be answered (Chapter 1). Table 6.1 summarizes the three types of assessments that can be conducted in the social and economic evaluation of artificial reefs: (1) *monitoring* to determine reef utilization patterns; (2) *impact assessment* to understand the social and economic significance of reef utilization in the local area; and (3) *efficiency analysis* to determine the cost-effectiveness or net benefits of the reef. The complexity of evaluation varies with each type of assessment, as does the ability to gauge different objectives (as shown in Figure 6.2). Consequently, reef managers should carefully choose an assessment approach that balances the resources available for project evaluation with the objectives of the research agenda.

Monitoring can help determine whether an artificial reef is meeting the design criteria and whether the target user group is actually using the habitat. This type of evaluation is useful for evaluating broad project goals such as: (1) increase the number of shore-accessible recreational fishing sites in a coastal community; (2) provide nearshore fishing sites for small scale coastal fishermen in a coastal bay; (3) provide separate habitat sites for recreational divers interested in either spearfishing or photography; or (4) replace damaged natural reef sites with artificial reefs. In addition, biologists may be interested in the human effects on fish populations; i.e., do species preferences or catch-per-unit effort (CPUE) differ for those using reefs compared to other water resources?

Impact assessment is used to evaluate more specific project objectives that seek desired changes in economic activity or social structures. This type of evaluation focuses on the changes caused by the project and seeks to determine whether these changes have met the specific objectives. It is particularly important to determine whether the artificial reef project has produced more of the desired effect than would have occurred "naturally," without the site development. For example, a project objective to "increase nonresident fishing trips and economic activity in a coastal community by xx percent" could be evaluated with an expenditure and economic impact analysis that measures nonresident activity both before and after a habitat development project. Similarly, the evaluation of an objective "to increase small-scale commercial fishing sales from a port by xx percent" could compare sales levels after the project is initiated with preproject sales levels. The most commonly used methods for economic impact evaluations are economic base and input/output analysis. Impacts on social structures are evaluated with social impact analysis or importance/performance

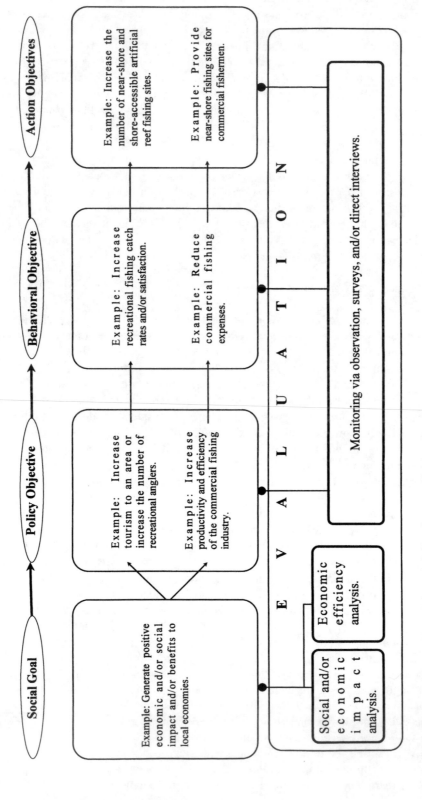

Figure 6.2 Example of a goal and objective definition for an artificial reef project.

Table 6.1 Types of Socioeconomic Assessment

Type One — Monitoring

Questions to ask:
  Who uses the artificial reef and its resources?
  When does use occur?
  Where does use occur?
  Why does use occur?

Techniques to use:
  Data collection and analysis from site observation, interviews, and/or surveys

Type Two — Impact Assessment

Questions to ask:
  What changes, if any, are measurable in economic or social activity due to artificial reef development and
    utilization?
  When do the changes occur?
  Where do the changes occur?
  Why do the changes occur?

Techniques to use:
  Economic base analysis
  Input/output analysis
  Social impact analysis

Type Three — Efficiency Analysis

Questions to ask:
  Are the project objectives being met at the least possible cost?
  Does the monetized value of project benefits exceed the project costs?

Techniques to use:
  Cost–effectiveness analysis
  Cost–benefit analysis

analysis which gives feedback on the level of relative importance participants place on specific characteristics and the degree to which the reef is providing those characteristics.

Efficiency analysis is another type of evaluation that is appropriate for objectives related to the economic performance of a project. Efficiency analyses are usually classified as either cost–effectiveness or cost–benefit evaluations. Cost–effectiveness analysis determines whether a project has produced (or could produce) the desired impact at least cost while cost–benefit analysis determines whether the monetized value of a project's benefits exceeds the costs. Both cost–effectiveness and cost–benefit analyses can be used to compare the performance of several artificial habitat projects, and the results can be compared to efficiency analyses of other types of enhancement projects. In addition, both types of analyses can be initiated in the planning phase of artificial development to make a preliminary evaluation of whether a project is a reasonable economic investment. Thus, a project objective to "increase the exploitable biomass of species yy in a coastal bay by xx percent using the least cost alternative" could be evaluated with a cost–effectiveness analysis. Alternatively, a project objective to "provide an artificial reef site to produce a positive economic return" could be assessed with a cost–benefit analysis.

6.3 SOCIAL AND ECONOMIC DIMENSIONS OF ARTIFICIAL REEFS

This section provides a discussion of the linkages between artificial reef goals, user interests, and the legal and institutional issues that influence reef development and assessment.

Table 6.2   Artificial Reef Stakeholder Groups and Institutional Frameworks

Stakeholder Group	Institutional Framework		
	Private	Communal	Public
Sport Anglers			X
Sport Divers			X
Artisanal or Commercial Fishers	X	X	X
Commercial Divers	X		X
Resource Managers and Scientists	X		X
Environmental Groups	X		X

6.3.1   User Interests and Institutional Elements

There are many human activities affected by artificial reef development. These activities foster a variety of stakeholder groups whose interests may be expressed in several institutional frameworks. Table 6.2 presents a listing of potential artificial reef stakeholder groups and the possible institutional frameworks that influence these stakeholder interests. In most situations, the majority of users benefit directly from artificial reefs as a public resource developed and managed by government agencies or communal groups. This is because, individually, each stakeholder group may not have the resources or the incentives to invest in an artificial reef project solely to meet its purposes. Commercial or communal fishing and diving groups may be the exceptions to the extent that they are able to finance and protect artificial reefs for their own private gain. However, when there are a multitude of potential reef users in an area, it becomes difficult to effectively restrict reef usage. In these cases, resource use congestion may occur eventually leading to user conflicts (Samples 1989) and/or resource degradation. Socioeconomic monitoring and evaluation can be used to identify incompatible or unintended uses and the potential source(s) of conflict at artificial reef sites, such as spearfishing at sites with heavy hook-and-line fishing pressure. With this information, reef managers can design regulations or institutions to reduce user conflicts (Milon 1991).

It is important to note that the term "stakeholders" includes not only groups (e.g., fishermen) that expect to benefit from, and thereby support, an artificial reef proposal, but also those who may actively oppose such development (e.g., environmental groups). The latter may question the goals and objectives of an artificial reef project and for this reason *stakeholder analysis* — which identifies relevant stakeholder groups, their expectations, and likely behavior — may be a necessary step to identify the concerns of opponents (Pickering 1997a). Stakeholder analysis can be thought of as a logical extension of the proposal made by a number of researchers (Ditton 1981; Graefe 1981; Bohnsack and Sutherland 1985; Milon 1991) that socioeconomic research should identify all of the likely winners and losers of a reef project. This information can be used to ascertain each group's power to influence project development, and efforts can be designed to address opposition.

An illustration of the role of stakeholder groups in artificial reef development is provided by the debate surrounding the decommissioning of North Sea oil platforms (Aabel et al. 1997). In operation, these structures act as *de facto* artificial reefs and the likelihood is that the "reef effect" of fish aggregation would continue if they were decommissioned. A number of decommissioning scenarios have been explored. Two options for platforms which (due to their size and water depths) are not required to be completely removed are: (1) to topple the inert platforms *in situ* or (2) to move parts of the jackets of platforms and group them in clusters at a limited number of pre-determined coastal or offshore sites. While there appear to be *prima facie* grounds for exploring the economic feasibility of these options, some environmental groups regard the disposal of platforms at sea merely as "dumping," irrespective of potential benefits. The power of these groups to influence events has been demonstrated by their successful mobilization of public opinion against the deep water abandonment of the *Brent Spar* oil platform (Wright 1998), despite the favorable findings of independent environmental assessments.

6.3.2 Legal Elements

As with most marine and freshwater resources, artificial reefs are typically open-access because property rights to artificial reef territory or productive output are not well defined and protected. Sometimes, as noted in the previous section, a private group may be able to build artificial reefs and exclude others from using them so that property rights are held in common by the private group. In these cases, the ability to control access to the habitat may be due to direct private ownership of the aquatic system, isolation from other user groups, exclusive user rights granted by a government authority, or *de facto* tenurial relationships enforced by custom and/or violence. Kurien (1995) provides a discussion of several types of institutional arrangements for communal management based on the history of artificial reef development in Kerala State, India.

It is important to recognize the potential scope of artificial reef ownership (or rights to a reef's marine product) can vary internationally according to different territories' laws for marine resources. British law, for instance, permits the creation of exclusive rights (termed *several fisheries*) to a reef owner that can be applied to certain types of shellfish. In 1997 the law was extended to include lobsters and other crustaceans, a change that will undoubtedly affect the economic prospects for artificial reefs constructed for the purposes of lobster stock enhancement (Jensen and Collins 1997; Pickering 1997b; Whitmarsh 1997). In general, though, there is limited formal recognition of private or communal rights to artificial reef resources. For example, Pickering reviewed the property rights structures for artificial reefs in the nations of the European Union and concluded:

> At both the international and national level, most fishing regulation is ... usually defined in a way that would not include such indirect activities as artificial reefs. As a consequence, unless specific legislation is made to the contrary, anyone can fish around a reef (Pickering 1997b:218).

6.4 SOCIAL AND ECONOMIC ASSESSMENT CONCEPTS

This section presents a general introduction to artificial reef monitoring, social assessment, and economic assessment along with references to more detailed discussions. The data collection and measurement techniques available for these types of assessments are discussed in Section 6.5.

6.4.1 Monitoring and Description

Monitoring is the most fundamental type of socioeconomic assessment because it can provide information to answer a number of basic questions for artificial reef evaluation (see Table 6.1). These questions are usually descriptive in nature and focus on the dynamics of reef use. For example, reef managers can employ monitoring to determine who uses specific artificial reefs and the reasons why they are used. A preliminary step for monitoring then is to identify potential users. Researchers have relied on a variety of sources including boat license screening surveys, intercept site surveys, charter/party boat passenger lists, commercial fishing boat license records, scuba or angler clubs, and saltwater fishing license records.

The data collection and evaluation techniques used to monitor the primary aspects of artificial reef usage and performance can be divided into three general categories: (1) direct observation of activity at the site(s); (2) on-site or shore interviews; and (3) mail or telephone surveys. These techniques can be used individually or in combination, if necessary, to customize and/or cross-check the data collected. Successful monitoring depends on a valid data profile that is not unduly influenced by events on one or a few days during the data collection period (Selltiz et al. 1976; Finsterbusch et al. 1983; Babbie 1997). Therefore, monitoring should not be done on a one-time basis and/or for a short interval. Regardless of the monitoring techniques employed, observation or interview schedules should be structured to systematically record information on specific aspects of user behavior associated with reefs. After monitoring data are collected, various types of frequency and/or statistical tests should be applied to summarize the important results of the study.

The primary considerations in determining the type of monitoring undertaken are: (1) the desired reliability of the data; (2) the resources available; (3) the research skill of the evaluator(s); and (4) the physical characteristics of the habitat site(s). Section 6.5 provides greater detail on the three general socioeconomic monitoring techniques (observation, interview, and survey) and the factors to consider when choosing among them. A brief review of previous studies illustrates the range of issues that can be addressed with socioeconomic monitoring. Readers interested in more information on individual studies can consult the study reports directly or detailed summaries in Milon (1989, 1991) and Roe (1995).

Ditton and Graefe (1978) used a mail survey to solicit information from recreational boat owners and charter/party boat captains. The survey was designed to obtain data on the use patterns for tire reefs, petroleum platforms, and sunken ships in coastal Texas, U.S. and the socioeconomic characteristics of the users. A stratified sample for the survey was generated from boat owner lists and charter/party boat advertisements in the areas surrounding Houston and Galveston.

Murray and Betz (1991) surveyed the attitudes of American sport anglers, commercial anglers, sport divers, and environmentalists about artificial reefs off the coasts of North Carolina, Florida, and Texas. Survey questions were designed to gauge artificial reef awareness among potential user groups and to assess the relative acceptance of various objectives for reef programs. Opinions regarding different management measures were also collected as well as information on the willingness to pay for artificial reef programs. A large part of the questioning effort was intended to provide feedback on measures such as time-of-day access limitations to minimize conflict around reef sites. The sample for the mail survey was selected from membership lists for various clubs and associations thought to have members who were knowledgeable about artificial reefs. This sample selection criterion places a greater emphasis on the selection of "heavy" artificial reef users and may not represent other potential reef users.

Ditton et al. (1995) conducted a mail survey of the charter fishing and diving industry in Texas to determine its use of artificial reef structures. This study is unique in that the survey included the entire population of the charter fishing and diving boats, not just a predetermined sample. The survey instrument requested detailed information on reef activities such as the rate and frequency of artificial reef use, the number of offshore trips taken, and the distances traveled to reef sites. There were also questions to elicit preferences for artificial reef materials, siting, and management options. Answers to these types of questions about users' preferences provide valuable input to the reef planning process. For example, the authors reported overwhelming agreement that there should be more state-sponsored reefs and that the most preferred material by captains of all boat types are petroleum structures, ships, and barges. Also, some existing sites were not used because of the distance from shore. Artificial reef program managers in Texas can use this monitoring information to justify the development and promotion of new reefs made from materials demanded by charter captains and to locate reefs within a practical distance from shore. The authors noted, however, that if the Texas artificial reef program sought to meet the diversity of user interests, similar surveys needed to be completed with other user groups (Ditton et al. 1995:24).

6.4.2 Social Assessment

Social assessment is used to isolate and measure changes that (could) occur in established social relationships, social structures, and normative systems when a policy or project is established (Vanderpool 1987). Before social system changes can be assessed, however, the conditions prior to project or policy implementation must be defined. These prior or baseline conditions have several dimensions: historical, cultural, demographic, social, and economic/ecological (Leistritz and Murdock 1981). A review of the first four dimensions can establish a baseline for social assessment, whereas the last dimension can be reviewed to portray preimpact conditions for a more detailed economic assessment (see Section 6.4.3).

Important background on the historical dimensions of artificial reefs might include information on commercial and recreational fishing patterns around a reef, past institutional arrangements for reef management, business cycles of reef-related industries and products, and past political conflicts over reef-fishing rights. The cultural dimension involves norms that define patterns of (perceived) ownership in artificial reef property, norms of reciprocity that define relationships on vessels or local reef-related businesses, and the extent to which conflict over reef resources can be attributed to divergent "worldviews" (Vanderpool 1987:481) of different stakeholder groups. Demographic dimensions of artificial reefs concern data on stakeholder characteristics (i.e., number, age, ethnicity, and income). Other aspects of the social dimension can deal with interactions between stakeholder groups (e.g., vessel owners and captains), changing patterns of ownership in reef-related industries, and power structures in communities affected by artificial reef development.

Baseline assessment of these social dimensions requires systematic and comparative monitoring of social variables that are consistent with the objectives of a project. Social variables can be based on changes in amount of use, types of use, types of fish targeted, type of social group utilizing a site, and/or level of satisfaction. Satisfaction can be used as a "global" measure (e.g., are you satisfied with the artificial reef you visited today?) or a specific measure of trip characteristics (e.g., was the reef close enough to shore, easy to find, did you catch the fish you expected, satisfied with the size of fish, or was the reef crowded?). Specific satisfaction questions provide more useful information that may assist managers in modifying a problematic situation. Details on constructing social impact or satisfaction questions can be found in Ditton and Graefe (1978); Ditton et al. (1981); Finsterbusch et al. (1983); Fedler (1984); and Milon (1988).

The primary approach for a social assessment analysis is to compare social variable monitoring data during more than one time period (pre- and postproject periods when available) using statistical tests (see Chapter 2). Direct observations, interviews, and mail and telephone surveys can be used. Since social assessment requires a comparison, however, it is important that the data collection process is carefully planned so valid statistical comparisons can be made. This means that the samples are properly selected and the same measurement variables are collected (see Chapter 2). Planning is necessary whether the comparisons are based on before-and-after measurements or a cross-section analysis using statistical controls. Detailed guidelines for social assessment and impact analysis are available in Finsterbusch et al. (1983); and Burdge (1996).

In many situations, the most desirable way to accurately evaluate a project's social impact is to measure the appropriate indicator variable before and after the habitat is established. For example, a project objective to reduce the distance traveled by inshore fishermen should not be evaluated by comparing the distance traveled by inshore habitat users with the distance traveled by offshore fishermen or with the distance traveled by other inshore fishermen who do not use the habitat. In the first comparison, the distance traveled by offshore fishermen is irrelevant since inshore fishermen would travel shorter distances regardless of whether the habitat exists. In the second comparison, the fishermen who selected the artificial reef may already travel less than the nonusers so that the effect of the project may be only to relocate or concentrate their fishing activity. The proper comparison is the distance traveled by habitat users before and after the project is established so that the net change in distance traveled can be measured.

While before-and-after measurement is the most desirable procedure for impact evaluation, it may not be practical. Data for fishing and diving activities may be limited so a special effort may be necessary to establish baseline measurements before the project begins. Alternatively, it may be possible to establish a control group of potential users who are restricted from actually using the project and measure their activity concurrently with the activity of users.

If both preproject measurements and control groups are not feasible, the only alternative is careful statistical analysis of the data to identify the influence of confounding effects on the outcome measure. This type of statistical control requires further information on appropriate factors that may influence the outcome. For example, in the evaluation of distance traveled by inshore fishermen after an artificial reef is established, data on boat length and target species may be used as controls to compare the distances traveled by reef users and nonusers. These factors may influence the choice of fishing site and hence the distance traveled. If the distance traveled by users with boats of the same length and the same target species choice as nonusers is lower, the conclusion that the habitat actually reduced distance traveled is more plausible than an evaluation based solely on distance traveled by all inshore users and nonusers. This type of statistical control depends on the evaluator's prior knowledge of likely confounding factors. Appropriate statistical procedures for this type of evaluation are described in Chapter 2 and in many standard statistics texts (e.g., Wright 1979).

To document the relative importance of various management-related characteristics of a reef (distance from shore, ease of location, quantity or quality of fish, etc.), importance/performance analysis can be used. Once a listing of management-related site characteristics has been identified, survey participants are asked to rate the importance of each characteristic (using a predetermined scale) and to describe how satisfied they are with that characteristic of a specific site. For example, a participant might report that it is very important for a reef to be "easy to find," but report that a particular reef was "not very easy to find" (low performance). By plotting the various importance/performance scores on a grid, reef characteristics that are over-performing or under-performing can be identified based on the level of importance they are assigned. Details of importance/performance analysis can be found in Martilla and James (1977).

Most social assessments in the marine setting have concentrated on the impact of various fishing regulatory measures on commercial anglers (Vanderpool 1987). The few studies dealing with artificial reefs tend to focus on community changes (impacts) fostered by reef development. Simard (1997) reviewed the social effects of artificial reefs in Japan. The author describes artificial reef development in Japan and summarizes surveys of fishers' perceptions of the effects of artificial reefs on fishing productivity and related social changes. One study surveyed 40 of Japan's approximately 4000 Fishing Co-operative Associations (FCAs) to determine their use of artificial reefs and assess their views regarding the reefs' abilities to attract and increase the productivity of marine resources. Another study examined the role of artificial reefs in the experiences of 65 fishermen who experienced increased catches. Simard (1997) noted evidence from these studies suggesting that artificial reefs played a "reassuring" role in the lives of Japanese fishers. This reassuring effect had several important social consequences for fishers who use artificial reefs, including an increase in the time that fishers' wives were able to spend on activities at home (this is likened to the influence of technological improvements on the farming lifestyle). Artificial reefs were also cited as part of the large-scale plan developed by/for FCAs that "removes dependency on nature by introducing human operations before the catch."

Kurien (1995) reviewed the social significance of artificial reefs in the artisanal fishing culture of Kerala State, India. His analysis was based largely on first-hand experiences of the author and on anecdotal evidence. A rich history of shared knowledge on reef building and maintenance activities in this area was reported as a manifestation of the fishers' conservationist ethic for marine resources. Kurien believes that artificial reefs have served as a symbol of communal efforts to "green the sea" and as a mechanism for collecting and disseminating transgenerational knowledge. Also, artisanal fishers used artificial reefs to effectively "fence-off" their exclusive fishing zones (assigned by the State) against the incursion of trawlers. This has allowed for the recovery of nearshore areas damaged by previous "overfishing" and contributed to a greater sense of security for artisanal fishers. Kurien suggests that increased confidence in marine resources has, in turn, promoted positive social impacts for the artisanal fishing communities that are similar to the "reassuring effects" noted by Simard (1997) in Japan.

6.4.3 Economic Assessment

Economic assessment of artificial reef performance can (1) provide in.
or potential economic impacts of reef development; and/or (2) determine whethe
an efficient (public or private) investment. The first type of assessment is an econon.
analysis while the second is an economic efficiency analysis. Both types of analyses requi.
socioeconomic data to evaluate the changes in economic activity and the benefits and costs associated with artificial reef programs.

6.4.3.1 *Economic Impact Analysis*

Economic impact analysis focuses on changes in sales, income, and employment resulting from
a project. It is the appropriate evaluation method when the objective of a project relates to desired
changes in the economy of a coastal community or region. Three types of impact from a project
can be measured: (1) the *direct* impacts or the changes in local spending or final demand for goods
and services immediately related to artificial reef development; (2) the *indirect* impacts due to
changes in the purchase of inputs by businesses directly impacted (from 1); and (3) *induced* impacts
that occur when local employees of the businesses directly (from 1) and indirectly (from 2) impacted
change their purchases of goods and services. The indirect and induced impacts of a project are
usually referred to as the *secondary* impacts because they occur only as a product of the direct
impacts. Economic impact analysis measures the total economic impact of a project on sales,
income, and employment by summing the direct and secondary impacts.

The direct economic impacts of an artificial reef project are measured using information obtained
through data collection and monitoring. For example, data on (average) local expenditures on fishing
trips to artificial reefs are necessary to estimate the direct economic impact caused by the use of
artificial reefs by recreational anglers. These expenditure data can be utilized in conjunction with
information from reef usage monitoring to estimate the aggregate direct impacts of artificial reef use.

Estimates of the secondary impacts of a project can be calculated with multipliers. A *multiplier*
expresses the relationship between the direct impact of a project in terms of the resulting changes
in sales, income (wages and earnings), and employment within the impacted area. These are called
the sales or output multiplier, the income multiplier, and the employment multiplier. Other multipliers also can be developed for special evaluation purposes (Stevens and Lahr 1988). Each
multiplier is a useful measure of changes caused by an activity, but the appropriate choice for an
impact analysis should be linked to the project objectives. Many reef development efforts seek to
improve community income and the number of jobs so income and employment multipliers would
be most useful.

The two most commonly used approaches to compute multipliers are the economic base and
input/output methods. Each method embodies a different view of the economic development process
that can influence the outcome of an impact evaluation. Milon (1991) presents an introduction to
economic-base and input/output analysis for artificial habitat evaluation. The various methods used
to compute multipliers are detailed in reference works such as Leistritz and Murdock (1981);
Hewings (1985); Miller and Blair (1985); Richardson (1985); and Propst and Gavrilis (1987). These
publications describe models developed specifically for use in the U.S. In practice, multipliers used
in economic impact studies should be estimated specifically for the country and/or region where
the analysis is conducted.

Buchanan (1973) provided the first published study of the economic impact of an artificial
habitat. This study used a combined direct observation and mail survey of anglers in Murrells Inlet,
South Carolina, U.S. Since baseline data prior to the development of the habitat were not available,
responses to mail survey questions were used to divide tourist anglers into those who would or
would not return to Murrells Inlet if the artificial habitats were not present. The group who would
not return was classified as the change in fishing effort caused by the habitat. This group accounted

.6% of the fishing effort and 10% of the direct expenditures by tourist anglers. There was no .empt to compute the secondary impacts of these expenditures on the economy.

Rhodes et al. (1994) used annual and monthly mail survey instruments to collect data on artificial reef usage patterns and factors that influence the level of fishing activities at reef sites in South Carolina. Samples for both surveys were drawn randomly from the population of registered boat owners with powered vessels greater than or equal to 16 ft. The authors extrapolated the average number of days that respondents reported fishing exclusively over an artificial reef to approximate the total number of artificial reef fishing days and related expenditures occurring (in 1991) along the South Carolina coast. A simple multiplier was used to approximate the cumulative impact of the artificial reef expenditures on the South Carolina economy.

6.4.3.2 *Efficiency Analysis*

A more thorough understanding of artificial reef utilization can be gained by evaluating reef efficiency because, while impact evaluation measures the effects of a project on socioeconomic conditions, there is no specific criterion embodied in impact analysis to determine whether the project is beneficial or harmful. As such, impact analysis is "value-free" in the sense that no value judgments are imposed by the analytical methods. By contrast, efficiency evaluations impose a standard to evaluate whether a project is good or bad. The standards are then evaluated with data to document how a reef is (could be) performing and/or the economic costs and benefits. This section briefly describes two types of artificial reef efficiency analyses: cost–effectiveness analysis and cost–benefit analysis. A more thorough treatment of these types of analyses for artificial habitat and coastal resource evaluation can be found in Bockstael et al. (1985); Milon (1989, 1991); Penning-Rowsell et al. (1992); Lipton et al. (1995); and Whitmarsh (1997).

Cost–effectiveness and cost–benefit analyses impose a standard to evaluate whether a project is preferred or not. The standard imposed by cost–effectiveness analysis, however, is different than the standard imposed by cost–benefit analysis. Cost–effectiveness seeks the project alternative that produces the desired outcome at the least cost, whereas cost–benefit analysis determines whether the monetary value of the outcome justifies the project cost. The choice of an evaluation standard depends on the objectives of the project and the characteristics of the benefits produced by the project. As a general rule, cost–effectiveness analysis is most useful when a specific tangible outcome is desired that cannot be easily measured in monetary terms. For example, the use of alternative artificial reef designs to provide habitat for an endangered species could be evaluated by comparing the costs per unit of shelter volume. Assuming shelter volume is an appropriate indicator of habitat availability, the least cost design would be preferred. Cost–benefit analysis is more appropriate when the benefits of the project can be monetized and associated with specific user groups. For example, alternative artificial reef designs to increase the exploitable biomass of a species for recreational and commercial harvesting can be monetized (Table 6.3) and then the aggregate monetary benefits can be compared to the costs of each design.

Table 6.3 Potential Benefits of an Artificial Reef Project

Benefit	Economic Parameter	Available Measurement Techniques or Data Sources
Marine Habitat/Species Preservation	Preferences for marine habitat preservation (consumer surplus)	Contingent valuation
Diving and Snorkeling Site	Demand for diving and snorkeling sites (consumer surplus)	Travel cost Contingent valuation
Recreational Fishing Site	Demand for recreational fishing sites (consumer surplus)	Travel cost Contingent valuation
Commercial Fishing Site	Fishers' profit (producer surplus)	Surveys Industry records Productivity approaches

Another important distinction between cost–effectiveness and cost–benefit analyses is the data that are required to conduct each type of analysis. The economic benefits of artificial reefs are typically related to ongoing activities at the reef site and concentrated among the project users or businesses that provide services to the users (e.g., charter boats and restaurants). This means that special monitoring efforts involving controlled observations, surveys, and/or interviews are required to obtain specific information on the benefits (value) of artificial reef use. On the other hand, the site-related and off-site-related costs of an artificial reef project are typically available via archival or record retrievals from the governmental or private entity that sites and constructs the reef. Thus, different strategies may be necessary to obtain information on the benefits of artificial reef use than are necessary to collect data on the costs of reef deployment.

It is important to distinguish efficiency analysis from economic impact analysis because the two types of evaluation are commonly confused. As discussed previously, economic impact analysis focuses on the direct and secondary changes in sales, income, and employment caused by a project. These changes are usually considered to be beneficial for the local community but the changes may only result from an offsetting decline in economic activity elsewhere (Talhelm 1985). In addition, an impact evaluation does not consider the costs to implement the project. Cost–benefit analysis considers both sides of the ledger and counts only the direct impacts on income or user benefits that result from the project. Other categories of what might be considered beneficial impacts such as sales and employment growth are not counted because these effects are caused by the change in income. Focusing on the direct income changes in cost–benefit analysis avoids the problem of double-counting the benefits and provides a consistent basis to compare benefits and costs.

For an economic efficiency analysis, the *benefits* of an artificial reef project are the net outcomes expressed in monetary units. Table 6.3 lists the potential benefits of an artificial reef project with the respective economic parameters and applicable measurement techniques. These benefits can include tangible outcomes such as the change in commercial fishery harvests due to a habitat and intangible outcomes such as the enjoyment a sport diver experiences from photographing a habitat. Tangible effects are valued by measuring the amount of money gained from the outcome such as the change in commercial fishery profits or income. Intangible effects are valued by measuring the beneficiaries' net willingness to pay for the outcome. In some cases intangible effects may be too nebulous to measure, such as when an artificial reef is deployed to protect an endangered species or for use in a research project. In these situations, cost–effectiveness analysis may be a more reliable evaluation tool than cost–benefit analysis.

Artificial reefs generally are perceived as beneficial due to their ability to attract and concentrate marine resources and possibly increase localized biomass (Pickering and Whitmarsh 1996). The concentrated marine resources can be beneficial to human use in one or a combination of two ways: (1) harvested biomass or enjoyment from a trip to an artificial reef site is higher than a trip to a natural habitat site and trip costs are the same; or (2) harvested biomass or enjoyment is the same but trip costs are lower. Both commercial and sport artificial reef users can enjoy the benefits from these two effects; however, the methods used to measure the expected benefits for commercial and sport users are different.

Commercial users' benefits are measured by the change in profits or "producers' surplus" that occurs due to the presence of the artificial reef, that is, the difference in producers' profit with and without the new habitat. With information on the potential changes in fishery productivity expected from the reef, the estimation of changes in producer surplus is fairly straightforward (see Milon 1989, 1991). However, it is important to bear in mind that key assumptions regarding the behavior of certain economic variables may be necessary. For example, property rights for access to the artificial reef must be defined and expected market prices for marketable catch must be estimated for the planning horizon (e.g., Whitmarsh 1997).

Sport user benefits from an artificial reef are measured by the users' willingness to pay or "consumers' surplus" for the habitat site. Consumer surplus is the amount sport users are willing to pay for the habitat over and above the actual expenditures incurred in using the habitat. It can be measured for both resident and tourist sport users because each group can directly benefit from the project. This is an important distinction from economic impact evaluation that attributes beneficial effects to new tourist spending. Note also that this measure of benefits does not include trip expenditures since these expenditures could be incurred elsewhere. These sport user benefits can be measured with survey research methods to identify users' preferences for artificial reef characteristics and to elicit willingness to pay. Bockstael et al. (1985, 1986); Milon and Schmeid (1991); Penning-Rowsell et al. (1992); and Lipton et al. (1995) provide introductory discussions and reviews of studies using these benefit estimation methods for natural resource and coastal valuation; Milon (1991) reviews applications of benefit measurement for artificial marine habitat.

The potential benefits from an artificial reef also may be "tangible" but indirect such as the functional benefits that a reef may provide to support other economic activities (Pickering et al. 1999). For example, the presence of an artificial reef may serve to divert effort away from an established commercial fishery that is currently overexploited, thereby indirectly increasing biomass and productivity (Whitmarsh and Pickering 1997). An artificial reef may also facilitate the capture of dissolved nutrients (mainly phosphorus and nitrogen) created by fish farms, again with a positive but indirect influence on aquatic production (Laihonen et al. 1997). Additionally, artificial reefs may fulfill a coastal protection and biological support function such as the situation in certain Mediterranean countries where *Posidonia* seagrass beds (which are an important habitat for juvenile fish) are protected by submerged anti-trawling structures (Bombace 1997). Since these indirect effects typically involve a series of physical impacts culminating in a change in socially valuable output (e.g., marketable fish), the functional benefits of an artificial reef can be measured provided that data can be obtained on the cause-and-effect physical relationships and the monetary value (in the form of market or surrogate prices) of the resulting product. It needs to be recognized, however, that even where a productivity approach to valuation can be used in this way, this methodology may not fully account for the total economic benefits resulting from an artificial reef project. Such would be the case where, for example, an artificial reef mitigated the effects of eutrophication caused by aquaculture. The perceived improvement in water quality might, in this situation, be more appropriately measured by a direct valuation method such as contingent valuation.

An important element of artificial reef efficiency evaluation is proper accounting for the economic costs of a project. This is a deceptively simple element of evaluation because all project costs are not always considered, resource costs are not always fully reflected in their prices, and costs incurred over the life of the project are often neglected. The proper scope for the costs of an artificial reef project is the full range of resources used from the initial design and planning stages to the final stage, including the cost of removing the habitat from the water, if appropriate. The categories of costs listed in Table 6.4 are representative of the private and/or social costs commonly incurred in an artificial reef project. Most of the cost categories such as personnel, transportation, maintenance, and planning are based on engineering analyses and are fairly straightforward to estimate. In practice, however, estimates of artificial habitat costs have tended to focus only on material and transportation costs (e.g., Prince and Maughan 1978; Shomura and Matsumoto 1982) and neglect the personnel costs to plan, design, administer, and implement the project. In addition, maintenance and evaluation costs over the life of the project for marker buoys, repositioning, or other activities after the initial deployment are often overlooked. Maintenance and evaluation expenses are necessary parts of project development and should be reflected in project costs. Potential external costs caused by a project, such as material removal from a beach after a storm or damage to vessels, also should be considered as part of project costs in the planning process. While these external costs may not be known until the project is implemented and may be avoided by proper planning and siting, the recognition that these are legitimate project costs encourages artificial reef planners to consider the full costs of particular designs.

Some costs, such as building rent or dockage fees, are relatively easy to measure because they have a market price that (should) reflect the economic cost of using the resource. Other resources used in artificial reef development, however, may be imperfectly priced or not priced at all. For

Table 6.4   Potential Economic Costs of an Artificial Reef Project

Costs	Economic Parameter	Available Data Sources or Measurement Techniques
Personnel	Wages and salaries	Project cost estimates
Planning and administration	Expenses	Project cost estimates
Construction and reef materials (including foregone scrap values)	Expenses	Engineering estimates
Transportation	Expenses	Engineering estimates
Marker buoy positioning	Expenses	Engineering estimates
Material maintenance	Expenses	Operation cost estimates
Dismantling and removal	Expenses	Engineering estimates
Liability insurance	Expenses	Project cost estimates
Damage from "loose" materials	Repair costs	Repair cost estimates
Damage to vessels	Repair costs	Repair cost estimates
Ecosystem disruption	Foregone productivity or nonuse values	Ecosystem or habitat valuation technique
Stock externalities[1]	Loss in productive value, the cost of compensating equipment	Consult a resource economist
Congestion externalities[1]	Loss in productive value, the cost of compensating equipment	Consult a resource economist

[1] See Milon (1991) for a discussion of stock and congestion externalities in the context of artificial habitat management.

these resources, special care has to be taken to ensure that their true opportunity cost is counted in the project cost analysis. The opportunity cost of using any resource for an artificial reef is the foregone economic return (value) that the resource would have brought in an alternative use. For example, the U.S. Department of Defense considered the use of demilitarized combat tanks for an artificial reef project (PRC Environmental Management, Inc. 1994). Since the tanks were already owned by the U.S. government, there were no out-of-pocket costs for acquisition. However, because the tanks could be sold as scrap metal and the proceeds used to buy a prefabricated habitat structure, the tanks had an alternative use value. The opportunity cost of using the tanks for an artificial reef project was the value of the tanks as scrap and this cost was included in the total project costs. Similarly, the opportunity cost of an obsolete offshore oil and gas platform that might be used for an artificial reef is equal to the onshore salvage value minus dismantling and transportation costs (Stelzer 1989). From a social accounting perspective, these opportunity costs are the true value of the resources that should be counted as an acquisition cost in an efficiency analysis. Examples of other ways to measure opportunity costs are discussed in greater detail in texts on efficiency evaluation such as Thompson (1980).

It is also important that measures of the benefits and costs of an artificial reef project cover the full time period of the project. For example, it is necessary to count costs that are anticipated for maintenance, monitoring, and removal of the reef structure at the end of the project because once a reef is deployed these costs become unavoidable. A further complication arises from the fact that a cost–benefit incurred in the later years of a project is not as costly/valuable as an initial cost–benefit due to the time value of money. The effects of time on the value of money are counted in efficiency analysis by discounting future costs and benefits to the present value. Thus, if the known useful life of a project was 3 years, the removal costs were $10,000, and the cost of money was 10%, the present value of these future costs would be $7,513. More information on discounting can be found in cost–benefit manuals such as Thompson (1980); Penning-Rowsell et al. (1992); and Zerbe and Dively (1994).

Few studies exist that completely assess the efficiency of an artificial reef project. Most studies focus on either costs *or* benefits and do not consider both for a balanced analysis. Early attempts at efficiency evaluations for artificial habitat are reviewed in Milon (1989, 1991). Whitmarsh (1997) reported on a cost–benefit analysis of different construction materials in artificial reefs for European lobster (*Homarus gammarus*) production. The project benefits for the analysis were estimated as revenues from lobster sales, though other possible "external" project benefits like improved recreational and commercial fishing were noted. The estimation of revenues anticipated over the 100-year life of the project used a forecast of lobster production and their unit value with assumptions about the following key variables:

- Property rights to the reef and its harvest,
- Catch rates, weights, and controls,
- Lobster prices,
- Hatchery releases,
- Natural colonization,
- Entry into and exit from the fishery.

The estimation of anticipated annual project costs included construction, sea transport and placement expenses, and the operating expenses associated with harvesting and the cost of juvenile lobsters. These annual costs were subtracted from the stream of projected lobster revenues to compare the benefits minus the costs, or the net present value (NPV), of constructing an artificial reef with either stabilized blocks of power station ash or quarry rock. NPV curves were estimated to show the sensitivity of the analysis to variations in key assumptions, and cost–benefit ratios/internal rates of return were calculated for the various project options. Risk simulations showed a negative NPV for the ash blocks and a positive NPV for the quarry rock. Two key issues arise from Whitmarsh's (1997) analysis: (1) results from economic analysis are highly dependent on reliable data for the biological productivity of artificial reefs; and (2) clearly defined property rights to exclusively harvest the fishery products are necessary to ensure the economic returns from artificial reef development.

A more recent study (Whitmarsh et al. 1998) examined in closer detail the circumstances under which lobster stock enhancement based on artificial reef technology would be economically viable. In the U.K., low recapture rates of hatchery-reared juvenile lobsters released into the sea have raised questions about the economic efficiency of public lobster stock enhancement programs (Lee 1994). However, given that lobsters from the wild population colonize artificial reefs (Jensen and Collins 1997), the use of such structures as a ranching substrate offers a possible way to increase the efficiency of harvesting and to raise the recapture rate. The main focus of the study by Whitmarsh et al. was the costs of recapture and their implications for the economic viability of a reef-based stock enhancement program. In the analysis, harvesting of the reef was assumed to take place by commercial fishers using baited traps under a range of cost assumptions. The limiting case of zero incremental harvesting cost was deemed to be valid only in situations where an established lobster fishery already exists. In most circumstances, the expectation was that harvesting enhanced stocks would require an increase in fishing effort and the commitment of additional factor inputs. These were assessed using data obtained from a cost and earnings study of commercial U.K. lobster fishermen. The different cost assumptions used were that the marginal increase in fishing effort is measured by (1) running costs only; (2) variable costs; (3) total costs; or (4) total costs including the opportunity cost of capital. The study found that the NPV of an artificial reef project for lobster stock enhancement was sensitive to assumptions about harvesting costs. Specifically, the *breakeven recapture rate* (i.e., the rate above which the NPV becomes positive) was significantly higher once the baseline assumption of zero incremental cost was relaxed.

6.5 DATA COLLECTION AND MEASUREMENT TECHNIQUES

Many of the studies reviewed in Section 6.4 used mail surveys to obtain data to evaluate whether or not an artificial reef was meeting social and policy objectives. There are, however, other techniques available to collect socioeconomic data. This section describes mail and telephone survey methods in greater detail and introduces two additional data collection and measurement techniques for social and economic assessment.

6.5.1 Direct Observation

One approach to obtain socioeconomic information about artificial reefs is through direct observation at the site. An evaluator or observer can record the number and type of boats and type of activity at an artificial reef site during different time periods. It also may be possible to use direct observation to discern boat length, the number of people in each boat, type of fishing, and likely target species. These additional data may be hard to collect accurately, however, without confirmation via boat registration data or other surveys for the surrounding coastal areas. Research suggests that simply measuring boat length correctly can be difficult, even with confirmation aids (Ditton and Auyong 1984).

Direct observation may be desirable if the site is easy to monitor and the purpose of evaluation is to identify general usage patterns. However, there are several limitations to recognize when considering direct observation as an option to collect socioeconomic data for artificial reefs. First, direct observation can be expensive since it requires personnel on the site for extended periods of time, travel costs may be high, and the safety of the observers must be considered. If the study area is large, then direct observation may not be practical or economically feasible for the entire area. Second, direct observation usually cannot provide demographic, social, or economic information about individual users. This could be a serious drawback since information on the socioeconomic characteristics of reef users is usually of interest in reef management decisions. Third, direct observation cannot provide data on individuals or groups who do not directly use a study site(s). Finally, it may be difficult to collect a truly random sample from direct observation, especially for sites such as petroleum platforms that require the cooperation of private owners and/or industry. Consequently, researchers may have to rely on a "convenience sample" based on where and when they can get cooperation at an observation site. This potential lack of representative (random) samples can limit the validity of statistical tests of differences in use characteristics among geographic areas or stakeholder groups and restrict the ability to formulate models to predict use patterns (Ditton and Auyong 1984).

The procedures for systematic observation are fairly straightforward. Most of the technical guidance focuses on providing accurate and realistic instructions to the observers and other concerns common to all survey research. Further details on systematic observation techniques are available in Webb et al. (1966); Weick (1968); Whiting and Whiting (1970); Hogans (1978); Peine (1983); and Smith (1991).

A variation on the direct observation method that may be useful for large, dispersed sites is aerial photography (Deuell and Lillesand 1982). Observations can be scheduled for selected time periods and the photo record can be used to identify the number and types of users. In addition to the disadvantages noted for on-site observation, aerial photography can be difficult to interpret for use patterns at sites, such as petroleum structures, where users tend to collect under superstructures. Furthermore, cost can become a limiting factor where additional flights are necessary to collect observations for use patterns over an extended period of time. Also, usage data collected with aerial photography will likely represent the upper bounds on usage levels at a site because aerial surveys are typically limited to good weather days. Analysts conducting aggregate economic studies should consider that aerial survey data may overstate benefits, costs, or impacts from an artificial reef project (Daniel and Seward 1975).

6.5.2 Direct Interviews

An alternative site-based survey method that is useful for socioeconomic data collection is the direct interview. Interviews may be conducted at a site and combined with observational data, or the interview may be conducted at user access points such as marinas, ports, or entrance ways to shore-accessed habitat sites. The direct interview can be used to collect user (and/or nonuser) profile data on socioeconomic characteristics, expenditures, and attitudes, and it provides the opportunity for the interviewer to properly identify species in the kept catch. The drawbacks of direct interviews are primarily cost related. It may be expensive to arrange an interview schedule for a sufficient period of time to provide representative data, especially if the sites and/or access points are widely dispersed. Also, respondents may not be willing to stop their activity for an interview or to reveal personal information (e.g., age and income) to a stranger. In some cases interviews can be combined with other types of surveys to obtain data on future use or other information that may be too difficult or time-consuming for an interview.

Effective standardized interviews have several important key elements (Fowler 1995): (1) good definitions are provided for all critical terms; (2) questions are tested in advance; and (3) questions are asked in a logical and nonrepetitive order. The first element is essential to ensure that respondents have a clear understanding of the questions posed without having to interrupt the interview flow with extraneous questions. This is especially important when respondents are presented with detailed information, such as artificial reef locations and ecological characteristics. The second interview element, pretesting of questions, helps to ensure that questions can be easily read by interviewers and easily understood by respondents. Pretesting can range from simple practice interviews with observation to more involved focus group evaluations and can even include intensive or cognitive evaluation of respondents' perceptions. The degree of pretesting warranted will depend on the inherent complexity of the questions (see Fowler 1995:Chapter 5). The final element for effective interviews requires an understanding of the potential thought processes of the respondents so that questions can be asked in a logical order. For example, an angler who responded "yes" to a question whether she had fished at an artificial reef site in the past 2 months will probably also give the site location and/or name. In this case, there is no need to force the interviewer to ask and code a separate location question which could unduly interrupt the interview process. To avoid such repetitious questions, interviewers should be given flexibility to ask questions in a more conversational manner, thereby allowing the interview to flow naturally. Additional details on interview techniques can be found in Fowler and Mangione (1990); Salant and Dillman (1994); Fowler (1995); and Babbie (1997).

6.5.3 Mail and Telephone Surveys

The flexibility and relatively low cost make mail and telephone surveys a common method to obtain many types of socioeconomic information, including data for artificial reef users. These surveys can be coupled with direct observation or short interviews to obtain more detailed follow-up information than can be acquired on-site. When mail or telephone surveys are used independent of interviews, a sample frame of potential artificial reef users (or nonusers) can be identified from public record lists, such as recreational and commercial boat registrations, recreational and commercial fishing licenses, and sport fishing and diving club membership. For example, Ditton et al. (1995) conducted a thorough review of magazines, newspapers, telephone books, brochures, business cards, and Chamber of Commerce records to develop a mailing list for full-time charter and dive boat operators in Texas.

Samples can be selected from lists of potential artificial reef users in a variety of ways, ranging from methods based on convenience to more complicated multistage selection protocols (Chapter 2). A simple, yet effective way to ensure that all relevant groups are represented in the sample is to proportionally stratify the sample (Henry 1990). For example, Ditton and Graefe (1978) stratified a sample frame of registered boaters in Texas to produce a proportional sample from two groups, those with vessels under 26 ft and those with vessels over 26 ft. Sample frame and size selection is discussed briefly in Section 6.4 and in greater detail in Chapter 2.

Survey instruments can be designed to elicit information about usage and catch at an artificial reef site and other natural habitat sites. Additional user and nonuser profile information includes socioeconomic characteristics, expenditures, and attitudes. The principal drawbacks of mail and telephone surveys are that respondents may not be able to provide information about the use of a specific site, the timing and duration of site use over extended periods of time, or the proper identification of species in the catch. These problems can be partly overcome by limiting the duration for recall (maximum of 3 to 6 months), by sequencing distribution of the survey over major periods of use, such as fishing seasons, and by including visual aids such as maps and charts.

The recall problems related to survey lag also can be overcome with logbooks that enable study participants to directly record information at the time of occurrence. A logbook program was used in conjunction with a survey of recreational anglers to identify use patterns and catch effort data for offshore oil and gas platforms in coastal Louisiana (Stanley and Wilson 1989). The trip logbooks required anglers to list the number of anglers in the party, the structure fished, the fishing method used, and the species and number of fish caught.

Other data collection problems such as inadequate sample size and survey instrument bias can be addressed by following generally accepted procedures outlined in social survey reference works such as Dillman (1978); Rossi et al. (1983); Salant and Dillman (1994); Babbie (1997); and Rea and Parker (1997).

6.5.4 Quality Control

There are advantages and disadvantages to each data collection technique described in the previous sections. Table 6.5 defines selected survey research criteria and describes how mail and telephone surveys, interviews, and direct observation compare for each. Similarly, Table 6.6 compares the relative reliability or availability of information on selected socioeconomic variables from each data collection method. The figures and ratings provided are based on the authors' experience in conducting socioeconomic research in a variety of settings. It is important to note that all research projects, regardless of the method used, have site-specific features that inevitably influence the ways in which the criteria parameters listed can be evaluated and the scope of socioeconomic data that can be collected. There are, however, quality control procedures that can be taken throughout any research project that can influence compliance with the research criteria and control the availability and quality of data.

Socioeconomic data are usually time consuming and/or expensive to obtain, therefore, quality control is important. Careful attention to detail during each step of the research effort will maximize the potential for quality data to support policy-relevant conclusions. Important aspects of quality control for socioeconomic evaluation include inter-observer or inter-interviewer reliability, standardized questioning, sample frame and sample size, coding consistency, and regular data collection frequencies.

If the data are collected directly by observers or interviewers, the observers or interviewers should use standardized indicators or questions in making their observations (Fowler and Mangione 1990). Observers and interviewers should be trained in the importance of consistency, accuracy, conformance with the sample selection criteria, and unbiased questioning. Role-playing (as respondent and interviewer) is also valuable to determine appropriate responses and strategies for dealing with potential problems.

Decisions about the sampling frame influence the total error and cost of data collection (Henry 1990). Some data collection techniques do not require a sampling frame, such as random digit dialing for a telephone survey. Other techniques, such as mail surveys, require careful choices between different target populations. For example, to collect socioeconomic data for artificial reef users, the sampling frame could be the general population in a coastal state or specific stakeholder groups, such as members of fishing and diving clubs. While a general population mail survey may

Table 6.5   Comparison of Socioeconomic Data Collection Methods[1]

Research Criterion	Definition	Method			
		Mail Survey	Telephone Survey	Interview	Observation
Training/ preparation time	Typical time required to select the sample, prepare/pretest and print the survey instrument, and/or recruit/train interviewers before data collection can begin	6–8 weeks	6–8 weeks	8–10 weeks	4–6 weeks
Data collection time	Typical time after training/prep is through before data analysis and report preparation can begin	2–5 weeks	1 week	2–5 weeks	50+ hours[2]
Response rate	Percentage of potential respondents initially contacted who actually complete the survey	20–60%	40–70%	60–80%	NA
Anonymity	Ability to maintain the anonymity of respondents	high	high	high	high
Compliance with instructions	Ability to ensure that respondents are complying with instructions and answering questions as intended	low	moderate	high	NA
Interviewer bias	Potential for the interviewer to influence respondent answers	none	moderate– high	high	NA
Flexibility	Ability to probe for more detail, explain unclear questions, and use visual aids	low	moderate– high	high	NA
Complexity	Ability to obtain complex information by providing more detailed instructions and lengthy lists of alternative responses	low	moderate	high	NA
Mean cost per respondent[3]	Total cost of the survey divided by the number of completed responses (in U.S. dollars)	$15–$50	$15–$60	$40–$120	Varies widely

[1] Based on information provided in Rea and Parker (1997) and experiences of the authors.
[2] Includes time for mission execution, key development, and photo preparation and interpretation (Deuell and Lillesand 1982).
[3] Estimates in 1998 dollars based on achieving 300 to 400 responses; smaller surveys may cost more than larger surveys.

Table 6.6   Comparison of Information Reliability from Socioeconomic Data Collection Methods[1]

Information	Relative Reliability/Availability of Data[2]			
	Mail	Telephone	Interview	Observation
Demographic data	3–4	3–4	4–5	1–2
Avidity/mode of activity	3–4	3–4	5	4–5
Trip location(s)	3–4	3–4	5	5
Trip purpose(s)	3–4	3–4	4–5	2–3
Vessel/equipment data	4	4	5	4
Expenditures	3–4	3–4	4–5	1
Catch data	2–3	2–3	4–5	3–4
Preferences/attitudes	3–4	4–5	4–5	1
Resource valuation	3–4	3–4	4–5	1

[1] Ratings based on the authors' experiences.
[2] Scale from 1 to 5 where highly unreliable/unavailable equals 1; highly reliable/available equals 5.

provide more representative sample data than a survey of fishing and dive club members, the former probably would be more expensive and suffer from low response rates.

Once the sample frame (if necessary) is determined, a sample size must be selected. The sample size will determine the confidence limits of estimators (see Chapter 2). Standard references on survey research (e.g., Henry 1990) provide tables to determine necessary sample sizes for specific levels of precision.

An integral part of socioeconomic sampling is the development and use of a consistent coding scheme and accuracy checks. Protocols should be developed to track data collection and entry, especially when more than one interviewer or observer collects and records data. One simple method to track which tasks have been completed and who was responsible is to require that each data procedure be initialed (either in writing or electronically) by the individual handling the data (Van Kammen and Stouthamer-Loeber 1998).

Finally, in the context of the adaptive management framework for artificial reef research (Figure 6.1), socioeconomic data collection should be conducted periodically (at regular intervals if possible) to provide a longitudinal record and to inform policymakers, managers, and other scientists. Attention should be given to the various cycles (weekday/weekend, time of day, in-season/out-of-season, weather, species closures, etc.) that may influence artificial reef utilization. Ideally, before reef deployment, a baseline of socioeconomic data would be recorded. Then, data collection would occur at regular intervals of peak-use, normal-use, and low-use time periods. Regular, continuous data collection, such as monthly recording of boats fishing over reefs, when and where new reefs are created, tourist counts, number of charter/party boats, and local expenditure data, will provide quality information for socioeconomic evaluation.

6.6 CASE STUDY — BOAT RAMP SURVEYING TO COLLECT SOCIOECONOMIC DATA ON ARTIFICIAL REEF USERS

One of the most difficult aspects of socioeconomic evaluation is collecting high quality data for artificial reef users. Artificial reefs are often far from shore and dispersed over a wide area. This situation makes on-site interviewing over any length of time difficult and costly. However, personal interviews with artificial reef users are very desirable because detailed individual and fishing catch information can be collected that may not be possible with mail and telephone surveys. The case study described below encounters many elements that pertain to a variety of reef assessment problems in other locations.

6.6.1 Survey Design

To evaluate whether a relatively low-cost method of personal interviewing of artificial reef users could be developed, a pilot study was conducted in the Tampa Bay, Florida area in cooperation with the U.S. National Marine Fisheries Service (NMFS). The NMFS coordinates the Marine Recreational Fishing Statistics Survey (MRFSS), the only continuous survey of marine fishing for the Atlantic and the Gulf of Mexico coasts. The MRFSS consists of a telephone survey and an on-site interview component. Details on the NMFS sampling and survey methods, as well as the procedures used for the effort, catch, and participation estimates can be found in Van Voorhees et al. (1992). The telephone survey collects data on the presence and number of marine and recreational anglers in the household, and the number, mode, and primary location of fishing trips in a 2-month period. At the same time, intercept interviews of anglers returning from fishing trips are conducted at water-body access sites. The intercept interviews collect the following information from shore, charter boat, and private boat anglers:

- Number, weights, and lengths of fish caught by species,
- State and county of residence of the anglers,
- Avidity level of the anglers, i.e., the number of trips per year,
- Mode of fishing,
- Primary area of fishing.

To determine whether a boat angler was an artificial reef user, a question was added to the basic MRFSS intercept interview for the pilot study. Anglers were asked whether they had fished within 200 ft (60.96 m) of an artificial reef. If the response was positive, follow-up questions regarding artificial reefs were asked to provide additional information about:

- Specific artificial reef(s) where the angler was fishing,
- Catch from the artificial reef(s),
- Reasons why the angler decided to fish at an artificial reef.

The pilot study was conducted in a four-county region around Tampa Bay in the first half of 1992 because these counties have well-established artificial reef programs. Many of the individual reef sites are marked as "fish havens" on local nautical charts and are well known to local fishermen. To help the angler identify specific artificial reef sites, detailed maps showing the exact location and name of artificial reefs were used by MRFSS interviewers working in the study area. A total of 46 reef sites in the area were identified on the maps. All of the sites identified were accessible only by boat. Since this was a pilot study, the number of follow-up questions was limited and other important socioeconomic data, such as artificial reef fishing expenditures, were not collected. In future studies expenditure data could be collected as part of the intercept interviews or solicited later in follow-up mail or telephone surveys.

6.6.2 Results

During the survey period 2255 boat anglers were interviewed at boat ramps and marinas in the study area for the MRFSS. Of this total, 79 anglers fished at one or more artificial reef sites on the day of the interview. Follow-up surveys were completed with 67 individuals or 85% of the anglers who had fished at artificial reefs. The difference between the number who fished at artificial reefs and the number of follow-up surveys completed is due to: (1) lack of cooperation by the anglers; (2) a failure by the interviewer to administer the follow-up survey to artificial reef anglers; or (3) other factors not identified by the MRFSS interviewers.

Of the 46 sites in the study area, artificial reef anglers interviewed for the follow-up survey used 19 sites. The vast majority of these sites were within 10 mi (16 km) of shore. Table 6.7 summarizes the demographic characteristics (residency status, average age, gender, and prior fishing activity) of artificial reef users and other anglers identified in the MRFSS and the pilot study. The majority of both users and nonusers were Florida residents and were also primarily residents of

Table 6.7 Demographic Characteristics of Artificial Reef Users and Other Boat Anglers

Characteristics	Artificial Reef Users	Other Anglers
Resident (Florida)	75 (94.9%)	2072 (95.2%)
Nonresident	4 (5.1%)	104 (4.8%)
Age (years)	38	37.5
Gender: male	88.6%	87.9%
female	11.4%	12.1%
Fishing days in prior 2 months (mean)	7.8	7.3
Fishing days in prior 12 months (mean)	50.3	44.6

Table 6.8 Importance of Different Reasons for Fishing at an Artificial Reef

	Very Important				Not Important
Reason	1	2	3	4	5
Reef close to shore	42.4%	18.2%	12.1%	1.5%	25.8%
Targeting snappers and groupers	27.3%	15.2%	18.2%	12.1%	27.3%
Reef easy to locate	63.6%	15.2%	4.5%	3.0%	13.6%
Expected better catch	45.5%	10.6%	16.7%	7.6%	19.7%

the four-county area surrounding Tampa Bay. These results indicate that tourists were not a major part of recreational boat fishing effort in the study area. Other socioeconomic differences between artificial reef users and other anglers were relatively minor. Artificial reef users were the same age as other anglers and both groups were predominantly male. Both groups of anglers were fairly avid fishermen with artificial reef users averaging 7.8 days fishing during the past 2 months and 50.3 days during the prior 12 months. Other anglers averaged 7.3 and 44.6 days during the past 2 and 12 months, respectively.

Table 6.8 shows the most important reason for fishing at an artificial reef was that the reef site was easy to locate. Artificial reef users also stated that they expected to catch more fish than they would have caught at natural bottom sites, and it was important that a reef was close to shore. These results suggest that convenience and expectations about catch motivated artificial reef site choice decisions.

Table 6.9 presents a comparison of the six most popular primary target species for artificial reef users and other anglers. For both groups, the most frequently cited (more than one third of each group) response was no target species. Artificial reef users who did have a target species cited grouper (Serranidae), red drum (*Sciaenops ocellatus*), and Spanish mackerel (*Scomberomorus maculatus*) as their most popular targets. On the other hand, other anglers cited spotted sea trout (*Cynoscion nebulosus*), snook (*Centropomus undecimalis*), and red drum as species they were targeting. These results suggest that artificial reef users were more focused on reef community fishes than other anglers.

6.6.3 Discussion

The pilot study demonstrated that direct interviews with reef anglers at boat ramps were a feasible method of data collection and the costs were not prohibitive. Cost per completed interview was about U.S.$40 as compared to a standard direct interview cost of between U.S.$70 to $120 (Table 6.5). These cost savings illustrate the advantages of augmenting existing fisheries data collection efforts to include relevant artificial reef information.

Several other results from the pilot study should be noted. First, the fact that less than 5% of 2255 boat anglers interviewed in the Tampa Bay area fished at an artificial reef indicates that the population of artificial reef users in the area was relatively small. If the study area is representative of other coastal areas, this means that it may be difficult, and costly, to acquire detailed socioeco-

Table 6.9 Primary Target Species Identified by Artificial Reef Users and Other Anglers

Artificial Reef Users		Other Anglers	
None	46.8%	None	33.4%
Grouper	10.1%	Spotted Sea Trout	16.1%
Red Drum	8.9%	Snook	8.4%
Red Grouper	6.3%	Red Drum	7.5%
Spanish Mackerel	6.3%	Grouper	6.2%
Sheepshead	3.8%	Spanish Mackerel	5.4%

nomic data for a large group of users. Also, it is unlikely that a broad areal survey would provide a large number of observations for a specific artificial reef site.

Although it may be difficult to acquire data on a large group of artificial reef users with boat ramp surveys, this method has some distinct advantages over mail and telephone surveys. The low number of artificial reef users suggests that it would take a very large number of either mail or telephone surveys just to contact a few artificial reef users. It might be possible to increase the incidence of artificial reef users in a mail or telephone survey by using saltwater fishing license records to establish the sample frame. However, either mail or telephone surveys still would be limited in the type of site use and catch data that could be collected.

These considerations suggest that, if the objective of socioeconomic monitoring and assessment is to collect statistically reliable information about a specific artificial reef site, it may be necessary to interview users directly on the site. This is likely to be expensive if the interviews are conducted over a lengthy period of time. This may be the only way, however, to obtain adequate sample sizes for a specific artificial reef site for precise statistical estimates of socioeconomic variables.

6.7 ACKNOWLEDGMENTS

The authors thank Bill Seaman for his assistance and encouragement. We also thank Virginia Vail and the Florida Department of Environmental Protection for financial support of the pilot study and Robert Hiett, QuanTech, Inc., Arlington, Virginia for his assistance in the design and conduct of the pilot study. Finally, we thank David Carter, graduate student in the Food and Resource Economics Department, University of Florida, Gainesville for providing library research and editorial assistance.

REFERENCES

Aabel, J.P., S. Cripps, A.C. Jensen, and G. Picken. 1997. *Creating Artificial Reefs from Decommissioned Platforms in the North Sea: Review of Knowledge and Proposed Programme of Research.* Report prepared for the Offshore Decommissioning Communications Project (ODCP), London, 129 pp.

Babbie, E. 1997. *The Practice of Social Research.* 8th Ed. Wadsworth Publishing, Belmont, CA.

Bell, F.W., M.A. Bonn, and V.R. Leeworthy. 1998. Economic impact and importance of artificial reefs in northern Florida. Report to the Office of Fisheries Management, Florida Department of Environmental Protection, Tallahassee, FL, 389 pp.

Bockstael, N., A. Graefe, and I. Strand. 1985. Economic analysis of artificial reefs: an assessment of issues and methods. Artificial Reef Development Center Technical Report No. 5, Sport Fishing Institute, Washington, D.C.

Bockstael, N., A. Graefe, I. Strand, and L. Caldwell. 1986. Economic analysis of artificial reefs: a pilot study of selected valuation methodologies. Artificial Reef Development Center Technical Report No. 6, Sport Fishing Institute, Washington, D.C.

Bohnsack, J.A. and D.L. Sutherland. 1985. Artificial reef research: a review with recommendations for future priorities. *Bulletin of Marine Science* 37(1):11–39.

Bombace, G. 1997. Protection of artificial habitats by artificial reefs. Pages 1–15. In: A.C. Jensen, ed. *European Artificial Reef Research, Proceedings, First EARRN Conference,* Ancona, Italy, March 1996. Southampton Oceanography Centre, Southampton, England.

Buchanan, C.C. 1973. Effects of an artificial habitat on the marine sport fishery and economy of Murrells Inlet, South Carolina. *Marine Fisheries Review* 35(9):15–22.

Burdge, R.J. 1996. *A Conceptual Approach to Social Impact Assessment.* Social Ecology Press, Middleton, WI.

Daniel, D.D. and J.E. Seward. 1975. *Natural and Artificial Reefs in Mississippi Coastal Waters: Sport Fishing Pressure and Economic Considerations.* Bureau of Business Research, University of Southern Mississippi, Hattiesburg.

Deuell, R.L. and T.M. Lillesand. 1982. An aerial photographic procedure for estimating recreational boating use on inland lakes. *Photogrammetric Engineering and Remote Sensing* 48(11): 1713–1717.

Dillman, D.A. 1978. *Mail and Telephone Surveys — The Total Design Method.* John Wiley & Sons, New York.

Ditton, R.B. 1981. Social and economic considerations for artificial reef deployment and management. Pages 23–32. In: D.Y. Aska, ed. *Artificial Reefs: Conference Proceedings.* Florida Sea Grant College Program Report 41, University of Florida, Gainesville.

Ditton, R.B. and A.R. Graefe. 1978. *Recreational Fishing Use of Artificial Reefs on the Texas Gulf Coast.* Texas Agricultural Experiment Station, Texas A&M University, College Station.

Ditton, R.B. and J. Auyong. 1984. *Fishing Offshore Platforms Central Gulf of Mexico.* OCS Monograph MMS84-0006, Minerals Management Service, U.S. Department of Interior, Metairie, LA.

Ditton, R.B. and L.B. Burke. 1985. *Artificial Reef Development for Recreational Fishing: A Planning Guide.* Sport Fishing Institute, Washington, D.C.

Ditton, R.B., A.R. Graefe, and A.J. Fedler. 1981. Recreational satisfaction at Buffalo National River: some measurement concerns. Pages 9–17. In: *Some Recent Products of River Recreation Research.* USDA Forest Service Report NC-63, St. Paul, MN.

Ditton, R.B., L.D. Finkelstein, and J. Wilemon. 1995. *Use of Offshore Artificial Reefs by Texas Charter Fishing and Diving Boats.* Texas Parks and Wildlife Department, Austin.

Fedler, A.J. 1984. Elements of motivation and satisfaction in the marine recreational fishing experience. Pages 75–83. In: R. Stroud, ed. *Marine Recreational Fisheries.* Vol. 9. National Coalition for Marine Conservation, Savannah, GA.

Finsterbusch, K., L.G. Llewellyn, and C.P. Wolfe, eds. 1983. *Social Impact Assessment Methods.* Sage Publications, Beverly Hills, CA.

Fowler, F.J., Jr. 1995. *Improving Survey Questions: Design and Evaluation.* Sage Publications, Newbury Park, CA.

Fowler, F.J., Jr. and T.W. Mangione. 1990. *Standardized Survey Interviewing: Minimizing Interviewer-Related Error.* Sage Publications, Newbury Park, CA.

Graefe, R.B. 1981. Social and economic data needs for reef program assessment. Pages 152–166. In: D.Y. Aska, ed. *Artificial Reefs: Conference Proceedings.* Florida Sea Grant College Program Report 41, University of Florida, Gainesville.

Henry, G.T. 1990. *Practical Sampling.* Sage Publications, Newbury Park, CA.

Hewings, G.J.D. 1985. *Regional Input-Output Analysis.* Sage Publications, Beverly Hills, CA.

Hogans, M.L. 1978. Using photography for recreation research. U.S. Forest Service Research Note, PNW-327, U.S. Department of Agriculture, Portland, OR.

Holling. C.S. 1978. *Adaptive Environmental Assessment and Management.* John Wiley & Sons, New York.

Jensen, A.C. and K. Collins. 1997. The use of artificial reefs in crustacean fisheries enhancement. Pages 115–121. In: A.C. Jensen, ed. *European Artificial Reef Research, Proceedings, First EARRN Conference,* Ancona, Italy, March 1996. Southampton Oceanography Centre, Southampton, England.

Kurien, John. 1995. Collective action for common property resource rejuvenation: the case of people's artificial reefs in Kerala State, India. *Human Organization* 54(2):160–168.

Laihonen, P., J. Hanninen, J. Chojnacki, and I. Vuorinen. 1997. Some prospects of nutrient removal with artificial reefs. Pages 85–96. In: A.C. Jensen, ed. *Artificial Reef Research, European Proceedings, First EARRN Conference,* Ancona, Italy, March 1996. Southampton Oceanography Centre, Southampton, England.

Lee, D. 1994. The potential economic impact of lobster stock enhancement. M.Sc. dissertation, University of York, U.K. 27 pp.

Leistritz, L.F. and S.H. Murdock. 1981. The socioeconomic impact of resource development: methods for assessment. Westview Press, Boulder, CO.

Liao, D.S. and D.M. Cupka. 1979. Economic impacts and fishing success of offshore sport fishing over artificial reefs and natural habitats in South Carolina. South Carolina Marine Resources Center Technical Report 38, Charleston.

Lipton, D.W., K.F. Wellman, I.C. Sheifer, and R.F. Weiher. 1995. Economic valuation of natural resources — a handbook for coastal resource policymakers. NOAA Coastal Ocean Program Decision Analysis Series No. 5, NOAA Coastal Ocean Office, Silver Spring, MD.

Martilla, J.A. and J.C. James. 1977. Importance-performance analysis. *Journal of Marketing* 41:77–79.

Miller, R.E. and P.D. Blair. 1985. Input-Output Analysis: Foundations and Extensions. Prentice-Hall, Engle-wood Cliffs, NJ.

Milon, J.W. 1988. The economic benefits of artificial reefs: an analysis of the Dade County, Florida reef system. Florida Sea Grant College Program Report No. 90, University of Florida, Gainesville.

Milon, J.W. 1989. Economic evaluation of artificial habitat for fisheries: progress and challenges. *Bulletin of Marine Science* 44:831–843.

Milon, J.W. 1991. Social and economic evaluation of artificial aquatic habitats. Pages 237–270. In: W. Seaman, Jr. and L.M. Sprague, eds. *Artificial Habitats for Marine and Freshwater Fisheries,* Academic Press, San Diego.

Milon, J.W. and R. Schmeid. 1991. Identifying economic benefits of artificial reef habitat. Pages 53–57. In: J.G. Halusky, ed. *Artificial Reef Research Diver's Handbook.* Florida Sea Grant College Program, University of Florida, Gainesville.

Milon, J.W., C.F. Kiker, and D.J. Lee. 1997. Ecosystem management and the Florida Everglades: the role of the social scientist. *Journal of Agricultural and Applied Economics* 29(1):99–107.

Murray, J.D. and C.J. Betz. 1991. User views of artificial reef management in the southeast. University of North Carolina Sea Grant College Program Report 91-03, Raleigh.

Peine, J.D., ed. 1983. *Proceedings, Workshop on Unobtrusive Techniques to Study Social Behavior in Parks.* National Park Service, Atlanta, GA.

Penning-Rowsell, E.C., C.H. Green, P.M. Thompson, A.M. Coker, S.M. Tunstall, C. Richards, and D.J. Parker. 1992. *The Economics of Coastal Management: A Manual of Benefit Assessment Techniques.* Belhaven Press, London.

Pickering, H. 1997a. Stakeholder interests and decision criteria. Paper presented at EARRN Workshop 3: Socioeconomic and legal aspects of artificial reefs, Olhao, Portugal, July 1997.

Pickering, H. 1997b. Legal framework governing artificial reefs in the EU. Pages 195–232. In: A.C. Jensen, ed. *European Artificial Reef Research, Proceedings, First EARRN Conference,* Ancona, Italy, March 1996. Southhampton Oceanography Centre, Southampton, England.

Pickering, H. and D. Whitmarsh. 1996. Artificial reefs and fisheries exploitation: a review of the "attraction versus production" debate, the influence of design and its significance for policy. *Fisheries Research* 31(1–2):39–59.

Pickering, H., D. Whitmarsh, and A.C. Jensen. 1998. Artificial reefs as a tool to aid rehabilitation of coastal ecosystems: investigating the potential. *Marine Pollution Bulletin* 37:505–514.

PRC Environmental Management, Inc. 1994. Draft economic analysis of marine environmental enhancement project. Prepared for Defense Logistics Agency, Alexandria, VA.

Prince, E.D. and O.E. Maughan. 1978. Freshwater artificial reefs: biology and economics. *Fisheries* 3:5–9.

Propst, D.B. and D.G. Gavrilis. 1987. Role of economic impact assessment procedures in recreational fisheries management. *Transactions of the American Fisheries Society* 116:450–460.

Rea, L.M. and R.A. Parker. 1997. *Designing and Conducting Survey Research: A Comprehensive Guide.* Jossey-Bass Publishers, San Francisco.

Reggio, V., ed. *Petroleum Structures as Artificial Reefs: A Compendium.* Minerals Management Service Report No. 89-0021, U.S. Department of Interior, New Orleans, LA.

Rey, H. 1990. Toward the formulation of a method to assess the socio-economic impact of artificial reefs. Pages 295–302. In: *IPFC Symposium on Artificial Reefs and Fish Aggregating Devices as Tools for the Management and Enhancement of Marine Fishery Resources.* Colombo, Sri Lanka, May 12–17, 1990.

Rhodes, R.J., J.M. Bell, and D. Liao. 1994. Survey of recreational fishing use of South Carolina's marine artificial reefs by private boat anglers. Project No. F-50 Final Report, Office of Fisheries Management, South Carolina Wildlife and Marine Resources Department, Charleston.

Richardson, H.W. 1985. Input-output and economic base multipliers: looking backward and forward. *Journal of Regional Science* 25(4):607–61.

Roe, B. 1995. Economic valuation and impacts of artificial reefs: a summary of literature on marine artificial reefs in the United States. Prepared for the Artificial Reef Technical Committee, Atlantic States Marine Fisheries Commission, Alexandria, VA.

Rossi, P.H., J.D. Wright, and A.B. Anderson, eds. 1983. *Handbook of Survey Research.* Academic Press, New York.

Salant, P. and D. Dillman. 1994. *How to Conduct Your Own Survey.* John Wiley & Sons, New York.

Samples, K.C. 1989. Assessing recreational and commercial conflicts over artificial fishery habitat use: theory and use. *Bulletin of Marine Science* 44:844–852.

Selltiz, C., L.S. Wrightsman, and S.W. Cook. 1976. *Research Methods in Social Relations.* Holt, Reinhart & Winston, New York.

Shomura, R. and W. Matsumoto. 1982. Structured flotsam as aggregating devices. NOAA Technical Memorandum NOAA-TM-NMFS-SWFC-22. National Marine Fisheries Service, Southwest Fisheries Center, Honolulu, HI.

Simard, F. 1997. Socio-economic aspects of artificial reefs in Japan. Pages 233–240. In: A.C. Jensen, ed. *European Artificial Reef Research, Proceedings, First EARRN Conference,* Ancona, Italy, March 1996. Southampton Oceanography Centre, Southampton, England.

Smith, H. W. 1991. *Strategies of Social Research.* 3rd Ed. Holt, Reinhart & Winston, Orlando, FL.

Stanley, D.R. and C.A. Wilson. 1989. Utilization of offshore platforms by recreational fishermen and scuba divers off the Louisiana coast. Pages 11–24. In: V. C. Reggio, ed. *Petroleum Structures as Artificial Reefs: A Compendium.* Minerals Management Service Report No. 89-0021, U.S. Department of Interior, New Orleans, LA.

Stelzer, F., Jr. 1989. Rigs-to-Reefs as an alternative to platform salvage. Pages 143–154. In: V.C. Reggio, ed. *Petroleum Structures as Artificial Reefs: A Compendium.* Minerals Management Service Report No. 89-0021, U.S. Department of Interior, New Orleans, LA.

Stevens, B.H. and M.L. Lahr. 1988. Regional economic multipliers: definition, measurement, and application. *Economic Development Quarterly* 2:88–96.

Talhelm, D.R. 1985. The economic impact of artificial reefs on Great Lakes sport fisheries. Pages 537–543. In: F.M. D'Itri, ed. *Artificial Reefs: Marine and Freshwater Applications.* Lewis Publishers, Chelsea, MI.

Thompson, M.S. 1980. *Benefit-Cost Analysis for Program Evaluation.* Sage Publications, Beverly Hills, CA.

Van Kammen, W.B. and M. Stouthamer-Loeber. 1998. Practical aspects of interview data collection and data management. Pages 375–397. In: L. Brickman and D. J. Rog, eds. *Handbook of Applied Social Research Methods.* Sage Publications, Newbury Park, CA.

Van Voorhees, D.A., J.F. Witzig, M.F. Osborn, M.C. Holiday, and R.J. Essig. 1992. Marine recreational fishery statistics survey, Atlantic and Gulf coasts, 1991–1992. Current Fisheries Statistics Number 9204, National Marine Fisheries Service, Silver Spring, MD.

Vanderpool, C.K. 1987. Social impact assessment and fisheries. *Transactions of the American Fisheries Society* 116:479–485.

Walters, C.J. 1986. *Adaptive Management of Renewable Resources.* McGraw-Hill, New York.

Webb, E.J., D.T. Campbell, R.D. Schwartz, and L. Sechrest. 1966. *Unobtrusive Measures: Nonreactive Research in the Social Sciences.* Rand McNally & Co., Chicago.

Weick, K. E. 1968. Systematic observational methods. Pages 357–451. In: G. Lindzey and E. Aronson, eds. *The Handbook of Social Psychology.* 2nd Ed. Addison-Wesley, Reading, MA.

Whiting, B. and J. Whiting. 1970. Methods for observing and recording behavior. Pages 282–315. In: R. Naroll and R. Cohen, eds. *A Handbook of Method in Cultural Anthropology.* Columbia University Press, New York.

Whitmarsh, D. 1997. Cost-benefit analysis of artificial reefs. Pages 175–193. In: A.C. Jensen, ed. *European Artificial Reef Research, Proceedings, First EARRN Conference,* Ancona, Italy, March 1996. Southampton Oceanography Centre, Southampton, England.

Whitmarsh, D. and H. Pickering. 1997. Commercial exploitation of artificial reefs: economic opportunities and management imperatives. Research Paper No. 115, Centre for the Economics and Management of Aquatic Resources, University of Portsmouth, Portsmouth, U.K.

Whitmarsh, D., S. Pascoe, A.C. Jensen, and R.C.A. Bannister. 1998. Economic appraisal of lobster stock enhancement using artificial reef technology. Page 69. In: S. Pascoe, C. Robinson, and D. Whitmarsh, eds. *Proceedings, European Association of Fisheries Economists Bioeconomic Modeling Workshop,* December 17–18, 1997, Miscellaneous Publication No. 39, Centre for the Economics and Management of Aquatic Resources, University of Portsmouth, Portsmouth, U.K.

Willmann, R. 1990. Economic and social aspects of artificial reefs and fish aggregating devices. Pages 384–391. In: *IPFC Symposium on Artificial Reefs and Fish Aggregating Devices as Tools for the Management and Enhancement of Marine Fishery Resources.* Colombo, Sri Lanka, May 12–17, 1990.

Wright, A.D. 1998. Speech given in a debate in the House of Commons. Hansard No. 1788, London, May 11–14, 1998.

Wright, S.R. 1979. *Quantitative Methods and Statistics: A Guide to Social Research.* Sage Publications, London.

Zerbe, R.O. and D.D. Dively. 1994. *Benefit-Cost Analysis in Theory and Practice.* Harper Collins College Publishers, New York.

CHAPTER **7**

Integrating Evaluation into Reef Project Planning

William J. Lindberg and Giulio Relini

CONTENTS

7.1 SUMMARY

This chapter provides guidance for building effective evaluation into the overall planning of an artificial reef project. The second section presents the chapter purpose and then describes the general steps and issues encountered in the planning process. In the third section, those steps and considerations are illustrated through a variety of hypothetical examples. The fourth section discusses logical issues in the interpretation of results. The chapter ends with comments on the importance of credible artificial reef evaluation.

7.2 INTRODUCTION

To paraphrase the definition given in Chapter 1, evaluation is simply answering whether an artificial reef or reef system is satisfying the purposes for which it was built. Unfortunately, truly excellent examples of artificial reef evaluations are rare in comparison to the level of reef development worldwide. This may be so, in part, because evaluation is too often an afterthought rather than an integral part of the preparation and planning of a reef building project. Even where exemplary evaluation efforts are described (e.g., California State Lands Commission 1999), the reports convey what was or is being done, but not necessarily why the evaluation has taken the particular form that is described. In order to implement the guidance offered in earlier chapters, we need to understand the thought processes that precede a good evaluation. Despite many potential similarities among reef projects, each evaluation effort is unique, with its own situation, setting, and constraints. Rather than simply reviewing past examples of reef evaluations and assessments, this chapter emphasizes how one might think through the planning process.

7.2.1 Purpose

Previous chapters offered guidance on study designs, useful data, and preferred methods for physical, biological, and socioeconomic evaluations of artificial reef performance. This chapter brings together the general issues encountered when preparing to evaluate artificial reefs. Whereas various methods and protocols for data collection and analyses can be obtained from Chapters 3 through 6, here we focus on the initial reasoning that leads to effective evaluations.

This chapter, in conjunction with Chapter 2, offers guidance for answering whether an artificial reef or reef system is satisfying the purposes for which it was built. Given the limited human and financial resources generally available for reef evaluation, it is essential that evaluation projects be designed from the outset to answer this question. Although technically correct methods produce technically correct data, they are not by themselves sufficient. The data must be relevant to the question and conform to an adequate and appropriate sampling or study design. Otherwise conclusions could be invalid or useless, and the evaluation effort wasted. The challenge is to work within the real-world constraints of time and money to design and execute an evaluation strategy capable of yielding meaningful results.

Because every reef project is unique, a prescription for all evaluation needs is impossible. We can, however, offer a logical framework and examples to serve as guides for the planning and preparation of reef evaluation projects.

7.2.2 General Steps for Evaluation Planning

To be most effective, evaluation planning should be integrated into the general planning process from the very beginning of a reef development project. However, unlike the linear thinking of project planning outlines, evaluation thinking is necessarily nonlinear as depicted in Figure 7.1. To aid the reader, Figure 7.2 provides a map for where chapter subsections fit with respect to this nonlinear thought. To introduce the planning considerations, major tasks are grouped into five categories summarized in Sections 7.2.2.1 through 7.2.2.5. These considerations then are elaborated by the examples in Section 7.3. In the examples, subsection headings used with Figure 7.2 will further guide the reader. We contend that modest forethought about evaluation requirements prior

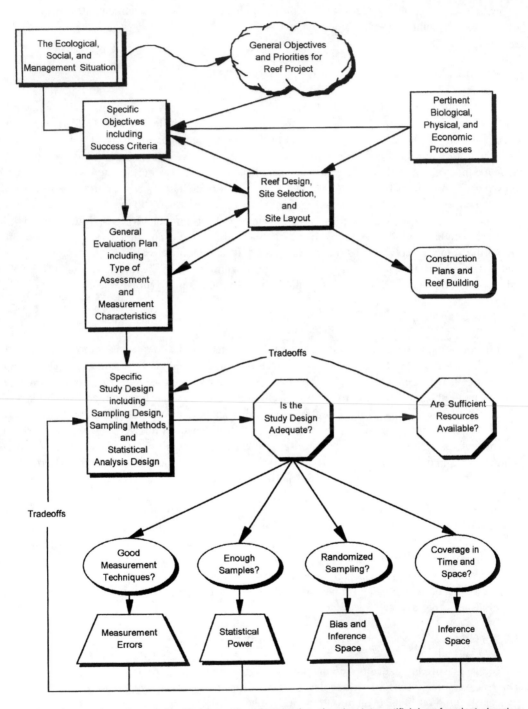

Figure 7.1 General flow diagram for the integration of evaluation planning into artificial reef project planning.

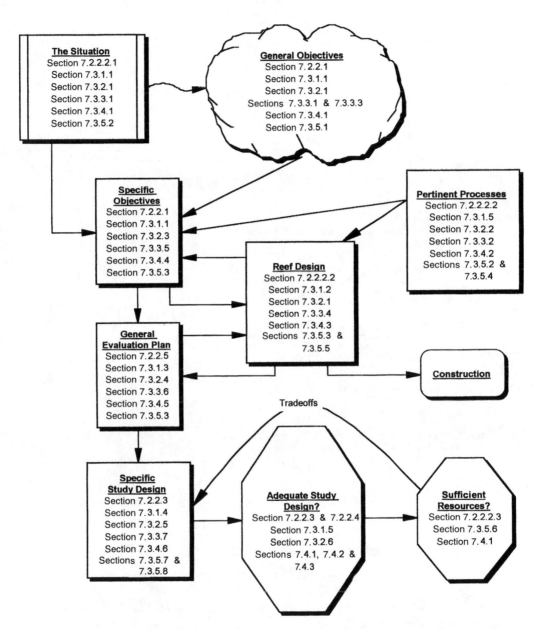

Figure 7.2 A map or reader's guide to where chapter subsections fit in the nonlinear thinking of evaluation planning. Sections 2.x have general discussions, while Sections 3.x.y refer to specific examples where the x is the example number and the y is the subsection within that example.

Table 7.1   Objectives for Artificial Reefs, Grouped with the Most General to the Left

---
To Promote Economic Development

To Enhance Seafood Production

Artisanal Production
Commercial Production
  Fisheries Systems (common property applications)
  Aquacultural Systems (private property applications)

To Enhance Recreational Activities

Consumptive Use of Fishery Resources
  Common Property Systems
    Public Angling
    Public Spearfishing
    Fishing Charters and Guides
  Private Property Systems
    Fishing Charters and Guides
    Personal Reef Locations
Nonconsumptive Use of Natural or "Archeological" Resources
  Common Property Systems
  Public-Access Scuba Diving
  Diving Charters and Guides
  Private Property Systems
  Underwater Submarine Tours
  Scuba Diving "Parks"

To Enhance or Manage Living Marine Resources

To Allocate or Protect Marine Resources

To Protect Adjacent Valued Habitat
To Enforce Separation of Gears in Fishing Conflicts

To Enhance or Conserve Fisheries Stocks

To Alleviate Bottlenecks in Early Life Histories
To Enhance Fish Habitat Quality (e.g., in marine reserves)
To Diminish Local Fishing Mortality

To Mitigate or Restore Marine Habitats

To Mitigate Habitat Loss
To Improve Water Quality

To Restore or Conserve Biodiversity

Note: The listing is not exhaustive, and a given reef project may have more than one objective.

to reef construction will return major dividends in the credibility of results and conclusions derived from the evaluation effort.

7.2.2.1   Define the Management Objectives and Success Criteria

In most cases, an artificial reef or reef system will have one primary and one or more secondary management objectives, which express the practical reasons for building a given reef. Several types of general objectives are outlined in Table 7.1. These need not be mutually exclusive, nor is this listing necessarily exhaustive. However, when multiple management objectives apply to a given project, it is important to define which is primary and perhaps to rank the remainder in order of importance. By focusing on the most important objectives, the planning decisions that involve tradeoffs can be made easier without jeopardizing the relevance of the evaluation project.

The objectives in Table 7.1, discussed briefly in Chapter 1, are merely potential starting points for the task of defining a management objective or objectives. The general objective, once identified, must be honed to the sharpest possible specificity. A specific management objective provides more direction to the planning process than do general or vague objectives. The specificity helps identify the elements of reef design or architecture that are most germane to the reasons a reef is to be built and supports choice among the available design options. The specificity also helps in clearly identifying the nature of the data most germane to whether the artificial reef is satisfying the purpose for which it was built. An integral part of a specific management objective is the success criteria statement introduced in Chapter 2, Section 2.3.1. The process of refining the general objective to the specific is entirely analogous to the thinking of a researcher who proceeds from a general question to a testable hypothesis, as outlined by Green (1979).

At this point, it is useful to distinguish between management and research objectives for artificial reef studies. Management objectives are the practical consequences expected from a reef structure, whereas research objectives focus on what is to be learned about the natural patterns and processes being altered by the reef structure. Obviously, the two are closely related and both can be part of the same project. Many published artificial reef studies have focused entirely on research objectives (e.g., *Bulletin of Marine Science,* Vols. 37[1], 44[2], and 55[2, 3]) rather than management objectives, and may not conform to the meaning of evaluation intended in this book.

Admittedly, many artificial reefs have been built without specific objectives to help guide their physical design. Nevertheless, identification and refinement of management objectives for a reef after the fact can still guide the planning of its evaluation.

7.2.2.2 Identify Key Features of the Situation

Artificial reefs are not developed and evaluated in a vacuum. Their performance, relative to stated objectives, depends on the context in which they are placed (i.e., ecological, social, and management) and their physical design and placement. Our ability to evaluate performance depends on the available human and financial resources, the technical skills or competence of the people involved, and the inevitable tradeoffs between what we would like to know or need to know and what we can realistically expect to accomplish. Whether we can measure everything desired or not, key features need to be identified in order to adequately plan any reef evaluation.

7.2.2.2.1 Ecological, Social, and Management Contexts — The practical benefits, or detriments, actually derived from an artificial reef result from that structure somehow altering the processes of the system into which it is placed. The physical structure of a reef alters biological and/or physical processes, presumably to enhance the "goods and services" (*sensu* Christensen et al. 1996) desired from an ecosystem. The presence of the reef might also alter social processes. From our perspective, artificial reefs are simply tools, and their consequences derive from how well those tools are being used, which in turn is determined by the management structure superimposed on the natural and social systems. The more one knows about each of these elements, the more likely one is to design and develop an effective artificial reef for the intended objectives, and the more likely one is to evaluate the appropriate characteristics.

The context in which an artificial reef is to function has several facets. The *ecological context* of a natural system consists of the biotic and abiotic components of the system and the natural processes that connect them. Of particular importance are the organisms considered products of the manipulated ecosystem. In fisheries, these are referred to as the target species. Target fishes and invertebrates, or in some cases plants, have complex life histories with ecological requirements that vary with life history stage. Often, only some stages are associated with particular reefs. As noted in Chapters 3 and 4, the physical environment (e.g., water circulation, flow, turbulence, temperature, and clarity) affects the distribution of nutrients, primary production, and the trophic status of systems, while as noted in Chapters 4 and 5, such factors can affect the behavior and production of target species. Animals are adapted to acquire their essential resources, namely food, shelter, and mates. A myriad of ecological processes and relationships determine how successful they are. The effects of habitat selection, food web dynamics, predator/prey relationships, competition, etc. on bioenergetics and population dynamics entail complex issues of spatial and temporal scale that researchers are just beginning to comprehend (e.g., landscapes, meta-populations, source-sink dynamics, etc.). It is clear that the more one understands how a reef functions ecologically, the better able one is to design a reef for a specific objective and then evaluate its biological effectiveness.

The *social context* for a reef refers to the customs and practices of people who stand to benefit from an artificial reef or those affected by its presence. Are the users primarily recreational or commercial fishers or recreational divers? Are there conflicts among local groups or might the presence of a reef impede the activities of others? What kind of boats, if any, will people use to access the reefs? Are the vessels owned privately by individuals, cooperatives, or corporations or chartered by tourists and occasional users? Are the users technically modern or dependent on artisanal methods? What kind of gear is traditionally used in the local fishery and is it compatible with the anticipated reef design? What is the duration and frequency of traditional fishing trips, and is that by choice or necessity? Answers to questions such as these help one to anticipate the effects that a proposed reef might have on people and the effects that people might have on a proposed reef. If the objectives for a reef involve economic development (see Table 7.1), then prior consideration of the social context directs attention to measurable characteristics for socioeconomic evaluations (Chapter 6). If the objectives involve some enhancement of living marine resources, then such insights can help guide reef and study designs to account for fishing pressure where appropriate in the evaluation process.

The *management context* of a reef includes the laws, regulations, policies, and practices that govern the use of marine resources at and near the reef. At the forefront here is the question of ownership. Are we dealing with common property, private property resources, or some level of communal property rights? How are artificial reefs dealt with in the fisheries management framework or environmental regulation of the associated state, province, or country? Are there special permitting and reporting requirements or provisions for controlling access according to the objectives asserted for a reef? If so, what level of enforcement exists and is it primarily legal or traditionally social? Of comparable importance are the issues of competing or incompatible uses for the sea floor and adjacent water column, and the hazards to navigation created by reef structures. The management context warrants special consideration in the definition of objectives, design of reefs, and evaluation planning because the ability to achieve desired objectives might be determined as much by management constraints as it is by ecological processes. In such cases, an objective evaluation of management practices might be a valuable piece of a reef evaluation plan.

7.2.2.2.2 *Rationale for Reef Design and Site Layout* — In the overall planning process, defining a rationale for the physical design of a reef and the layout of a reef site is necessarily interwoven with the refinement of specific management objectives and an understanding of the ecological, social, and management contexts. If the objectives are vague or general and the context poorly known or not explicitly considered, the resultant reef design and site layout will be weakly justified and give little direction to evaluation planning. By contrast, one can develop strong justifications that lead to effective evaluations by thinking through the ecological and social processes that might be manipulated to achieve desired benefits. This is a creative thought process in which various options and alternatives for how reef structure might alter key processes is explored. Obviously, this activity is limited by the scientific knowledge and understanding one has of the ecological and social processes related to the stated objectives. Here also it is important to distinguish between what is known, what is hypothesized, and what is assumed about key processes and the effects of reef structure. If the reef planning process were approached without these distinctions, the entire project might be based on gross misconceptions, unwittingly perpetuated by the subsequent evaluation.

One need not be a professional research scientist to undertake this exercise, but the thinking is entirely analogous to the conception of a manipulative field experiment. After all, the placement of an artificial reef into an ecosystem is inherently a manipulation of some portion of that system. Consequently, and as noted in Chapter 2, evaluation planning is served well during the design phase of a construction project by considering the possibilities and perhaps the need for relevant comparisons or experimental contrasts, appropriate controls, and adequate replication in the physical configuration of a reef system. With forethought, artificial reef projects can offer unique opportunities to put into place physical infrastructures for good experimental designs. But when the evaluation is planned after a reef or reef system is built, these options generally are no longer available.

Nevertheless, when planning an evaluation for existing reefs, it is useful to follow a similar creative thought process, i.e., given the general objectives and context, and also given the existing physical reef configuration, how might relevant ecological or social processes be affected? Again, this explores the various options and alternatives for how reef structure might alter key processes and allows the original general objectives to be refined to somewhat more specific objectives *post hoc*. The advantage of doing so is to identify potential comparisons and measurable characteristics that might become the focus of an evaluation plan. The study design might then become more comparative than purely descriptive, which in turn, could greatly affect the conclusions and practical utility derived from the results.

Before leaving the rationale for reef design and site layout, consider also the central role that routine predeployment site surveys and postdeployment documentation should play in the planning processes. When done well, predeployment surveys of a proposed reef site add detail and specific information about the ecological context or the landscape into which a reef will be placed. Such information obviously can be pertinent to the final reef configuration that would implement a design strategy for a given objective. Oftentimes predeployment surveys are minimally directed to address reef stability concerns and the requirements for authorizing permits. We suggest that they be conducted so as to provide a map of the ecological landscape. Likewise, postdeployment documentation confirms the extent to which a planned configuration was actually achieved and provides a baseline for future evaluations of reef stability. In cases where a specific configuration was not planned, thorough postdeployment documentation also can give details that might allow an evaluation planner to engage in the creative thinking described in the previous paragraph. Without an accurate reef map, that planning exercise would occur in a relative vacuum. In either case, reef maps are useful when deciding the details of sampling designs.

7.2.2.2.3 *Tradeoffs, Available Support, and Levels of Effort* — Reality sets in when desired objectives and favored plans have to be reconciled with the resources available to do the work. Time, money, and talented personnel are finite. This becomes the sobering feature of any planning situation. Yet tradeoffs are necessary and can be just as challenging and rewarding as refining the objectives, discerning the context, and developing a sound rationale.

For both the reef design and evaluation planning, one should work through the prior planning steps first without much regard for practical constraints, but always mindful that constraints will come. We suggest this to safeguard the creativity needed for earlier steps to be effective, but mostly to ensure that the project is being driven by its objectives. Specify the vision first and then figure out how to get there. Then one can take either of two approaches: set out to generate the resources needed to implement the vision, or take the resources at hand and pick accordingly from among alternative design options.

Tradeoffs are most easily decided when priorities are clear. Do first that which is most important. For the physical reef design, build in the characteristics that most assure the primary objective for the project, and only then modify the plans to accommodate secondary objectives where they do not jeopardize the primary objective. Likewise for the evaluation plan, make the primary objective the core of the study design, and only add elements pertinent to the secondary objectives as additional resources permit. There is a real cost to data, therefore, decide what data you absolutely must have (see Chapter 2, Sections 2.3.3 and 2.3.6.1). Remember the original question whether an artificial reef or reef system is satisfying the purposes for which it was built. When multiple objectives are involved, it is better to answer this question with confidence for one objective than to attempt to answer it for many and end up answering it for none.

The possibility also exists that available resources (i.e., time, money, and personnel, but mostly funding) might be inadequate for any credible evaluation of the planned study objectives. For example, costs might limit the number of samples or replicates to such a degree that no conclusions would be valid given the anticipated variation and low statistical power (see Chapter 2, Sections 2.3.5.4 and 2.3.5.5.5). In such cases, the study objectives should be scaled back accordingly or the evaluation project deferred until adequate resources can be generated.

7.2.2.3 Decide on Sampling Design

The sampling design follows from decisions about the type of assessment (i.e., descriptive, comparative, or associative/predictive; Chapter 1) that is possible, given the reefs in question and the measurable characteristics appropriate to "specific" management objectives. Decisions about sampling design determine the adequacy of the resultant data in several respects, and that adequacy determines the success and validity of an evaluation effort. To judge the adequacy one needs to answer several questions:

- *Are there enough samples and/or replicates?* If not, only very strong effects, or what might be called "sledgehammer" effects, will be detectable. Subtle yet consequential effects could actually exist but remain undetected with low sample sizes.
- *Are the replicates independent and appropriate to the specific objectives?* If inappropriate, the data obtained from the sampling effort will not help answer the questions being asked. The data may be technically correct, but they are nevertheless useless for the stated purpose.
- *Are the measurement techniques sufficiently accurate and precise?* If not, one has to contend with unwanted variation or "noise." If imprecise, a greater number of samples will be necessary to detect the same level of reef effect. If inaccurate, any estimates derived from the data will have minimal utility because their values will be suspect.
- *Are the samples adequately distributed in time and space?* If not, the inferences one can legitimately draw from the results will be limited to a restrictive range of conditions. This diminishes the utility of an evaluation to future decisionmaking.
- *Are the resources (time, money, and personnel) sufficient to execute the planned sampling design?* If not, the evaluation effort will fail. One must revisit the tradeoffs and assessment strategy, as discussed above, and decide how not to waste the evaluation resources.

7.2.2.4 Anticipate How Data Can Be Analyzed or Summarized

The eventual analysis of data will depend on the type of assessment (i.e., descriptive, comparative, associative/predictive), characteristics of the data and data set (e.g., discrete or continuous variables and missing data), and the logical organization of data imposed by the experimental and/or sampling design. One should refer to Chapter 2 and statistics textbooks for the various analytical procedures that might be appropriate for any given study. Of particular importance here is the need to recognize the assumptions of any analyses being considered, to determine whether the anticipated data conform to those assumptions, and to examine how robust the statistical procedures are to violations of their assumptions. Of course, one also must be confident that the analyses actually address the desired comparisons, contrasts, or associations, given the specific objectives that motivated the evaluation. If the evaluation planning up to this point has been thorough and done well, one could readily seek the recommendations of a statistician or knowledgeable colleague for appropriate statistical procedures. However, if the preceding planning steps have not been well executed, your statistical consultant first would have to help you identify and resolve the inadequacies before addressing the proper statistical analyses. Obviously, corrections are much easier to make before data are gathered than afterward.

7.2.2.5 *Recognize the Connection Between Study Design and Legitimate Conclusions*

Specific objectives should state or imply the criteria of success for a given reef project, as noted above in Section 7.2.2.1. A statistically valid analysis of appropriate data will allow one to judge whether those criteria have been met. Beyond that, however, the generalizations one draws from a particular study are only as valid as the logic and scope of the study design. How representative of reefs, in general, are the various elements of the situation under study? What is the population of reefs to which the study might legitimately apply? Will the data be representative of conditions across a broad or narrow range of space and time? One has to be careful not to overgeneralize. For example, if the ecological processes being altered by a reef depend largely on local conditions or specific time periods, then it would be a mistake to extrapolate inferences to geographic areas or time periods where key conditions differ. One can minimize the risk of overgeneralizing results by clarifying the intent of the study while it is being planned. Is the intent strictly limited to a particular reef, or is the intent to learn something more general that might be applied to future reef projects. If it is the latter, then particular attention should be paid to the tradeoffs that might limit or unintentionally restrict the inference space of the resulting study. At the very least, one should be objective and self-critical in the reporting of conclusions from an evaluation study and honest about the level of empirical support associated with any inferences that are asserted.

7.3 SELECTED HYPOTHETICAL EXAMPLES

By itself, a conceptual model of evaluation planning (e.g., Section 7.2.2) does not adequately convey the nonlinear thinking, or simultaneous thinking on several fronts, that is part of the planning process. We hope to convey this aspect of planning through hypothetical examples drawn from across the spectrum of general objectives listed in Table 7.1. We do not claim that the situations for these examples are representative (although some are based on real-world cases). Nor do we suggest that the reef designs and evaluation plans presented here should actually be adopted in lieu of alternatives that you, the reader, might conceive.

Our intent is to illustrate the thought processes involved in planning. To do this, we call attention in each example to places where the right or wrong decision can make the difference between an effective and ineffective reef evaluation plan. Some examples depict currently routine objectives and situations, while others are for more innovative applications of artificial reef technology. Once again, evaluation is done to answer the question whether an artificial reef or reef system is satisfying the purposes for which it was built.

Artificial reefs have been used throughout the world for a variety of general objectives (Table 7.1), although each global region seems to have its own distinct "flavor" (Figure 7.3). In North America, artificial reefs primarily have been used to enhance recreational activities, with some applications to artisanal fisheries (e.g., lobster casitas) and mitigation for habitat loss (e.g., kelp restoration). In Asia, particularly Japan, reef technology has served commercial seafood production almost exclusively, while in the Indo-West Pacific, simple reefs are part of artisanal fisheries. In Europe, reefs have been used to protect habitat and allocate marine resources, with only modest explorations of fisheries and aquacultural uses. Nowhere, as yet, have artificial reefs been developed explicitly to enhance or conserve fisheries stocks in the context of natural ecosystems, despite popular perceptions to the contrary. The hypothetical examples below are intended to reflect this breadth of global interests and potential applications. Throughout these examples we defer to Chapters 2 through 6 for the details and references associated with various methods.

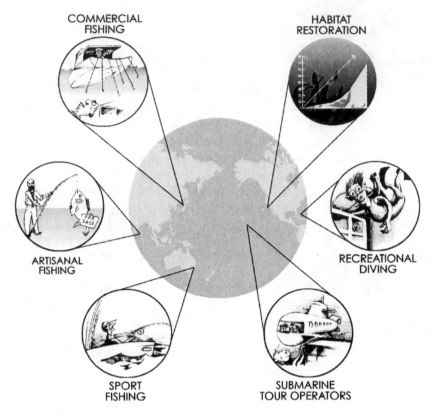

Figure 7.3 A diverse group of interests have need to evaluate the performance of artificial reefs for various economic, social, and ecological purposes.

Examples 1 and 2 deal with the enhancement of recreational and artisanal fisheries, respectively, and portray planning where the associated biological processes being manipulated are either ignored (Section 7.3.1) or not yet known (Section 7.3.2). Examples 3 and 4 also deal with straightforward objectives, i.e., protecting adjacent habitat and improving water quality. Example 5 assumes fisheries conservation as the primary objective and economic development as the secondary, and is the most complex example in the set. Example 5 also weaves research objectives into the evaluation of management objectives. The reader may use Figures 7.1 and 7.2 as guides to the nonlinear thinking conveyed by the subsections within each example.

7.3.1   Example 1: To Enhance Recreational Use of Common Property Resources

7.3.1.1   The Situation Leading to General and Specific Objectives

A regional governmental agency in an industrial nation built 20 fishing piers 2 years ago along the shores of its major estuaries to provide more recreational fishing opportunities for anglers without boats. The fishing piers, to some extent, dictated where anglers would be concentrated, which may not be where the favored fish occur, such as copper, quillback, yellowtail, and black rockfish, family Scorpaenidae (*Sebastes caurinus, S. maliger, S. flavidus,* and *S. melanops,* respectively).

The responsible ministry wants to evaluate the utility of building artificial reefs around fishing piers to enhance angler satisfaction. In this example, the general situation has given rise to a general management objective that has more to do with socioeconomic goals than with ecological goals. Attainment of the primary objective implicitly depends on the ability of artificial reefs to make rockfish accessible to anglers on the piers, but exactly how or why they do so is of little immediate concern to those responsible for the project. An assumption has been made that a shortage of suitable rock habitat in the immediate vicinity of some fishing piers is the reason why the catch of rockfish, and thus angler satisfaction, has apparently been lower at some piers than at others.

A more specific objective, therefore, is to elevate the rockfish catch-per-unit (CPUE) effort and user satisfaction at the least popular fishing piers to the levels currently enjoyed at the top 20% of the fishing piers. Not only is this objective more specific, but it also conveys the success criterion.

7.3.1.2 Consider the Reef Design

For this example, construction considerations and the fishing piers themselves are strictly determining the reef designs and site layouts, which is contrary to the planning protocol in Figure 7.1. Natural quarry rock will be used to create horseshoe-shaped reefs around the ends of selected piers. The granite boulders for this project will be approximately 0.5 m in diameter, to be deployed by front-end loaders directly from the pier decks into the water. Sufficient material will be deposited to make rock piles 10 m wide and 2 m high, with the horseshoes being approximately 60 m long. The reef materials will be deployed only around the ends of piers because of common knowledge that harvestable, non-spawning rockfish are not normally caught in the shallower water.

7.3.1.3 Develop a General Evaluation Plan

A general evaluation plan follows directly from the general situation and the general and specific objectives. In this case, the evaluation planning also directly contributes to the site selection for reef development. The specific management objective suggests a comparison between the four most popular and productive fishing piers (i.e., 20% of 20 piers) and some of the previously least popular and productive piers, after they have been enhanced by artificial reefs. The project planners realize that statistical analyses are easier with balanced study designs (i.e., equal sample sizes), therefore, four of the poorly performing piers will be chosen to receive reefs. Both the construction and evaluation must be completed within 18 months in order to prepare program budgets for the subsequent fiscal year; thus neither time nor money allows pre- and postconstruction comparisons.

An assumption is made, for the purpose of selecting reef sites, that pier usage accurately reflects the CPUE and user satisfaction of each pier. Descriptive statistics are already available for the number of anglers visiting each pier monthly. Such data had alerted recreation program managers to the discrepancies among fishing piers in the first place. However, a variety of uncontrolled factors also could be contributing to the variation in usage among piers (e.g., proximity to urban populations, convenience of parking, and/or locally depleted fisheries), hence the four sites for reef development will be drawn at random from the ten piers with the lowest average angler usage. With this done, construction will commence immediately.

From the considerations above, it is obvious that the reef evaluations are intended to serve decisions about future agency programs and budgets. Therefore, the specific study design should be a cost–benefit analysis (Chapter 6, Section 6.4.3.2), which would allow the success criterion to be evaluated while serving that intent. The costs for the original piers and for the artificial reef enhancements are or will be known precisely by the sponsoring ministry. A specific study design needs to include how data pertinent to the benefits will be gathered.

The measurement characteristics were generally indicated by the specific objective, i.e., CPUE for rockfish and angler satisfaction. Direct interviews of departing anglers would be the best methodology for this study, given the needed information and the small number of distinct study sites (i.e., four reef-enhanced piers and four "criterion" piers). The survey instrument and interview procedures must be professionally developed and implemented to avoid biases. At the very least, interviewers must generate the following types of information: how long did each angler fish, how many fishing lines did they use, the number of fish (e.g., rockfish by species) landed vs. the numbers kept, and metrics for angler satisfaction with the fishing trip to the particular pier being sampled. In order to assess the money value of benefits to anglers, questions also should be included to reveal the worth of the experience, such as how much might they be willing to pay. Of course, additional fisheries and economic data could and should be gathered at the same time, considering the fixed costs of conducting such interviews. Given a very limited agency budget for the evaluation and the relatively high costs of direct interviews, particular attention also must be paid to the sampling design, its adequacy, and the tradeoffs that may be necessary.

7.3.1.4 Develop a Specific Study Design

Although the locations for sampling are already set by the objective and general study design, the timing of sampling is not. It is likely that angler usage of fishing piers is nonrandom in time for reasons unrelated to fishing success or satisfaction (e.g., anglers' work schedules). Therefore, existing usage data need to be examined for temporal patterns on daily, weekly, and monthly time scales. Then, a stratified sampling design can be planned to ensure adequate coverage in time, with the number of interviews per time period (time periods = strata) being proportional to the fraction of total variance in pier usage represented in each strata. This should maximize the precision obtained for the overall estimates of CPUE and angler satisfaction. Maximizing the precision is important in this case because the number of replicates (i.e., piers) per treatment (i.e., reef-enhanced piers vs. "criterion" piers) is nominal (i.e., n = 4). (The greater the precision, the greater will be the confidence in any conclusions from the analysis.) Within each time strata, the individual anglers to be interviewed will be systematically selected from among those departing during that time period. The first interviewee will be randomly selected from among the first five anglers, and then every fourth departing angler thereafter will be interviewed until the targeted sample size for the strata is attained or the time period passes. Although not strictly random, this procedure should have no obvious bias and is much more practical in this context than attempting complete randomization.

Two types of analyses are anticipated for these data. First, the CPUE data and angler satisfaction metrics for each pier can be averaged across the time strata for the entire study period, and the resultant grand means can be statistically compared using a simple t-test, if the assumptions of normality and variance are reasonable. If not, simple nonparametric procedures (e.g., a Wilcoxon rank-sum test) can be used.

For this example, it is important to recognize that success, according to the criterion in the specific objective, would be inferred if no statistical difference existed between the two pier types or if the reef-enhanced piers outperformed the top 20% "criterion" piers. For this reason, a one-tailed statistical test should be used (i.e., Hypothesis: "criterion" piers > reef-enhanced piers), and particular attention should be paid to the probability of a Type II error. If the "criterion" piers performed significantly better than the reef-enhanced piers, then the enhancement could be judged a failure and yet the enhanced piers might, nevertheless, have improved in their performance. Unfortunately, preconstruction sampling in this example would have precluded drawing such a conclusion, or allowed one to estimate the gain that had actually been achieved.

Furthermore, the cost–benefit analysis likely would be undertaken only if the outcome of the previous analysis had indicated success. For such an analysis, ratios of benefits (monetized and not monetized) to actual costs (pier + reef) could be obtained for each replicate pier, and those ratios could be statistically analyzed in a manner similar to that described above, with similar caveats about one-tailed tests and Type II errors.

7.3.1.5 Critically Examine Limitations of the Project

A brief critique of this example can be very instructive. First, the failure to incorporate pertinent biological and physical processes into the refinement of the general objective to specific objectives, and into the design of the reef itself (see Figure 7.1), greatly undermines the potential for success. It also limits the inferences that can be drawn from the evaluation study as proposed. For example, rockfish can be long-lived and very site-attached, readily leading to localized overfishing. How the artificial reefs might be colonized by rockfish and how the local populations might be sustained despite directed fishing are major factors in the long-term success of this management practice. Furthermore, the time course for those processes was not considered in setting the schedule for the evaluation project.

If colonization of the artificial reefs by harvestable fish occurred within the time frame of the project, but was simply the result of redistribution from adjacent habitat, then the proposed evaluation study might indicate success, but that positive effect could be temporary. In this scenario, decisions to expand the management practice could be technically supported by the study results yet still be logically untenable because of unrecognized assumptions. Conversely, if colonization occurred by the settlement of larval fish or small juveniles, which must grow to harvestable size, then results of the planned evaluation study could indicate failure when, in fact, the potential for longer-term success might be real. Unfortunately, the failure to consider such processes in the reef design would also undermine this second scenario.

Rockfish are also highly predatory and some species will cannibalize small juveniles. All size-classes depend on shelter space scaled to body size for survival. By building the reefs with just one size of boulder, the size range of created cavity spaces might be too narrow for the requirements across the full-size range of fishes. This could inadvertently create a shelter bottleneck in local rockfish populations and defeat the purpose for building these reefs in the first place. It would have been far better to synthesize the biology of the targeted species along with the ecological processes that might determine their abundance and distribution, and then to propose the alternative hypotheses as secondary specific objectives for the evaluation study and reef design.

7.3.2 Example 2: To Enhance Artisanal Seafood Production

7.3.2.1 The Situation Leading to General Objectives and General Reef Design

Along the coast of the Indian Ocean, artisanal fisherfolk have traditionally used tree branches anchored with rocks as artificial reefs, or more properly perhaps, as fish aggregating devices (FADs) to enhance their catch. The practice has been refined through experience over generations to the point where particular species of trees are preferred for their attractiveness to fish, a response that may be chemically mediated (Sanjeeva Raj 1996). However, total fisheries landings increased substantially with the advent and expansion of mechanized fishing, particularly trawling, which began offshore but has moved to inshore waters (d'Cruz et al. 1994). Associated declines in catch-per-unit effort for artisanal fishers are an ever growing concern, especially for the majority who are sustenance fishers using hooks and hand lines from canoes in nearshore waters. Nongovernmental organizations (NGOs) working with local fishing communities are attempting to ensure the livelihoods of artisanal fisherfolks through the development of artificial reefs.

Several problems are inherent in the traditional artisanal reef practices. Tree branches are temporary reef structures that represent a recurrent expense for fishers to deploy. The continuing demand for suitable and preferred trees is causing a depletion of terrestrial vegetation having other economic and ecological values. To the extent that the tree branches act as attractants rather than productive fish habitat, the traditional artisanal practice may actually exacerbate fisheries depletion. Consequently, the general objectives for the NGO reef programs are to enhance artisanal seafood production and to enhance or manage local living marine resources, with the economic objective being the higher priority (see Table 7.1). In other words, a sustainable artisanal fishery is desired.

In this hypothetical example, an NGO is working with several artisanal fishing villages and the mechanized fishers of a region to refine the general objectives into a reef plan. The general strategy is to deploy a reef tract paralleling the coastline just beyond the range of artisanal fishers, and then to deploy numerous small patch reefs inshore from that reef tract within easy reach of the artisanal fishers. The intent is for the reef tract to separate trawlers from the enhanced artisanal fishing grounds and to serve as a protected source of fishes and shellfish to colonize the more accessible nearshore patch reefs. Of course, the practicality of this strategy depends on cooperation from the mechanized fishing fleet and perhaps governmental regulation, as was done for a similar situation in Malaysia (Polovina 1991).

7.3.2.2 *Consider Pertinent Biological, Physical, and Socioeconomic Processes*

Implicit in the general strategy are specific objectives and assumptions about the situation (Figure 7.1). It is assumed that social systems, mediated by the NGO and fisheries management system, can effectively close an artificial reef tract to mechanized commercial fishing. (With only slight rephrasing, this assumption can be stated as a specific project objective to be evaluated.) Uncertainty about the extent to which social and regulatory systems alone can exclude mechanized fishing from both the offshore reef tract and inshore artisanal grounds further suggests that the reef designs in this application need to include passive barriers to mechanized fishing, e.g., snags against trawls and long lines. From the combination of social and legal constraints and the physical barriers, one might expect measurable trends over time in prohibited commercial fishing activity. Initially, compliance with the closure might be high but nevertheless incomplete. Compliance might then decline for a period, resulting in an increase in gear loss to physical snags. But, as the costs for noncompliance accumulate in the commercial fleet, one might expect a longer-term trend toward more complete voluntary compliance.

Also implicit in the general strategy above are assumptions about pertinent biological processes (Figure 7.1). The offshore artificial reef tract is intended to be a source for fishes and invertebrates to colonize the inshore patch reefs. But, this is assumed in the absence of adequate life history and ecological knowledge of the harvested species. Will the inshore patch reefs be colonized predominantly by planktonic propagules or by mobile juveniles and adults? Is the direction of larval transport, or movement patterns of later life history stages, offshore-to-inshore or perhaps vice versa, or even alongshore? Are the fisheries populations spatially structured in any way that might affect the resupply of the artisanal fishing grounds? Where and when does reproduction occur? Are there distinct nursery habitats or ontogenetic shifts in habitat requirements? Are the harvested species highly resident or transient and is residency density dependent? What is the rate of colonization and the anticipated time course for development of fishable stocks on the artisanal reefs? Is the anticipated rate of resupply adequate to offset the harvest rate from artisanal fishing? Many such questions are pertinent to effective reef design and evaluation planning, but in cases like this we do not start with the answers. Therefore, it would be helpful in the planning process to hypothesize ways such a reef system might function ecologically (i.e., alternative models), and then design an evaluation study enabling us to judge which of the alternatives is apparently true.

At least three simplistic ecological models are implied by the questions above. The first assumes no significant spatial structuring of the reef fish populations. Dispersal of planktonic larvae would be nondirectional within and between the offshore reef tract and inshore artisanal patch reefs, with little postsettlement movement by juvenile and adult reef fishes (i.e., a simple bipartite reef fish life history). The second model assumes spatially structured populations in which reproductive adults reside and spawn on the protected offshore reef tract, planktonic larvae settle on structured habitat inshore, and postsettlement juveniles reside and grow on the inshore artisanal patch reefs until they approach maturity and emigrate to the offshore reef tract. A third model assumes that spawning stocks actually reside outside the region encompassed by the offshore reef tract, and that both the reef tract and inshore patch reefs must be colonized by larval fish transported by ocean currents from beyond the constructed (and protected) reef tract.

The efficacy of the NGO's reef development strategy depends on which ecological model is closest to the truth. If the first, then one might expect sustainable artisanal harvests inshore due to protected spawning stocks and closure of the life cycle on reef tracts offshore. If the second, then one might expect sustainable artisanal harvests inshore only if the intensity of fishing allows an adequate number of juveniles to escape harvest and join the protected spawning stock on offshore reefs. If the third model is correct, then mechanized fishing elsewhere might continue to deplete reproductive stocks, although protection of juveniles on the offshore reef tracts might offset that loss to some unknown degree. Obviously, a biological as well as socioeconomic evaluation is needed to judge if this reef system can contribute to a sustainable artisanal fishery.

7.3.2.3 *Refine the General Objectives to Specific Objectives*

To reiterate, the general objective here is to enhance artisanal seafood production and to enhance or manage local living marine resources. To help accomplish this, the two-part strategy is (1) to create numerous small fishing reefs accessible to artisanal fishers inshore; and (2) to establish an enhanced marine protected area and boundary between mechanized and artisanal fishing by creating an offshore reef tract incompatible with mechanized fishing. The general objective and considerations above lead to specific objectives that include success criteria (Figure 7.1):

Objective 1 — To stabilize the CPUE of artisanal fishers in the participating coastal communities at or above their present-day levels of CPUE.

Objective 2 — To eliminate mechanized fishing from the area delineated by the offshore artificial reef tract and the reef-enhanced artisanal fishing grounds inshore.

Objective 3 — To complete the life cycle of important fishery species entirely within this regional system of offshore reef tract and inshore artisanal fishing reefs.

7.3.2.4 *Develop a General Evaluation Plan*

Given the specific objectives and scale of the proposed reef system, the general evaluation plan must be for a descriptive study, rather than a comparative or predictive one (see Chapter 2). All these objectives will require monitoring over time, but only the first really warrants an ongoing monitoring program to detect changes in artisanal fishery trends. Obviously, measurements of artisanal catch and effort will be needed to evaluate Objective 1. For Objective 2, lost fishing gear entangled by artificial reefs can be quantified to evaluate compliance by mechanized fishers. For Objective 3, the occurrence of each life history stage of harvested species can be evaluated by sampling at the inshore and offshore artificial reefs.

7.3.2.5 *Develop a Specific Study Design*

The details of specific sampling designs, sampling methods, and statistical analyses (Figure 7.1) have to be decided by answering who, what, where, how, and when for each objective. One trained NGO staff person can be dedicated part-time to Objective 1 for at least the first 5 years of monitoring. Given current illiteracy rates, voluntary reporting by artisanal fishers would not be practical, but the intent is to develop basic recordkeeping capabilities within each village during this start-up period, thereby educating the client group and possibly freeing NGO resources for other needs. Simple catch-and-effort data will be gathered by observing and interviewing artisanal fishers as they return from fishing at each participating village (i.e., a creel survey). The simplest measures of fishing effort will be recorded — the number of returning boats observed per village landing site, the number of fishers per returning boat, and the duration of the fishing trip for each boat as reported during interviews.

Likewise, simple measures of catch will be recorded — the number of fish per boat, total weight of fish per boat, and a species list. Direct observation and courteous questioning of the fishers should provide everything but the weight. Because scales or balances are not common or always reliable in such villages, the NGO must provide an appropriate transportable balance for its staff. For this objective, the question of when to sample and how many fishers to sample requires the greatest thought in planning the specific study design.

To establish trends in artisanal CPUE, the sampling for Objective 1 should begin immediately, well in advance of reef construction, and then continue uninterrupted into the future. Although the time series of CPUE data should be continuous, one can anticipate adjustments in sampling effort (e.g., frequency of sampling and number of fishers) as more is learned about trends and variability in such data. One can anticipate variation among fishers, among fishing days, months, seasons, and years, and among fishing villages, therefore, such information must be included in each data record. However, what is of consequence are the trends in CPUE over time. The period between onset of monitoring and reef construction can be used to establish a baseline for the criterion of success and to examine variability in CPUE in order to finalize the sampling design for when reefs are built. Initially, the NGO staff person, with additional help if necessary, would spend 1 day per week at each fishing village. All returning boats will be surveyed in the order of arrival. Sources of variation in these initial catch-and-effort data would be analyzed and used to determine sample sizes (i.e., number of fishers per village) and sampling frequency (e.g., weekly, monthly, or seasonally) needed for the subsequent ongoing monitoring. Statistical trend analyses can be routinely updated by the NGO central staff on a quarterly basis for the first 5 years of monitoring and then perhaps annually thereafter if the artisanal fishery is stable.

Success for Objective 1 would be indicated by the CPUE of artisanal fishers remaining at or above the level determined just prior to reef construction. However, continued monitoring would be advisable to detect changes in the system imposed by such factors as a deteriorating natural system, depletion of spawning stocks outside the study area, or perhaps increases in the number of artisanal fishers.

For Objective 2, an NGO-leased vessel towing an underwater video camera will transect the offshore reef tract semiannually to survey for lost nets and long lines. The data to be recorded are simply the number of entangled nets or longlines encountered per unit of sampling effort, e.g., per hour of video transect at a constant speed. When suspected lost gear is observed on the video monitor, the vessel can stop, record the location, and deploy scuba divers to confirm the sighting and either recover the gear or tag it for identification during future surveys. The divers may be either NGO staff or qualified volunteers. Three transects will be surveyed along the length of the reef tract, one along the seaward edge, one along the midline, and one along the landward edge of the reef tract. Occurrences of mechanized fishing directly on the inshore artisanal fishing grounds can be recorded during interviews of artisanal fishers routinely conducted for Objective 1. Similarly, any occurrences of nets and long lines entangled on inshore artisanal fishing reefs can be recorded during the fisheries-independent sampling for Objective 3. As data accumulate from the video transects, they can be plotted as total new sightings of lost fishing gear or average new sightings per unit area of reef tract vs. the sampling periods (i.e., time as the x-axis). The same can be done for the data from artisanal fishing grounds. Success for Objective 2 would be indicated by a decline in lost fishing gear over time, asymptotically approaching zero.

Objective 3, which is to complete the life cycle of important fishery species entirely within the regional system, will require fisheries-dependent and fisheries-independent sampling by appropriately skilled biologists. Both types of sampling will be conducted monthly over a 2-year period, beginning when the artificial reefs are 3 years old. The delayed start should allow time for colonization of the new reefs and the maturation of resident fishes. The fisheries-dependent sampling will be limited to inspecting the catch of artisanal fishers and recording the sizes and reproductive condition (e.g., running ripe eggs and milt, stage of gonadal development, gonadal-somatic indices, etc.) of major fisheries species. This can be done in conjunction with the observations and interviews for Objective 1.

The fisheries-independent sampling will involve spearing juvenile and adult-sized fish on reefs by divers, ichthyoplankton surveys, and diver inspections of reefs for recently settled fishes. The speared fish will be measured and inspected for reproductive condition in the same manner as for the fisheries-dependent sampling. A maximum of 15 fish per species will be speared during any given sampling period. These fish will be distributed to charities serving the artisanal fishing villages once the data are gathered. Ichthyoplankton surveys will involve appropriately sized plankton nets towed across the offshore reef tract, the inshore artisanal fishing grounds, and the midpoint between those two areas. Three replicate tows will be conducted at each sampling area during each month. The plankton samples from each tow will be concentrated and preserved onboard the NGO's vessel and subsequently sorted in the laboratory to identify and enumerate larvae of major fishery species. To document the occurrence of recently settled fisheries species, divers already trained to recognize the early stages of important reef fishes will search systematically across the surfaces and in the cavities of the offshore reef tract and inshore artisanal patch reefs. Voucher specimens will be collected by hand nets to confirm the species identifications, but no attempt will be made to quantify the abundance of early settlers.

Success for Objective 3 will be indicated, but not confirmed, by the common occurrence of all life history stages for target species within the entire reef system.

7.3.2.6 Consider Limitations of the Planned Study

It is important to acknowledge that, even if all life history stages are found, the proposed level of sampling and analysis for Objective 3 is inadequate to allow a conclusion of self-sustaining local reef fish populations. However, a persistent absence of major life history stages within the regional reef system would seriously cast that possibility into doubt. At best, the spatial and temporal patterns of life history stages within the reef system would give an indication as to which of the three ecological models summarized above is most likely true and has consequence for judging the potential for sustainability.

7.3.3 Example 3: To Protect Adjacent Valued Habitat[1]

7.3.3.1 Define the General Management Objectives and Situation

Government officials and conservation leaders in a moderately sized traditional fishing community on the Mediterranean coast are proposing to build a number of artificial reefs. One of their main goals is to prevent illegal otter-trawl fishing in order to protect biological resources, natural communities, and nursery habitats in shallow coastal waters. Most of the nonartisanal fishing boats in this region are using otter trawls throughout the year at depths up to 800 m. The damage caused by otter trawls is due to mesh size (selectivity of the net, in particular the cod-end) and to the structure of the gear itself. In particular, the footrope scrapes the bottom, while two otter boards (the function of which is to pull the net mouth open) plough the bottom leaving two parallel furrows far apart. It is not difficult to imagine substantial damage caused by repeated trawling, especially where seagrasses and biogenic hardgrounds occur. Various national and European Union laws forbid cod-end mesh sizes smaller than 40 mm (opening) as well as bottom trawling inside the 50-m depth contour. Because such activity is still quite common, it has become necessary to interfere with illegal trawling by means of obstacles. Artificial reefs are the tools of choice because they offer additional potential benefits.

7.3.3.2 Identify the Local Ecological Context and Pertinent Processes

The sea floor within 25 km of this particular coastal fishing community is a landscape of various substrata and habitat types, e.g., photophilic algae assemblages, coralligenous bottom, small bed

[1] In some places, artificial reefs have been built specifically to protect adjacent habitat or sensitive biotopes (Lefevre et al. 1983; Relini and Moretti 1986; Assoc. Monégasque pour la Protéction de la Nature 1995; Sanchez-Jerez and Ramos-Esplà 1995; Badalamenti and D'Anna 1997; Bombace 1997; Charbonnel et al. 1997; Harmelin and Bellan-Santini 1997; Moreno 1997). This hypothetical example, like the others, draws from such experience to illustrate the thinking entailed in project planning.

rocks and seagrass meadows of *Cymodocea nodosa* up to 15 m depth and of *Posidonia oceanica* up to 35 to 40 m.

Seagrass beds of *P. oceanica* located inshore from 2 to 40 m depth, and especially the area from 8 to 30 m, are of particular concern. Seagrass habitat serves as spawning grounds, nursery habitats, natural baffles protecting shorelines from high wave energy, habitat for a very high number of algae and animals, production of oxygen, export of production, etc. The importance of these seagrass beds derives from their ecological role and quite high primary production.

The plant, 80 to 90 cm high, with 5 to 6 leaves per shoot (each leaf is about 1 cm wide) attached to the rhizome effectively absorbs the energy of swells and currents, retains sediment, and thus protects shorelines. The living leaves absorb currents; the dead leaves accumulate on beaches and modify or reduce the effects of waves. The rhizomes ("matte") maintain the sediment. Their interlacing mats form an elastic and rigid structure that reduces the effects of wave surge.

A density of 4000 to 8000 leaves/m^2 represents a nice trap for sediment and a good protection for deposited material. If the sedimentation is between 5 to 70 mm/year the complex of rhizomes ("matte") can grow up. Conversely, a meadow can disappear due to excessive sedimentation or to erosion by wave action, with insufficient deposition of sediment. The growth of the "matte" can last thousands of years. The dead leaves contribute organic content to the sediments and, in part, are exported toward the beach or to other ecosystems. At considerable depths, leaf detritus is an important source of food, as on the bottom at 500 to 800 m where red shrimps are fished.

Numbers can give an idea of the production per square meter and year — roots: 80 g dry weight; rhizomes: 20 to 40 g d w; leaves: 700 to 2000 g d w; epiphytes on leaves. 500 to 1500 g d w. The Leaf Area Index (LAI) is 10 to 20 m^2/m^2; the biomass of leaves is 400 to 1200 g d w/m^2; biomass of epiphytes on leaves is 10 to 320 g d w/m^2; biomass of epiphytes on rhizomes is 1 to 10 g d w/m^2; biomass of rhizomes is 1500 to 4000 g d w/m^2; biomass of fauna (fish and cephalopods excluded) is 500 g d w/m^2; and total flora biomass is 3500 to 6000 g d w/m^2. As mentioned above, a considerable part of this production is exported.

In the bay of Calvi (Corsica Island, France), where half of the bottom surface (about 20 km^2) is covered by *P. oceanica,* the activity of the seagrass bed in May is responsible for nearly all the diel variation of both dissolved oxygen and inorganic carbon. Its zone of influence covers the whole bay. The balance of net photosynthesis is –61 μM inorganic C per gram of dried decalcified leaves when estimated from pH and alkalinity measurements, and +79 μM O_2 per gram of dried decalcified leaves when estimated from dissolved oxygen analysis. The photosynthetic quotient is 1.09 (Frankignoulle et al. 1984).

The ichthyofauna community in *P. oceanica* is very important. For example, Harmelin-Vivien (1982) has studied fishes in the Regional Park of Corsica, between 15 and 40 m deep. This community included 41 fish species belonging to 19 families, which were classified as resident (56%), as transient (22%), and as occasional species (22%). The most important families were Labridae, Scorpaenidae, Serranidae, and Centracanthidae, together representing 41% of the total number of species and 87% of the total biomass. Important species also belonged to Congridae, Ophidiidae, Gadidae, Syngnathidae, Sparidae, Gobiidae, and Bothidae.

The nocturnal fishes of *Posidonia* beds were more abundant and more diversified than the diurnal fishes. At night, the number of species increased as well as number of individuals and biomass. Several species were caught only at night, some nocturnal carnivores (Congridae, Gadidae, and Ophidiidae) as they moved out of their shelter and some diurnal zooplankton feeders (Centra-canthidae and Pomacentridae) as they moved into the grass canopy to rest. *Posidonia* fish fauna presented high diversity values (3.11 < H' < 4.09). Species composition and quantitative structure of the *Posidonia* fish fauna studied in Corsica was very similar to that observed in the seagrass beds in the Marseilles Gulf or around Port-Cros Island. Strong homogeneity occurred in the fish fauna of *Posidonia* seagrass beds in the northwestern Mediterranean Sea, despite minor local variations.

Ardizzone and Pelusi (1984) documented damage by illegal trawling inside *Posidonia* beds, and confirmed the importance of protecting this habitat. The trawled, low density *Posidonia* meadow had a monotonous catch composition, with *Octopus vulgaris* being dominant. Other species often caught, but of low economic interest, belonged mainly to the families Labridae and Scorpaenidae. The limited number of valuable permanent fish species is related both to the reduction of *Posidonia* density and associated fauna, and to the overfishing of slow-growing fish species. In fact, the continuous trawling of a *Posidonia* bed, after a few years of extremely rich harvests of valuable species (such as *Dicentrarchus labrax, Litognathus mormyrus, Pagellus erythrinus, Diplodus sargus, D. vulgaris, D. puntazzo, Sparus pagrus, Mullus surmuletus*), leads to a gradual decline of these stocks. Because of the high exploitation through unselective trawls (mesh size 16 mm at cod-end), this fishing activity also often damages juveniles of these species.

7.3.3.3 Refine the General Objectives

Given the priority of protecting the *Posidonia* habitat, the primary objective for this project is to eliminate any trawling and anchoring on the meadows. Continuous anchoring by pleasure boats or fishing boats can have deleterious effects on the meadows because anchors and, in part, chains destroy plants and sometimes eradicate rhizomes. The best way to avoid this damage is to prepare small mooring buoys attached to the blocks of artificial reefs to provide boats an alternative to anchoring. In so doing, care also must be taken that the reefs themselves do not inflict habitat damage. A precise map of the meadows is necessary to prepare for the placement of reef blocks. For the purposes of this example, secondary objectives include enhancing biodiversity and artisanal fishing, but complete evaluation plans will not be developed for these.

7.3.3.4 Develop the Reef Design and Site Layout

The design of modular reef units and their physical layout on the sea floor will likely determine how well they stop illegal trawling and anchoring. To be effective for secondary goals, such as enhancing benthic organisms and fishes, the modules also must have some characteristics detailed in Examples 4 and 5 and also in Chapters 4 and 5. On the basis of experiences to date, anti-trawling reefs should have the following characteristics:

1) Be lasting, but not polluting,
2) Be heavy and strong, so as not to be broken or moved by large otter trawls,
3) Have as many different-sized holes as possible, while maintaining structural integrity,
4) Have some surfaces protected against siltation or sedimentation,
5) Have links for ropes or chains useful for additional mariculture operation (such as mussels or oysters hung in special cages) or for mooring boats.

In some places, concrete perforated blocks ($2 \times 2 \times 2$ m cubes with holes of 20 to 40 cm diameter) have been particularly suitable when arranged as pyramids of 5 or 9 cubes each (Bombace 1981; Relini and Relini Orsi 1989). Elsewhere, anti-trawling devices have been made of concrete blocks with protruding iron bars or hooks. These are effective against large, commercial trawls but are also harmful to artisanal gear such as gillnets, trammel nets, and hand lines. Given the essential small-scale fisheries of the community that requested habitat protection, such devices would be inappropriate. For this project, simple concrete perforated blocks arranged as 5-cube pyramids will be built as the anti-trawling reefs.

A second point to consider is how to distribute these reefs to protect the largest possible area with a minimum number of modules. This is a balance between construction costs and anti-trawling effectiveness. Corridors smaller than 50×250 m between reefs should generally exclude commercial otter trawls. For the landscape in this example, a latticework pattern of reefs will be deployed around the *Posidonia* bed and inside areas that are not covered by plants. In some small areas, single blocks are also useful for protection and as refuges. A map of biological communities at and around the individual construction sites would help to avoid damage to the habitats being protected by artificial reefs. Particular attention must be paid to seagrasses (e.g., *Posidonia oceanica*) and to coralligenous-like assemblages. Such a map would also aid the planning of exact placement points for a better integration of the natural and artificial habitats (e.g., nearby rocky substrata or wrecks may be sources for the colonization of newly immersed blocks). In fact, benthic organisms (adults or larvae) and fishes will be attracted by new surfaces and refuges available on the concrete blocks.

The presence of the blocks and mooring systems, while avoiding eradication of *Posidonia*, are predicted to have a positive effect on seagrass, not only by stopping the decrease of meadow area, but also by allowing a recovery to ensue. The recovery may be slow but important, and after some years, it should be possible to notice growth of plants, increased numbers of shoots, and a greater area of sea floor covered by the seagrass.

A third point to consider is the stability of the modules and the prevention of sinking. This very important point is sometimes underestimated, with deleterious consequences. There are cases of boulders placed as artificial reefs that completely disappeared into the soft sediment after some years. Knowledge of the bottom type is essential for project design. For this example, the sediment is assumed to be soft and deep, and the bottom currents occasionally strong. Therefore, it will be necessary to prepare a "bed" of small stones or pebbles, on which to arrange the block pyramids to better ensure stability of the structures. From an ecological perspective, the increased structural complexity and added microhabitats should also increase species diversity associated with the reefs. To the extent that small stones prevent erosion around the big blocks, potential physical effects on the surrounding natural habitat are expected to be minimal.

Another important role of these small stones is to facilitate the natural transplanting of broken pieces of *Posidonia* transported by current and the waves. In some places artificial transplanting by scuba divers could be tested.

7.3.3.5 *Refine Specific Objectives to Include Success Criteria*

As a consequence of considering which physical and biological processes are important in the reef design, the general objective can be refined to five somewhat more specific objectives:

Objective 1 — To eliminate from the protected area all physical disturbance of *Posidonia* habitat associated with anchoring and mainly with otter trawling, especially scars on the sea floor from trawl doors and broken or mowed seagrasses and associated organisms from footropes and trawl bags.

Objective 2 — To provide stable anti-trawling reefs that do not alter the physical structure or biological composition of the adjacent habitat that the reefs are intended to protect.

Objective 3 — To provide stable anti-trawling reefs that cannot be displaced by trawls or storms, and that do not scour or settle into the sand substratum at a rate faster than would ensure at least a 25-year effective lifespan.

Objective 4 — To provide stable mooring systems for boats on the reef or on the *Posidonia* bed.

Objective 5 — To increase the local diversity of benthic invertebrates and demersal fishes by adding taxa associated with hard substrata at various stages of their life histories (e.g., eggs, larvae, juveniles, or adults).

7.3.3.6 *Decide on the General Evaluation Plan*

An experimental approach that entails full replication of protected and unprotected *Posidonia* beds is not practical, given the scale of construction efforts and the geographic range that would be required. However, comparisons of data collected before and after the construction of anti-trawling reefs can be used to evaluate the two primary objectives, i.e., Objectives 1 and 2 above. A time series of monitoring data on broad-scale physical habitat damage from trawls, both before and after construction, would allow quantitative comparisons for Objective 1. Likewise, a time series from finer scale physical and biological sampling in protected habitat immediately adjacent to the artificial reef sites would be needed to evaluate Objective 2. To evaluate Objectives 3, 4, and 5, the monitoring would need to begin just after reef construction is completed. For Objectives 3 and 4, the data must represent the physical condition of the reefs themselves, including measures of scour, settling, and/or burial, and be sufficient for estimating the life expectancy of the reef based on engineering calculations. For Objective 5, the data must represent the biological communities directly associated with the artificial reefs, and be certain to include taxa of greatest interest to the local fishing community. Because time and money will limit sampling, greater monitoring effort should be directed to the primary objectives, 1 and 2, than to the secondary objectives, 3, 4, and 5.

7.3.3.7 *Develop the Specific Study Design*

To estimate the extent of trawl damage within the area to be protected and to detect changes following construction of the anti-trawling reefs (i.e., to evaluate Objective 1), a systematic transect sampling plan will span the entire area of protected *Posidonia* bed. These will be video transects that employ wide-angle lenses on either a remotely operated vehicle (ROV) or a towed, downward-looking camera, or direct examination by scuba divers, depending on costs and available funding. The aim of these observations is to determine if there are still signs of trawling activity or anchoring revealed by furrows, holes, broken plants, etc. Seasonally, surveys by ROV or towed camera would be along a grid of parallel and perpendicular transects 100 or 200 m apart, depending on the area to be surveyed. Along with the ROV or towed camera, it will be useful to plan some targeted surveys by divers in sites of particular interest.

To detect changes in the protected habitat potentially induced by the reefs themselves (i.e., to evaluate Objective 2), permanent monitoring stations will be established along transects going 100 to 200 m away from the anti-trawling reef sites nearest the protected *Posidonia* beds and also inside the protected area. Trained scuba divers will swim each transect during each sampling period, and employ the same methods at each monitoring station along its length. Obviously, these transects will be less numerous than those used for Objective 1, but the resolution of data will be greater. Along each transect some m^2 quadrats will be chosen randomly or at 50 m intervals. In each quadrat, once a year, the number of shoots and length of leaves will be measured to evaluate an increase in the growth and in surface coverage of *Posidonia*. This work is not difficult but it is time consuming; therefore, the number of stations (quadrats) chosen will be in relation to the manpower or money available. At least 20 stations per year must be surveyed. In a dense meadow, more than 700 shoots/m^2 can occur, while in a poor, sparse meadow, the number of shoots can be less than 50 shoots/m^2. A more sophisticated evaluation can consider primary production following classical methods described in Chapter 4. Also important are any qualitative and quantitative changes of fishes.

To evaluate the stability of the anti-trawling reefs and mooring systems (i.e., Objectives 3 and 4), physical data must be collected as described in Chapter 3, but for the structural integrity of the pyramids, a visual inspection by scuba divers, ROV, or camera is sufficient. If the cubes are not properly positioned, it means that hydrodynamics or mechanical action by man has affected the pyramids. Erosion of corners indicates that the concrete was not of good quality.

To evaluate the biota associated the anti-trawling reefs (i.e., Objective 5), especially the fauna of fisheries interest, limited surveys of the hard-bottom benthos and fishes will be conducted annually. As previously noted, a full evaluation plan for these secondary objectives will not be developed for this example, yet it is instructive to consider how they might be approached.

Three primary methods might be used to describe and quantify the flora and fauna closely associated with surfaces of the artificial reefs. These are (1) nondestructive methods (i.e., visual census by scuba divers by quadrats or transects, including photos or video records); (2) destructive methods (removing by scraping organisms within quadrats after photo or video records); and

(3) immersion and removal of panels or other ad hoc substrata. Each of these methods has limits and biases, but the last one has yielded good results (Relini 1976; Relini et al. 1994) because it is possible to take a piece of substratum into the laboratory with all the organisms intact. Thus, positions of individuals are maintained allowing densities, sizes, degree of maturity, and coverage by each taxon to be accurately assessed. If the substrata are immersed on the reef for increasing periods of time (e.g., 1, 3, 6, 9, and 12 months), it is then possible to track settlement periods of the main organisms and the development of the fouling community.

An assessment of fish can be made using different sampling techniques, including destructive methods (e.g., fishing nets, traps, hooks, long lines, and spearfishing) or conservative methods (e.g., visual census by scuba divers, audio and video recording, and ROVs) (see Chapter 5; Charbonnel et al. 1997, Harmelin and Bellan-Santini 1997). Descriptions of the fish community can be generated in terms of the number of individuals, species, biomass, and various community indices. However, the sampling design employing such methods should be adequate to detect rhythms or patterns on various time scales (e.g., seasonality), as well as to describe the development of the fish assemblage in concert with the benthos. If possible, the sampling design should also allow comparisons to be made with natural hard-bottom habitats of the region.

For this example, the biological or ecological aspects of the natural habitat and artificial reefs might be of greatest interest to the experts qualified to evaluate them. Nevertheless, it is important to concentrate attention during the evaluation planning and execution directly on the first two or three objectives. These mostly involve physical disturbance and the ecological consequences of disturbance, and will have the greatest bearing on judging whether or not the anti-trawling reefs are fulfilling the purpose for which they were built.

7.3.4   Example 4: To Improve Water Quality

7.3.4.1   Identify the Situation and Define General Management Objectives

The watershed surrounding a picturesque, small coastal bay (9 km^2 with a coastline of about 11 km) has been altered by agricultural and urban runoff to such an extent that water quality in the bay has deteriorated.

The agricultural pollutants arrive through very small streams while urban sewage is discharged mainly by a pipeline, 2 km into the middle of the bay at a depth of 12 m. Authorities believe that anthropogenic eutrophication from nonpoint sources has greatly increased phytoplankton production and suspended particulates in the bay. Changes in water clarity and color have been documented and are of concern to local citizens and business people, who derive value from the aesthetics of the bay. Recreational sailing, swimming, power boating, and general tourism are important to the local economy, whereas fishing activity within the bay yields minimal profits because of the decline of valuable fish. At times, a considerable increase of low-value small pelagic fishes occurs.

Headlands at the mouth of the bay support dense mats of mussels, characteristic of rocky intertidal and upper infralittoral shorelines, however, the interior of the bay has very little hard substratum. The perimeter is mostly sandy beach with only a few exposed geological features. The bay bottom is almost entirely sand/mud or mud overlying bedrock. Historically, seagrass beds occupied the center of this relatively shallow bay (sloping to 25 m depth), but seagrass is now sparse and only in the outer part of the bay, presumably because of insufficient light penetration and/or too much siltation of materials with high organic content.

Artificial reefs have been proposed as a way to sufficiently increase the population of mussels to filter the bay waters and thereby restore water clarity and color. Today, the bay has an average Secchi depth of 2 m and a distinctive brownish-green cast (extinction coefficient more than 1.0; more than 80% of total radiation is absorbed within the uppermost 1 m). Historically, visibility was often top-to-bottom, up to 15- to 18-m depth, with the clear water reflecting skyblue hues.

Although citizens and businesses would like to restore that condition, the responsible authorities are less optimistic in their expectations.

7.3.4.2 Identify the Pertinent Biological and Physical Processes

The construction of artificial reefs to enhance the settlement of filter feeders has the potential to improve water quality in areas with high eutrophication (nutrients, salts, and organic matter in suspension) and high concentrations of suspended particles, living (phyto and zooplankton) or not (tripton). The impact of filter-feeding organisms on the ambient water increases with rising population density. Beds of mussels (e.g., *Mytilus edulis* in the Atlantic; *M. galloprovincialis* in the Mediterranean) represent by far the highest population densities reached by any suspension-feeding organisms. Densities of about 1 kg dry body mass per m^2 are typical of mussel beds from various types of habitat, both from the tidal zone and from sublittoral habitats (Barker Jørgensen 1990). Oysters (*Ostrea edulis, Crassostrea* spp.) also could be successful in filtering water.

In the Adriatic Sea, 80 to 120 kg/m^2 of mussels (5 to 6 cm shells) are harvested per year from artificial reefs or gas extraction platforms (Bombace 1981; Relini et al. 1998). In this situation, the dry body mass can reach 25 kg/m^2 with densities of mussels up to 834 ind/dm^2 and 1117g/dm^2 or 438 ind/dm^2 and 1553g/dm^2 wet weight (Relini et al. 1998).

The rates at which mussel beds process the surrounding water can be estimated from the size frequency distribution of mussels on the beds and the relation between size and clearance. Rates of water processing amounting to about 10 m^3 m^{-2} h^{-1} seem to be typical, as quoted by Barker Jørgensen (1990). The impact of such rates may be assessed if the efficiency with which the mussels retain the various constituents, suspended or dissolved, is known. Besides phytoplankton, detritus, and silt, these constituents include bacterioplankton and small organic molecules, particularly amino acids, as well as oxygen.

Dense populations of suspension-feeding bivalves may process the ambient water at such high rates that phytoplankton production can be affected and even controlled by the populations of bivalves. Cloern (1982) provided evidence for such control in south San Francisco Bay. This shallow area receives large inputs of nutrients, however, despite eutrophication, the standing biomass of phytoplankton remained low. The grazing by zooplankton accounted for only a small reduction in the net rate of phytoplankton growth. But calculations based on the biomass and size distribution of benthic suspension-feeding bivalves and realistic rates of water filtration showed that bivalve populations filtered a volume of water daily that exceeded the total volume of the shallow waters.

A further example of such impact was reported from the Potomac River, Maryland, where Cohen et al. (1984) found that a reduction in phytoplankton concentration was inversely related to the density of the clam *Corbicula fluminea*. Determinations of filtration rates showed that the densest *Corbicula* populations would process a volume of water equivalent to the entire water column in 3 to 4 days. This high filtration of the water can decrease the turbidity of water itself. However, there are problems with biodeposition because feces and pseudofeces produced by bivalves can change rheological properties of sediments. Increased amounts of sediments and biodeposits also may become suspended by water turbulence in connection with periodic tidal currents and wave action, particularly during storms. In shallow waters with strong resuspension of sediments or high concentrations of particulate matter from sewage outfalls, the concentration of seston can run as high as 100 mg l^{-1} or more.

In such places, suspension-feeding bivalves are exposed to much higher particle loads than can be ingested, and surplus matter is expelled as pseudofeces. Mussels and oysters, which may inhabit highly turbid waters, begin to produce pseudofeces at seston concentrations of a few mg l^{-1}. Moreover, seston in such habitats mostly consists of silt and detritus of little or no nutritional value, and most of the ingested matter is egested as true feces. Both feces and pseudofeces are deposited as sediment more rapidly than the suspended matter from which they originate. Thus, the water-processing activity of suspension-feeding bivalves enhances the rate of deposition of the seston.

In some bivalves the rates of biodeposition have been measured or calculated. For *Crassostrea virginica* exposed to a mean seston concentration of about 9 mg l^{-1}, deposition was from 4 to 9 g per g dry body mass per month. For *C. gigas*, biodeposition was 9 to 10 g per g dry body mass when the seston was 42 mg l^{-1} in July and 21 mg l^{-1} in September.

In the Baltic Sea, it was estimated that the mussel population (*Mytilus edulis*) increased the annual total deposition of C, N, and P by 10%, and that the population circulated and regenerated 12 and 22%, respectively, of the annual N and P demands for pelagic primary production. Thus bivalves can play an important role linking the pelagic and benthic compartments of an ecosystem.

7.3.4.3 Decide on Reef Design and Site Selection

The availability of hard substrata can limit the abundance of sessile filter-feeders, and artificial reefs might alleviate that limitation. In Pomerian Bay (Poland) concrete pipes were used for an artificial reef experiment to remove nutrients. The annual filtration potential of dominant species, mainly mussels (*Mytilus edulis*) and barnacles (*Balanus improvisus*), was estimated to exceed 5500 m^3 of water per m^2 surface area (Chojnacki et al. 1993; Chojnacki 1994). Some preliminary results indicated that the reefs had an effect on water quality, e.g., a declining trend in nitrate (NNO$_3$) and an increasing trend in visibility (Laihonen et al. 1997).

Given the size of the bay to be restored and existing estimates of filtration and assimilation rates, one should be able to estimate the additional coverage and biomass of mussels needed to effect a discernible improvement in water clarity and color. Authorities and local managers, with the agreement of fishermen and environmentalists of NGOs, decided to build artificial reefs in the bay with the main goal of increasing the filtering biomass of mussels, but also to increase shelters for fish. For these purposes, pyramids of five cubic blocks (2 m side) were chosen. Four blocks were put on a bed of small stones or pebbles, 80 cm apart from each other, with the fifth block on top of the four. The top blocks bear an iron bar to fix two nylon ropes suspended between two close pyramids. The ropes are very useful for suspending cylindrical nylon netting for mussels removed from natural hard substrata. This operation facilitates the larval release and then colonization of substrata. Pyramids are positioned 20 m apart from one another and about 30 m from the sewage outfalls, and around the pipe. For heavy settlement of mussels it is better for the substrata to be less than 18 m deep. The main settlement occurs in May, and one year later, a biomass of 50 to 100 kg/m^2 and densities of more than 10,000 ind/m^2 can be reached. The bay has a volume of 66 million m^3 of water, but the part influenced by pollution is 33 million m^3. To filter this volume of water per day (24 h), at a rate of 10 m^3 per h per m^2 of mussels (Barker Jørgensen 1990), it would be necessary to cover a surface of about 137,500 m^2 with mussels, or at least double that for filtration twice a day. It may be that under conditions of 10,000 mussels/m^2 the filtration rate could be higher. One pyramid has about 90 m^2 for mussel settlement, which increases to 150 m^2, with mussels hanging from ropes and on the surface of the bed of pebbles. Therefore, at least 916 pyramids are needed. In order to reduce the number of pyramids, it is possible to increase the surface of each pyramid by adding concrete pipes among the pyramids. Another option is to filter water every 2 or 3 days, in which case half to one third the number of pyramids are needed. All these possibilities must be tested in relation to the load from sewage outfalls and small streams and to the turnover rate of phytoplankton seasonally dominant in the bay. A denser mussel population can increase the rate of filtration per square meter.

In any event, the actual parameter values must be determined from field data, and one must be very careful when predicting a change in the quality of water because such systems are very complicated. For example, nutrient loading coming from the land into the bay likely varies or might change according to other controls.

7.3.4.4 Refine General Objectives to Specific Objectives, Including Success Criteria

The general objective was to increase the population of mussels sufficiently to filter the bay waters and thereby restore water clarity and color to historical conditions. However, given the estimates of filtering capacity per unit of mussels and the level of reef construction that is practical, the specific objectives are as follows:

(1) To increase the overall standing stock of mussels in the bay to an average of 80 kg m^{-2} over the next 5 years, and thereby,
(2) To increase the average Secchi depth of the bay from 2 to 8–10 m, and to decrease the average chlorophyll *a* concentrations by 70%.

7.3.4.5 Develop a General Evaluation Plan

In this example, the effectiveness of the artificial reefs in improving water quality can be evaluated by monitoring the parameters we expect to see improve. These would include water clarity (e.g., Secchi depth), chlorophyll *a* concentrations, etc. Criteria for success were stated as part of the specific objectives (Section 7.3.4.4).

However, with this simple monitoring approach, one could not technically conclude that the reefs caused the favorable changes. Confidence in ascribing changes in water quality to the mussel reefs would be improved by also estimating the biomass of filter feeders and their biodeposits, and determining if these measurements were sufficient to account for observed changes in water quality of the bay. Of course, the sampling designs for all these variables would have to be adequate for both the spatial and temporal variation in the bay.

In an evaluation plan, cost–benefit is also important. In this case, an additional benefit is linked to possible harvesting of mussels and fishes. In particular, the high biomass production of mussels may allow for its harvest and sale after appropriate sanitary control (sometimes bivalve mollusks need to be treated in special depuration plants).

7.3.4.6 Develop a Specific Study Design

To determine the positive influence of filtration by mussels, it is necessary to have some data before the construction of artificial reefs and a specific study design for after their deployment. The following data are needed: the organic and nutrient (N and P compounds) load of different runoffs, nutrient concentrations at different sites and depths in the bay, Secchi depths, chlorophyll *a* concentrations, the dominant species of phytoplankton in different seasons (in particular the ratio of diatoms/dinoflagellates), and some information about the bottom (granulometry, organic matter content, and redox). Also, a visual census and/or sampling by trammel net of fishes would be useful, as would a map of mussels and other filter-feeding organisms in the bay. All this information has to be collected before and after the construction of artificial reefs at least seasonally, at different stations and at different depths (water column). The sampling stations have to be placed at different distances from outfalls. Comparison of data before and after the reef construction can give information about success of the initiative and how to improve the positive influence on quality of the water.

7.3.5 Example 5: To Alleviate Life History Bottlenecks

7.3.5.1 Define the General Management Objectives

Business leaders in a small, rural coastal community want to stimulate general economic development through eco-tourism (sometimes referred to as nature-based tourism). They envision recreational fishing and diving as important reasons visitors would come to their community. However, having seen environmental problems created elsewhere by development, these leaders want to stimulate business activity in a manner that conserves rather than threatens the ecological basis of the local economy. Artificial reefs were assumed to be compatible with their interests, thus on referral from others, they enlisted the help of an experienced university researcher. After several discussions of what was known and not known about artificial reefs, about species of local fisheries interest, and about natural ecological processes, it was decided that fisheries conservation would be the primary general objective for their reef program, and that economic development would be the secondary general objective.

The decision to emphasize conservation over direct economic returns was derived from an assumption that sustainable fisheries would support long-term economic growth and stability, whereas depleting the fisheries would yield only short-term profits. These general objectives clarify the values and priorities of those wanting the reefs built. But the general objectives lack specificity, which can only be achieved by considering key features of the situation and by developing scientific rationales for reef designs and site layouts.

7.3.5.2 Identify Key Features of the Ecological Situation

The most valued reef fishery species of the region is gag grouper, *Mycteroperca microlepis* (family Serranidae), a protogynous hermaphrodite (i.e., female-first sex reversal) with a spatially structured, complex life history. For gag, spawning occurs in late winter to early spring well offshore in deepwater aggregations of many reproductively active females and a few large males. The planktonic larvae are transported inshore by seasonal currents where they settle in structured shallow-water habitats, predominantly seagrass beds. Expansive seagrass beds in this area are considered to be among the most pristine and productive nursery habitats for gag throughout its range. However in this region, when young-of-the-year gag emigrate from seagrass habitat in the fall of each year, they must traverse a broad expanse (i.e., > 30 km) of shallow continental shelf sand-bottom and sand-veneer habitat before encountering patchy rock outcroppings with sufficient complexity to provide their preferred shelter. Even at that point, suitable natural reef habitat is, by all accounts, sparse.

Recent research in the same general region, using artificial reefs for experimental purposes, has confirmed that juvenile-to-adult gag are seasonally gregarious and stay resident on small patch reefs for an average of 10 months, with some staying upwards of 2 years. They are highly site-attached with home ranges of no more than 500 m in diameter. Gag also exhibit a capacity to home when displaced to distances of 3 km. The studies revealed that juvenile-to-adult gag select habitat primarily on the basis of available shelter and only secondarily on the basis of food resources, and it appears that they do so in a density-dependent manner. However, their growth rates and condition are significantly greater on small, widely scattered artificial reefs than on large or closely spaced artificial reefs or natural rock outcroppings of the region. These findings are consistent with the processes of density-dependent habitat selection, coupling to off-reef prey sources, and an interaction between shelter and food usage. Nevertheless, when experimental reefs were opened to public recreational fishing, the loss far exceeded the production of gag biomass from even the best reef types.

Several lines of evidence suggest that a demographic bottleneck regulates the transition of juvenile gag into the reproductive stock. Considerable variation in size at age has been documented for the gag population as a whole (e.g., the total length of 4-year-old females can differ by as much as 25 cm), and as experimentally shown, growth and condition vary with patch reef size and spacing. The preferred and most productive reef habitat types are apparently sparse across the area transitioned by juvenile-to-adult gag, despite the existence of premium nursery habitat inshore. To the extent that it exists, a demographic bottleneck could result from direct mortality associated with inadequate shelter and from diminished growth and condition. In fish, size and condition, often measured as relative weight (W_r), are highly correlated with fecundity and survival.

Given this ecological situation, the general primary objective of fisheries conservation can be refined, namely, to alleviate a demographic bottleneck in the life history transition of juvenile-to-adult gag grouper. However, this bottleneck is strongly suspected, but not yet scientifically known to exist. Thus, an ancillary objective emerges to test whether or not the suspected bottleneck actually exists. This ancillary objective simply acknowledges an opportunity to learn more about natural ecological processes and how artificial reefs might be used as fisheries management tools. It also clarifies a desire to do the evaluation in a manner that would have relevance beyond just the local coastal community and immediate reef project. Although now refined, the primary objective is still not specific enough to convey a criterion of success, which in this case is implied by the ancillary objective. To get more specific, one must consider what measurable characteristics are relevant, and what reef designs and site layouts have the greatest potential to alleviate the bottleneck, if it exists.

Relevant measurable characteristics would pertain to the demographic parameters of mortality, growth, and fecundity. Mortality cannot be measured directly, but in proper comparisons, the abundance of resident gag could be taken as a proxy for survival, the inverse of mortality. Growth or growth rates can be measured directly as changes in length (total length, fork length, and standard length) between the marking and recapture of fish, but in an open population such as gag, this would require a tremendous logistical effort. The alternative is to estimate growth from analyses of otoliths, although this would require some sampling without replacement on reef sites. Fecundity, or the number of eggs produced by a female, is not directly relevant because females in the area of the shallow continental shelf, where the bottleneck is suspected to occur, are either immature or prereproductive. Gonadal development apparently does not occur until after they depart for spawning grounds offshore. However, given the strong relationship in fish between condition factors and subsequent fecundity, it would be reasonable to use relative weight (W_r), a derived variable, as a proxy for potential fecundity (with the specific empirical relationship for gag to be established by other studies).

One would expect that if a bottleneck of the type hypothesized here were to be alleviated, then the abundance of juvenile-to-adult gag would increase, as would their growth rates and relative weights. However, a descriptive study of a reef or reef system, no matter how quantitative, would not allow one to determine whether the suspected bottleneck actually existed or was alleviated by the artificial reefs, i.e., whether the reefs had fulfilled the primary purpose for which they were built. At the very least, quantitative comparisons are needed.

The very nature of a demographic bottleneck is that it restricts the flow from one life history stage to the next or subsequent stages. Because of spatial structuring in the gag population, i.e., life history transitions from inshore to well offshore, the suspected bottleneck can be evaluated by comparisons of measurable characteristics across the shallow continental shelf. Before considering what comparisons to make and how to make them, we need to consider the physical design and layout of artificial reefs that might alleviate such a bottleneck for gag.

7.3.5.3 Design and Layout the Reefs, and Refine to Specific Primary Objectives

The general strategy for a reef design follows directly from the recent gag and reef research summarized above. Site selection and site layout are dealt with below. Moderately small patch reefs should be used (e.g., 1-m high and 2-m × 2-m footprint), with cavity spaces scaled to the body sizes of aggregated juvenile-to-adult gag (e.g., 50 cm diameters). The cavities should provide an overhead environment similar to small rocky ledge overhangs or small, shallow caverns in which gag naturally hide. These patch reefs should be widely spaced (i.e., ≥ 225 m apart) over a broad area and be positioned so as to complement any natural rock outcroppings in the immediate area. These design elements offer the greatest potential, based on current knowledge, for enhancing growth and condition of gag during the life history phase of interest here.

However, substantial loss of legal-sized, prereproductive females to fishing mortality is a major concern and would be contrary to the primary objective. Therefore, the patch reef architecture also should include snags to fishing tackle that could passively diminish the CPUE of recreational anglers. Additional strategies for diminishing potential fishing mortality are discussed below with respect to key features of the social and management situation, and of course the secondary objective of stimulating recreational activities for economic development.

Traditionally, artificial reefs and reef systems have been very localized structures relative to the geographic scale of fish populations. To accomplish the objectives of this project and to anticipate measurable effects on demographic parameters, a great many small, widely scattered patch reefs, as described above, would have to be distributed across the broad continental shelf between expansive seagrass beds inshore and naturally occurring ledge systems well offshore. Furthermore, because emigration routes from nursery habitats of the region are not yet known, one margin of the reef system should parallel the seagrass beds and be long enough for juvenile gag to have a high probability of encountering the reefs, regardless of their specific routes. For these reasons, and because only so much area can be adequately covered by predeployment site surveys, a 256 km^2 area is planned for development. Figure 7.4 depicts several geographical considerations, which are discussed below.

Descriptive and anecdotal accounts for artificial reefs in 6 m of water suggest that juvenile-to-adult gag do not occupy the shallowest reefs as readily as they do slightly deeper reefs (i.e., 13 m). Therefore, the inshore margin of the reef system or reef site will be along the 9-m depth contour, approximately 16 to 24 km offshore and 8 to 10 km from the seaward margin of the seagrass beds. (This decision also eliminates the need for state artificial reef permits in addition to the federal

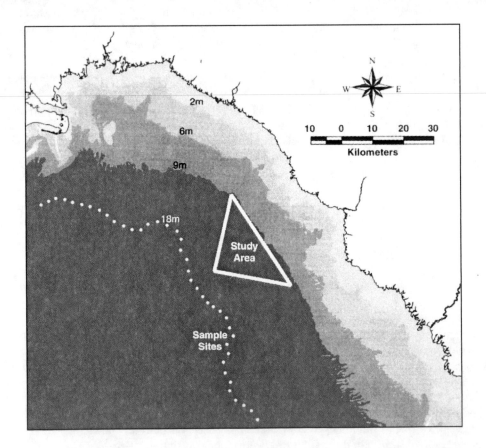

Figure 7.4   Vicinity map for the fishery conservation reef tract and its evaluation reefs, with geographic positions of key ecological features indicated.

permits required of all artificial reefs.) The inshore margin of the reef system will be 33 km long, centered on the mouth of the principal river of the area and bracketing pristine seagrass beds just to the north and south.

According to anecdotal reports from fishermen, natural rock outcroppings and ledges are most common seaward of the 15-m depth contour. Given the shallow slope of the bottom in this region, a triangular area that has a 33-km base and covers 256 km^2 would also have its apex at approximately the 15-m depth contour 16 km further offshore. Although seemingly arbitrary, the triangular shape will actually aid the evaluation of this reef system's primary and ancillary objectives.

The suspected bottleneck is best evaluated by comparing what comes out of the reef system to what went into it. Therefore, a line of 40 small, isolated evaluation reefs will also be built along the 18-m depth contour offshore from the reef system and extending 60 km north and south of the western apex of the reef system (Figure 7.4). These evaluation reefs will have the same architecture as the basic units within the reef system, but will be spaced at intervals of 3 to 4 km. The evaluation reefs will be built prior to the reef system itself in order to obtain a predevelopment baseline. Several predictions and relationships can be tested and quantitatively described with this arrangement.

Initially, the abundance, growth (or size at age), and condition (W_r) of gag should not differ among evaluation reefs regardless of their proximity to the reef system. However, if the suspected bottleneck actually exists and if the reef system alleviates that bottleneck, then as the reef system becomes fully colonized and maturing gag pass through it toward offshore spawning grounds, each of these measurement characteristics should increase on the evaluation reefs closest to the artificial reef system. A significant relationship should emerge over time, such that the abundance, growth, and condition of gag will become an inverse function of the distance of evaluation reefs from the western apex of the reef system. The exact shape of that function must be determined empirically (see below). Based on prior research in the region with gag and another large-scale reef system, the time for this pattern to emerge is expected to be 4 to 6 years, roughly corresponding to the time required for gag to become reproductively active. However, several assumptions underlie these predictions.

It is assumed that all of the evaluation reefs will be colonized initially by preexisting juvenile-to-adult gag that immigrated to offshore habitats prior to the placement of the reef system, and that these fish originated from the inshore nursery habitat of the region. If this assumption is false, it is of little consequence so long as there is no geographic pattern or bias that would confound the expected patterns. It is assumed that the production of young-of-the-year gag from seagrass beds inshore of the entire length of evaluation reefs is spatially variable, but not biased in favor of the nursery grounds immediately inshore from the primary reef system. This assumption can and should be tested by appropriately sampling nursery habitat inshore from the line of evaluation reefs. By adding appropriate sampling within the reef system, it also should be possible to establish the quantitative, empirical relationship between year-class strength in nursery habitat and subsequent recruitment to the adult stock with and without the bottleneck alleviated. This information bears directly on the effect of the bottleneck on the population dynamics of the regional stock and the potential fisheries management benefits of expanding such reef development across a broader geographic range. However, given the stochastic nature of settlement and recruitment processes, this phase of evaluation might require 5 to 10 years to establish an adequate time series of data.

At this point in the planning process, the reef design, site layout, and type of assessment for the primary objective has been essentially set. Architectural details for the reefs themselves remain to be specified, but are mostly irrelevant to our discussion of evaluation planning. The predictions developed above constitute the specific primary objectives for the reef project and evaluation study that inherently include the criteria for success. By testing the explicit assumptions mentioned above and establishing the empirical relationship associated with them, the evaluation study would be relevant to issues far broader than just the specific performance of one local reef project. The evaluation plan for the primary objective and its ancillary objective will not be complete until details of the sampling design (from Chapter 2) and sampling methods (from Chapter 5) have been specified. However, before doing so, we need to consider more carefully the key features of the social and management situation and how to accomplish and evaluate the secondary objective of economic development.

7.3.5.4 Identify Key Features of the Socioeconomic and Management Situation

The community is closely tied to water-related activities. Local residents are involved in a variety of service sector businesses, including a few small hotels and resorts, a few restaurants and fish houses, several small marinas and bait and tackle shops, and some individual guide services. Commercial fisheries still persist, largely for nearshore stone crabs (*Menippe mercenaria*) and blue crabs (*Callinectes sapidus*) and for offshore grouper (family Serranidae). Crabs are trapped and grouper are caught by hook-and-line. A handful of commercial fish trappers target black seabass (*Centropristis striata*) in federal waters offshore from the planned reef site. Inshore net fisheries have been virtually eliminated by the combination of fisheries regulations and a state constitutional amendment that banned most nets. Consequently, mullet (*Mugil cephalus*), spotted seatrout (*Cynoscion nebulosus*) and redfish (*Sciaenops ocellatus*) are no longer as important commercially as they once were. Increasingly, weekend and vacation homes are being purchased or developed by people residing elsewhere. Recreational fisheries are dominated seasonally by bay scallops (*Argopectin irradians*), spotted seatrout, and redfish. To a much lesser extent, offshore fishing and diving on natural and artificial reefs are being pursued by weekend visitors and tourists with their own boats. The portion of service sector businesses associated with these offshore activities is most likely to expand in direct response to reef development, although the diversification of accessible water-related activities is expected to stimulate the local economy as a whole. Unfortunately, a substantial recreational fishing effort directed to the planned reef system, whether by angling or spearfishing, can potentially prevent these reefs from accomplishing their primary objective.

Within the typical fisheries management situation for this locality, the location of every artificial reef is publicized and the only restrictions are those applicable to any given fishery, e.g., seasonal closures, size limits, and bag limits. Furthermore, although the snags to fishing tackle on each patch reef might reduce angling mortality somewhat, they will not affect mortality due to fish traps or spearfishing, which are commercial and recreational activities, respectively.

Two adjustments of the typical management framework are needed to safeguard the primary objective of fisheries conservation. First, the actual locations of the patch reefs should not be advertised or made public. Public angling on these reefs is not to be prohibited, it is simply not to be promoted. (In fact, local public education programs should be instituted to foster practical fisheries conservation practices such as catch-and-release, distributing fishing effort so as not to "fish out" sites, and promote the benefits of not telling others about discovered reef sites.) Second, the body that has authority over the reef site in question should designate the entire reef site a Special Management Zone (SMZ) to exclude both fish traps and spearfishing. For compliance and enforcement, it may be necessary to configure the SMZ as a rectangle using lines of latitude and longitude around the reef site. The mechanism and precedent exist for obtaining SMZ status, and it is best pursued at the same time that federal permits for reef construction are being sought.

Among the key features of the management situation, the need to withhold specific reef locations from the general public is perhaps the most problematic because of conflicting public policy. In particular, the federal policies and procedures for charting artificial reefs on nautical charts do not presently accommodate the direct application of reef technology to fisheries management as anticipated here. Although such issues do not bear directly on evaluation planning, they are central to overall project planning and the ultimate success of the project to be documented by an evaluation. In some cases, the policy impediments may be sufficiently severe to decide against continuing with a planned project, in which case further evaluation planning becomes irrelevant.

7.3.5.5 Design Reef Modifications and Refine the Secondary Objective

If the management provisions above can be implemented, then modifications to the basic reef design and site plan can be made to better accommodate the secondary objective of economic development. For example, a small fraction of the concealed patch reefs, i.e., no more than 10%, can be built without angler snags, thereby providing slightly greater angler satisfaction to those who discover them. As these become known to fishing guides, the guides might incorporate visits to such reefs in their strategies for successful charters. Also, two to four designated fishing reefs could be built within the system, at some distance from sanctuary reefs. The locations of fishing reefs could be marked by buoys and posted at marinas primarily to direct tourists renting boats, etc. Of course, from the perspective of fisheries ecology and source-sink dynamics, these few fishing reefs might best be considered "sink" habitats drawing their harvestable fishes from many surrounding "source" habitats that have not been promoted for public fishing. In a similar fashion, one or two specific reef locations within the overall site could be developed exclusively for divers, and these could be set aside in the SMZ to exclude angling along with spearfishing and fish traps. The designated dive reefs could be designed for human aesthetics by their shapes and layout, while the complexity and composition of the structures could, by design, favor a variety and abundance of local marine life, including otherwise exploited fishes. Locations of dive reefs could be marked by mooring buoys and posted in local dive shops. The key to accommodating consumptive and nonconsumptive recreational activities in the overall reef project plan is to maintain a conservative balance between the primary and secondary objectives.

Consideration of the social and management situation, and the adjustments to reef plans discussed above suggest several measurable characteristics pertinent to an economic impact analysis as outlined in Chapter 6. The economic activities of several types of service sector businesses (e.g., marinas, fishing guides, hotels and restaurants, and dive shops) are expected to increase as a direct result of increased recreational activities associated with the artificial reefs. The levels of economic activity can be evaluated before and at intervals after reef development. Concurrently, if study resources permit, the same evaluations could be done for other small, rural coastal communities to the north and south. These would serve as control communities. By reference to Figure 6.2 (Chapter 6), one can readily discern the specific social goals, policy objectives, behavioral objectives, and action objectives in the preceding paragraphs. Given the complexity and potential for bias in such a study, it would be wise to contract with an experienced professional for the overall economic impact analysis. Nevertheless, it will be instructive to consider how such a study might be conducted, which we will do after first considering the sampling design needed for an evaluation of the specific primary objective, i.e., alleviating the suspected life history bottleneck for gag.

7.3.5.6 Consider Available Support

Before committing further to evaluation planning, a pragmatist would judge the likelihood of obtaining resources needed for a project of such magnitude. In this example, private support, university and resource agency expertise, and local government participation have combined for the reef planning and evaluation planning and to obtain the essential federal permits for reef construction. Construction funding for the reef system itself is anticipated from several sources. The private sector is committed to building the designated fishing and diving reefs, under public supervision. An application for one reef construction grant will be submitted to the state's Artificial Reef Program to put in place the 40 small evaluation reefs offshore, but a combination of state construction grants, private foundation grants, and public works programs will be needed for the bulk of the small, widely scattered patch reefs. The availability of funds will obviously determine the timetable for construction and the start-up of the evaluation. Beyond that, to actually conduct the postconstruction ecological and economic evaluations, participating university faculty and staff will pursue, in cooperation with local governments, some combination state reef monitoring con-tracts and federal research grants. It is anticipated that three to four federal grant programs will be receptive to such proposals. The ecological and economic proposals can be submitted and funded separately, but in a complementary manner. The exact costs in those proposals will depend on details of the specific study designs. Of course, funding constraints imposed by sponsoring agencies may require tradeoffs in some details even after proposals have been approved.

7.3.5.7 Decide on Sampling Designs and Sampling Methods

The general evaluation plans detailed above were developed for both the primary and secondary objectives. All that remains for this example are the specifics of the study designs. But as an old adage says, "The devil is in the details." We will address the ecological evaluation first and the economic evaluation second. Review the predictions, empirical relationships, and testable assump-tions in Section 7.3.5.3. These convey the specific objectives, success criteria, types of assessments, and measurement characteristics for the primary general objective (see Figure 7.1), all of which need to be kept firmly in mind while planning the sampling design and sampling methods.

As conceived in the general evaluation plan, the primary objective will require data from which estimates of abundance, growth, and condition can be obtained for the gag occupying evaluation reefs, inshore seagrass beds, and the reef system itself. The evaluation reefs should be occupied primarily by young, adult female gag throughout the year, with the possibility of some fish moving offshore toward spawning grounds during the autumn and winter. For that reason, and given prevailing working conditions offshore, sampling of the evaluation reefs for gag abundance should be done during summer months. In the region of interest, larval gag settle into seagrass beds during late spring and grow rapidly there until emigrating in early fall. Therefore, sampling of nursery habitat for all measurement characteristics also must be done during the summer. Young-of-year (YOY) gag first occupy patchy reef habitat during the fall, and juvenile-to-adult gag occupy such habitat year round. Given the sampling already required during summer and the advantages of documenting colonization by YOY, sampling of the reef system itself would best be done in the fall. At this time, the data pertinent to abundance, growth, and condition of gag on area reefs can be gathered concurrently. Energy budgets of juvenile-to-adult gag apparently exhibit an annual cycle (i.e., summer growth in length shifting to autumn weight gain), thus autumn is the best time to capture both recent growth and maximum condition. For this reason, the sampling of evaluation reefs for gag growth and especially gag condition also must be done during the same time frame. This answers the question of when to sample on a broad scale.

Decisions on where to sample and how to sample need to be made together. For the artificial reefs, individual patch reefs will be the sampling units. Given the small size of the patch reefs to be built, the large size of the target species, and the observation that gag do not become diver-shy in the absence of spearfishing, the preferred method to obtain accurate abundance data is by total visual counts by scuba divers. However, because visual estimates of fish size by divers lack precision and accuracy, the visual census of gag should use only broad-size categories for enumerating the fish (e.g., 20 cm intervals in total length), and divers should be aided by T-bar meter sticks (Chapter 5). A standardized counting protocol should be used for all sampled patch reefs, whereby a wide perimeter count is done as the diver first approaches the patch reef, followed by a close inspection of cavities. Throughout the count, every effort must be made to avoid double-counting individuals and to include gag that might have moved to the periphery of the visual range when first disturbed. Given the possibility of diurnal activity patterns in the association of gag with reef structure, visual fish counts also should be standardized to the same time of day (e.g., 09:30 to 15:30) to minimize variance from that source.

An alternative sampling method is advanced hydroacoustics (Mason et al. 1999) for estimating gag numbers, sizes, and biomass. Gag may be more amenable to hydroacoustic sampling than other reef fish and demersal species, given the habitat of gag to aggregate above reef structure oriented into the current, and then to shift upward in the water column when a boat passes slowly overhead (L. Kellogg, personal communication). However, hydroacoustics also would require much more expensive equipment and technical competence, but the advantage might be fewer field days required. Of course, for this method, sampling might need to be restricted to daytime flood or ebb tides and not slack-tide periods. Prior to choosing hydroacoustics over visual counts, a direct comparison of sampling efficiency should be done specifically for gag in the context of such artificial reefs.

To obtain data for growth and condition, specimens must be physically captured, then measured (i.e., total length, fork length, and standard length), weighed, and have otoliths extracted for subsequent analysis. Again, the patch reef is the sampling unit and the measurements from several fish must be combined (e.g., averaged) to represent each sampling unit. For efficiency in the field, baited fish traps on 24-h soak times will be used under state and federal scientific collecting permits as the primary means of capturing fish. Hook-and-line fishing can supplement the catch when necessary. These are merely collecting procedures in this example, and are not intended as quantitative sampling methods.

At least 10, but preferably 15, gag per patch reef will comprise a sample for any given sampling period, and sampling will be with partial replacement. A subset of five randomly selected fish from each sample will be sacrificed for otolith analyses. The remainder will be returned alive to their point of capture after measurements have been made. Fish lengths will be measured using a standard fish board. Fish weights will be measured on-site by using an electronic balance coupled to a laptop computer for averaging many values over short time intervals. Relative weight, W_r, will be later calculated as a derived variable for condition based on fish weight adjusted for fish length. Otoliths will be processed in the laboratory for reading on a computerized video image analysis system, with two people reading each otolith independently to avoid bias and minimize errors. Careful data logs and specimen tags will be used to track specimens and otoliths throughout the boat and laboratory processing, such that all data from individual fish can be combined as necessary for size-at-age and incremental growth estimates.

The sampling design for evaluation reefs is straightforward. Every evaluation reef (i.e., 40 isolated patch reefs) will be sampled by the methods above once each summer for gag abundance, and once each fall for condition and growth. For the primary reef system, a representative sample of 40 patch reefs will be selected each year by a stratified random selection procedure. The strata will be equal areas of the reef site geographically positioned relative to axes paralleling the coast and perpendicular to it. This is to account for potential gag patterns resulting from adjacent nursery habitats and depth. A 5- to 10-year time series should be generated for testing the predictions and empirically describing the spatial and temporal relationships.

The sampling design for YOY gag in the inshore seagrass habitat will be most constrained by logistics and available support. Therefore, both high and low budget options are planned. In the low budget option, one would merely test the assumption that the production of YOY gag is variable along the coastline, but not biased in favor of those nursery grounds inshore of the primary reef system. This would involve a one-time sampling of a modest number of stations throughout the region. In the high budget option, many systematically selected, fixed stations would be repeatedly sampled over 5 to 10 years to correspond with the reef data sets. The systematic selection will ensure placement of stations directly inshore of evaluation reefs across their full range. In either case, the number and locations of sampling stations would be determined in cooperation with state fishery biologists and university researchers engaged in fisheries-independent monitoring. Regardless of the option chosen, the sampling procedures would be the same at each station. Absolute densities of juvenile gag will be estimated for each station by using standardized trawl sampling and the Jolly–Seber mark and recapture method for open populations (Koenig and Coleman 1998). Again, details of the sampling protocol would be decided in cooperation with others in the region already involved with fisheries-independent monitoring programs.

Now our attention turns to economic evaluation for the secondary objective, i.e., to stimulate economic development through increased recreational activities. Please review and bear in mind the strategies in Section 7.3.5.5 for increasing recreational use and user satisfaction and the types of specific objectives referenced in Chapter 6, Figure 6.2.

The secondary objective requires that the type of assessment be an economic impact analysis (see Chapter 6) and that data be collected from two sources, the recreational users of the resource (i.e., fishers and divers) and the businesses that serve those users. To ensure the quality of data, direct interviews of users will be conducted at boat ramps and marinas, while the service providers will be directly interviewed at their places of business (e.g., bait and tackle shops, marinas, motels, restaurants, dive shops, and charter boats).

Direct interviews are practical, in this example, because of the small number of access points and businesses that are concentrated in isolated, rural towns of the region. Mail and telephone surveys would not be as practical given the sparse local population relative to a geographically diffuse tourist population, and the low return rates normally associated with such surveys. The direct interviews would have to be conducted before and at intervals after the construction of the reef system because the secondary objective and economic impact analysis focus on changes in sales, income, and employment resulting from the reef project.

By including control communities, changes in the local economy due to artificial reefs could be distinguished with greater confidence from background economic changes due to other causes. Here again, the economic portion of the overall project evaluation would best be contracted with an experienced professional.

The direct interviews of users will employ a stratified selection process, whereby the samples get distributed across seasons and times of the week expected to differ in levels of recreational fishing and diving. Furthermore, the intercepts at access points such as boat ramps and marinas will be scheduled for periods of the day when traffic is high, e.g., early morning and late afternoon. Within sampling periods, the choice of which people to interview from among those returning needs to be guarded against bias by the interviewer. Therefore, in this case, a systematic selection process will randomly identify the first party to be interviewed and then interview every kth party thereafter (e.g., randomly draw a number between 1 and 10, say 7, interview the driver of the seventh boat to arrive, and then interview every fifth boater thereafter). To determine k consider how long an average interview is expected to take, the rate at which boaters typically arrive at ramps and marinas of the region, and the sample size needed for accurate estimates in such studies.

In addition to questions about whether people fished or dived, where the activity took place, what fish were caught, and user satisfaction, questions should be included to discover where the people are from (i.e., local, regional, or out-of-state), the expenditures made, and the monetized value of the experience. For this project, given the intent to attract visitors and tourists from outside the area, the Marine Recreational Fishing Statistics Survey described in Chapter 6 may be too biased toward local residents, so trained evaluation project personnel would have to conduct the interviews.

Direct interviews of service-providing businesses will be conducted annually just after the close of their fiscal years. In that way, accurate economic information about the entire year should be most readily available. The appointment for the first interview will be made at the very start of the reef project when a rapport and trust would have to be established with each proprietor, and confidentiality assured. For instance, when reporting the data, aggregate statistics will be used to describe each business type (e.g., marinas or bait and tackle shops) with at least three businesses sampled. Otherwise, broader aggregates may have to be used to ensure individual privacy. Again, details of the interviews need to be defined by a professional, but at the very least, the following information is needed: gross revenues, net profits, and the number of employees by income bracket.

7.3.5.8 Anticipate How to Analyze the Data

At this point in planning a complex project such as this, a professional statistician must be consulted to identify the analytical procedures most appropriate for the specific objectives, study design, and anticipated data. It is possible that preferred statistical procedures might warrant further modifications to the study design.

For data from the evaluation reefs, ecological patterns are predicted to develop over time among gag measurements (i.e., abundance, growth, and condition) and between these measurements and distance from the primary reef system. These relationships might be tested with a repeated measures analysis of covariance (ANCOVA), with distance as the covariate. An alternative might be a mixed model analysis of variance (ANOVA) in which distance is a fixed factor and the variance from time (i.e., changes over years) is modeled as a linear or nonlinear trend. However, it is likely that having only two reefs at each distance (n = 2), which is a physical constraint of controlling for depth, might be inadequate for either procedure. A less sophisticated approach would be first to do a regression analysis for each year (i.e., gag measurements vs. distance), and then do a regression of the slopes from the annual regression lines vs. sampling year.

Under the low budget option, the assumption of no biasing patterns in YOY gag across seagrass beds of the region could be evaluated by plotting and inspecting simple descriptive statistics from the sampling stations. Under the higher budget option, multiple/multivariate regression procedures could be used to relate the measures of gag on evaluation reefs to those same measures from the primary reefs and to the abundance of YOY in seagrass beds inshore from these artificial reefs. Appropriate time lags corresponding to life history transitions would need to be included. Such statistical modeling would require an adequate time series of data, perhaps 10 or more years.

The low cost option for economic evaluation could use multiple/multivariate regression procedures to relate expected increases in business activities, by business type, to changes in the level of recreational activity and user satisfaction associated with the artificial reefs. The higher cost option might entail using mixed model ANOVA to test the target community vs. control communities (fixed factors) while modeling trends in business activities over time.

For both the ecological and economic evaluation, a consulting statistician might well recommend that pilot studies be conducted or data be gleaned from the literature to run trial analyses and to determine the sample sizes needed to detect changes of various magnitudes. Of course, a well-planned evaluation study can anticipate the statistical procedures, but the final analysis cannot be decided until the data are in hand and the assumptions of the statistical procedures can be evaluated.

7.4 VALID INFERENCES AND CONCLUSIONS

In Section 7.2.1, we noted that technically correct methods produce technically correct data, which are not by themselves sufficient. The data also must be relevant to the question being asked and conform to an adequate sampling design. In Section 7.2.2.1, we explained that generalizations drawn from any particular study are only as valid as the logic and scope of its study design. Here we expand on that point with reference to the examples above.

7.4.1 Limited Study Designs for the Objectives

We have emphasized the utility of including "success criteria" in the specific objectives for a project, which greatly helps when defining the general evaluation plan and subsequent details of a study design (Figure 7.1). But a study design can be greatly constrained by time and money relative to the scope of the objectives. We see this in Example 7.3.2 concerning reefs deployed for a sustainable artisanal fishery. The scale of that reef system had to be large in order to be consistent with the ecological and socioeconomic situation. But that precluded having replicate reef systems for comparisons, or sampling the single reef system with enough intensity to thoroughly describe either the life history patterns of targeted fishes or the extent to which mechanized fishing was excluded from the offshore reef tract. The fisheries-dependent sampling of catch and effort was adequate to describe trends over time (i.e., if the artisanal harvest was stable or continuing to decline), but if changes were documented, then the remainder of the study would only give indications of why the reefs were not working as planned. The cause for such a failure might be suggested or inferred from the anticipated data, but it would be wrong to assert a reasonable inference as a hard-and-fast conclusion. As noted in that example, the study design was simply inadequate for that level of confidence.

By contrast, a single, large reef system was also the focus in Example 7.3.5. But in this case, it was hypothesized that a specific life history bottleneck existed for a particular species. The artificial reefs and evaluation study, as planned, constituted a direct test of an ecological process hypothesized *a priori* to be affected by appropriately placed reef structures. This was not simply a quantitative description of patterns. Clear predictions of what was anticipated were part of the specific objectives, and the study design was adequate to determine if those predictions materialized. Such studies allow conclusions about cause-and-effect to be asserted with greater confidence.

7.4.2 Explicit and Implicit Assumptions

Assumptions are an inevitable part of reasoning toward a conclusion and allow knowledge and understanding to be drawn from appropriate data and analyses. Explicit assumptions are preferable to implicit ones, because the validity of assumptions can only be judged or tested if they are first recognized. In Example 7.3.1, implicit assumptions were unwittingly made about the ecological processes that might make targeted fishes more readily accessible to the anglers. The reefs and the evaluation were planned as if rockfish populations were absolutely limited by habitat of the kind being constructed. For this example, the consequences of making an implicit assumption out of ignorance were discussed at length in Section 7.3.1.5. By contrast, a similar lack of ecological knowledge was part of Example 7.3.2, but here the assumptions about alternative life histories and habits of targeted reef fishes were made explicit. Consequently, plans could be made for limited sampling as part of the evaluation study to help judge which of the assumptions were most reasonable. As pointed out in Section 7.4.1, that portion of the planned study for Example 7.3.2 would be inadequate to assert definitive conclusions, but any resulting inferences about the functioning of the reef system could at least be accepted as reasonable. In many management situations that will be the best we can expect because higher degrees of certainty are unattainable given the inevitable tradeoffs and constraints.

7.4.3 Alternative Explanations

If one is an advocate for a particular position or course of action, then one organizes information (e.g., a study design and resultant data) to make that case, to reinforce the preferred position, and to diminish competing perspectives. However, if one is interested in knowing what artificial reefs can and cannot accomplish, then one weighs the alternatives objectively. For instance, in Example 5, we suggested that a bottleneck related to available shelter might exist in the life history of gag grouper; then again it might not. If it exists, then the planned reefs should alleviate the bottleneck and yield a predictable pattern of grouper abundance along a line of evaluation reefs. However, the same patterns offshore might merely reflect settlement patterns along nursery habitat inshore and have nothing to do with the artificial reefs. That would be an alternative explanation for the expected results. Therefore, as part of the evaluation planning, a sampling design was proposed for those inshore nursery habitats to evaluate that recognized alternative. By contrast, in Example 4, artificial reefs to support dense mussel beds were proposed to improve water quality in a bay with eutrophication. The planned monitoring of water quality characteristics could document improvements in the bay following construction and colonization of the reefs, but other reasons for the improvement could exist. Perhaps advanced sewage treatment and better controls on nonpoint source pollution were implemented at the same time the reefs were being built. How would one know what part of the improved water quality was attributable to which management practice?

The lesson reinforced here is that evaluation of artificial reefs, although focused on purposes for which reefs are built, cannot be planned in a vacuum. Whenever possible, alternative explanations for the expected outcomes should be considered as part of the planning process and dealt with as competing hypotheses or explicit assumptions, both of which entail more objectivity than simply ignoring the alternatives.

7.5 CLOSING REMARKS

This chapter, and indeed this book, were introduced with a guiding question for reef evaluation of whether an artificial reef or reef system is satisfying the purposes for which it was built. We have emphasized the need to define those purposes explicitly to refine the general or vague objectives into specific and testable ones. We have illustrated how specific objectives can lead to study designs capable of answering the question, but answering that question in a narrow sense is not enough.

Artificial reefs are merely tools, used by man to alter naturally occurring processes in order to yield some desired outcomes. Any tool used properly does good, but it has the potential to be misused, or used for its intended purpose but still do unintentional harm. The possibility of harm is increased by ignorance and reduced by knowledge. We need to embrace legitimate challenges to reef development as valuable opportunities for learning (e.g., the attraction/production issue, see Grossman et al. 1997; Lindberg 1997). By scientifically evaluating artificial reef projects, and reporting the results, we should all learn to better use these tools appropriately for specific benefits while avoiding their potential harm.

Ultimately, public and private sponsors of artificial reefs have a right to know their return on investments, and they have a need to know the attendant risks. Those of us involved in reef development must be accountable to our sponsors and simultaneously responsible toward the natural resources. Responsible reef planning would therefore consider any desired benefits specifically, as we have promoted throughout this book, but would also consider potential costs much more broadly. Costs go beyond direct expenditures on reef construction to include what economists call opportunity costs, or the replacement of one group's values by another's. Responsible and effective planning for artificial reef projects does not occur in a vacuum or with blinders on. Wisdom comes from asking the right questions and then heeding the answers. Therein lies the enduring benefit of objective, effective, and thorough artificial reef evaluations.

7.6 ACKNOWLEDGMENTS

The authors thank William Seaman, Jr. for organizing this volume, for providing us an opportunity to participate, and for patiently encouraging the preparation of this chapter. Our chapter drew heavily from Chapters 2 through 6, and we thank all our fellow authors for their contributions. Special thanks are due to Kenneth Portier for many stimulating discussions over the past few years concerning study design and how to communicate the underlying thinking and principles.

REFERENCES

Ardizzone, G.D. and P. Pelusi. 1984. Yield and damage evaluation of bottom trawling on *Posidonia* meadow. Pages 63–72. In: C.F. Bourdouresque, A. Jeudy de Grissac, and J. Olivier, eds. *International Workshop on* Posidonia oceanica *Beds.* GIS Posidonie Publication, France.

Assoc. Monégasque pour la Protéction de la Nature. 1995. *XX Ans au Service de la Nature.* Published by ECG, Monaco: 190 pp.

Badalamenti, F. and G. D'Anna. 1997. Monitoring techniques for zoobenthic communities: influence of the artificial reef on the surrounding infaunal community. Pages 347–358. In: A.C. Jensen, ed. *European Artificial Reef Research, Proceedings, First EARRN Conference,* Ancona, Italy, March 1996. Southampton Oceanography Centre, Southampton, England.

Barker Jorgenson, C. 1990. *Bivalve Filter Feeding: Hydrodynamics, Bioenergetics, Physiology and Ecology.* Olsen & Olsen Ed., Denmark, 140 pp.

Bombace, G. 1981. Note on experiments in artificial reefs in Italy. *Studies and Reviews (GFCM-FAO)* 58:309–324.

Bombace, G. 1997. Protection of biological habitats by artificial reefs. Pages 1–15. In: A.C. Jensen, ed. *European Artificial Reef Research, Proceedings, First EARRN Conference,* Ancona, Italy, March 1996. Southampton Oceanography Centre, Southampton, England.

Bulletin of Marine Science. 1985. *Third International Artificial Reef Conference.* Vol. 37, No. 1, Pages 1–402. University of Miami, Miami.

Bulletin of Marine Science. 1989. *Fourth International Conference on Artificial Habitats for Fisheries.* Vol. 44, No. 2, Pages 527–1081. University of Miami, Miami.

Bulletin of Marine Science. 1994. *Fifth International Conference on Aquatic Habitat Enhancement.* Vol. 55, No. 2 and 3, Pages 265–1360. University of Miami, Miami.

California State Lands Commission. 1999. *Final Program Environmental Impact Report for the Construction and Management of an Artificial Reef in the Pacific Ocean Near San Clemente, California.* State Clearing House No. 9803127. Prepared by Resource Insights, Sacramento, CA.

Charbonnel, E., P. Francour, and J.G. Harmelin. 1997. Finfish population assessment techniques on artificial reefs: a review in the European Union. Pages 261–277. In: A.C. Jensen, ed. *European Artificial Reef Research, Proceedings, First EARRN Conference,* Ancona, Italy, March 1996. Southampton Oceanography Centre, Southampton, England.

Chojnacki, J.C. 1994. Artificial reefs in the estuary river Odra as medium of revitalisation of marine environment. Vol. 1, Pages 40–48. In: *Proceedings, Third International Scientific Conference — Problems of Hydrodynamics and Water Management of River Outlets with a Special Regard to Odra River Outlet,* Szczecin, Poland.

Chojnacki, J.C., E. Ceronik, and T. Perkowki. 1993. Artificial reefs — an environmental experiment. *Baltic Bulletin* 1:19–20.

Christensen, N.L., A.M. Bartuska, J.H. Brown, S. Carpenter, C. D'Antonio, R. Francis, J.F. Franklin, J. A. McMahon, R.F. Noss, D.J. Parsons, C.H. Peterson, M.G. Turner, and R. G. Woodmansee. 1996. The report of the Ecological Society of America Committee on the Scientific Basis for Ecosystem Management. *Ecological Applications* 6(3):665–691.

Cloern, J.E. 1982. Does the benthos control phytoplankton biomass in South San Francisco Bay? *Marine Ecology Progress Series* 9:191–202.

Cohen, R.R.H., P.V. Dresler, E.J.P. Phillips, and R.L. Cory. 1984. The effect of the Asiatic clam, *Corbicula fulminea,* on phytoplankton of the Potomac River, Maryland. Limnology and Oceanography. 26:170–180.

d'Cruz, T., S. Creech, and J. Fernandez. 1994. Comparison of catch rates and species composition from artificial and natural reefs in Kerala, India. *Bulletin of Marine Science* 55(2):1029–1037.

Frankignoulle, M., J.-M. Bouquegneau, E. Ernst, R. Biondo, M. Rigo, and D. Bay. 1984. Contribution de l'activite de l'herbier de Posidonies au metabolisme global de la Baie de Calvi Premiers resultats. Pages 277–282. In: C.F. Bourdouresque, A. Jeudy de Grissac, and J. Olivier, eds. *International Workshop on* Posidonia oceanica *Beds.* GIS Posidonie Publication, France.

Green, R.H. 1979. *Sampling Design and Statistical Methods for Environmental Biologists.* Wiley, New York.

Grossman, G.D., G.P. Jones, and W. Seaman, Jr. 1997. Do artificial reefs increase regional fish production? A review of existing data. *Fisheries* 22:17–23.

Harmelin, J.G. and D. Bellan-Santini. 1997. Assessment of biomass and production of artificial reef communities. Pages 305–322. In: A.C. Jensen, ed. *European Artificial Reef Research, Proceedings, First EARRN Conference*, Ancona, Italy, March 1996. Southampton Oceanography Centre, Southampton, England.

Harmelin-Vivien, M.L. 1982. Ichtyofaune des herbières de Posidonies du parc national de Port-Cros. I: composition et variation spatio-temporelles. *Travaux Scientifiques du Parc National de Port-Cros.* 8:69–92.

Koenig, C.C. and F.C. Coleman. 1998. Absolute abundance and survival of juvenile gags in sea grass beds of the northeastern Gulf of Mexico. *Transactions of the American Fisheries Society* 127:44–55.

Laihonen, P., J. Hanninen, J. Chojnacki, and I. Vourinen. 1997. Some prospects of nutrient removal with artificial reefs. Pages 85–96. In: A.C. Jensen, ed. *European Artificial Reef Research, Proceedings, First EARRN Conference*, Ancona, Italy, March 1996. Southampton Oceanography Centre, Southampton, England.

Lefevre, J.R., J. Duclerc, A. Meinesz, and M. Ragazzi. 1983. Les récifs artificiels des établissementes de pêche de Golfe Juan ed de Beaulieu-Sur-Mer, Alpes Maritimes, France. Pages 109–111. In: *Journée Etudes Récifs Artificiels et Mariculture Suspendue.* Cannes, Commission Internationale pour l'Exploration Scientifique de la Mer Méditerranée.

Lindberg, W.J. 1997. Can science resolve the attraction-production issue? *Fisheries* 22:10–13.

Mason, D.M., A.P. Goyke, S.B. Brandt, and J.M. Jech. In press. Acoustic fish stock assessment in the Laurentian Great Lakes. In: M. Munawar, ed. *Great Lakes of the World.* Ecovision World Monograph Series. Backhuys, Leiden, The Netherlands.

Moreno, I. 1997. Monitoring epifaunal colonization. Pages 279–291. In: A.C. Jensen, ed. *European Artificial Reef Research, Proceedings, First EARRN Conference*, Ancona, Italy, March 1996. Southampton Oceanography Centre, Southampton, England.

Polovina, J.J. 1991. Fisheries applications and biological impacts of artificial reefs. Pages 153–176. In: W. Seaman, Jr. and L.M. Sprague, eds. *Artificial Habitats for Marine and Freshwater Fisheries.* Academic Press, San Diego.

Relini, G. 1977. Le metodiche per lo studio del fouling nell'indagine di alcuni ecosistemi marini. *Bolletino di Zoologia* 44:97–112.

Relini, G. and E. Moretti. 1986. Artificial reef and posidonia bed protection off Loano (West Ligurian Riviera). FAO Fisheries Report. 357:104–108.

Relini, G. and L. Relini Orsi. 1989. The artificial reefs in the Ligurian Sea (N-W Mediterranean): aims and results. *Bulletin of Marine Science* 44(2):743–751.

Relini, G., N. Zamboni, F. Tixi, and G. Torchia. 1994. Patterns of sessile macrobenthos community development in an artificial reef in the Gulf of Genoa (NW-Mediterranean). *Bulletin of Marine Science* 55(2):747–773.

Relini, G., F. Tixi, M. Relini, and G. Torchia. 1998. The macrofouling on offshore platforms at Ravenna. International Biodeterioration and Biodegradation, 41:41–55.

Relini, M., G. Relini, and G. Torchia. 1994. Seasonal variation of fish assemblages in the Loano artificial reef (Ligurian Sea NW-Mediterranean). *Bulletin of Marine Science* 55(2):401–417.

Sanchez-Jerez, P. and A. Ramos-Esplà. 1995. Influence of spatial arrangement of artificial reefs on *Posidonia oceanica* fish assemblages in the West Mediterranean Sea: importance of distance among blocks. Pages 646–651. In: Proceedings, *International Conference on Ecological System Enhancement Technology for Aquatic Environments*, Tokyo, Japan International Marine Science and Technology Federation.

Sanjeeva Raj, P.J. 1996. Artificial reefs for a sustainable coastal ecosystem in India, involving fisherfolk participation. *Bulletin of the Central Marine Fisheries Institute* 48:1–3.

Index

A

Artificial reefs
  abiotic factors (*see* fish and
      macroinvertebrates/assessment)
  Acanthuridae, family, 130
  accepted fact hypothesis (*see* experimental studies)
  Acoustic Doppler Current Profiler (ADCP) (*see*
      terminology)
  Actinopterygii, class, 130
  alternative hypothesis (*see* experimental studies)
  ambient environment, effects of (*see* reef engineering)
  Anarhichadidae, family, 131
  Anguilliformes, order, 130
  Antennariidae, family, 144
  aquaculture (*see* purposes of artificial reefs)
  Argopectin irradians, 226
  ascidians, 98
    assemblages (*see* definitions/fish and
      macroinvertebrates)
  assemblage structure and dynamics (*see* fish and
      macroinvertebrates)
  assessment, 13–15, 64–75 (also *see* fish and
      macroinvertebrates; trophic resources;
      socioeconomics; integration of evaluation
      and reef-project planning)
    approaches, 13–15
    basic types, 65–67, 83
    description, 14
    integration of evaluation and reef-project planning,
      204–231
    interaction with adjacent environment, 14–15, 71–73,
      86
    fish and macroinvertebrates, 134–156
      methods, 135–145
      sampling scale, 145–146
        space, 145–146
        time, 145
      strategies, 134–135
      techniques, 146–156
        destructive, 146–148
        nondestructive, 148–156
    postdeployment, 67
    predeployment, 65–67
    reef engineering, 76–83
    sample design, 73–75

  socioeconomics
    economic, 177
    economic impact analysis, 177–178
    monitoring and description, 173–174
    social assessment, 174–176
  study design, 27
  trophic resources, 104–118
    framework and approach, 106–107
    primary productivity, 115–118
      biomass changes, 115
      fate of primary production and fouling
        organisms, 117–118
      free water methods, 115–116
      incubation methods, 116–117
      quality control, 117
    sampling methods, 107–118
      chemical analysis, 112–113
      nutrients, 107–108
      plant and fouling invertebrates, 113–114
      primary producers and biofouling community,
        108–110
      quality control, 113
      sediments, 108
      water column constituents, 110–112
    types, 104–106
  types, 64–73
    existence and screening, 67–68
    postdeployment, 67
    predeployment, 65–67
  variables, 76–83
    currents, 79–80
    depth, 78
    global positioning, 76
    light, 81
    reef structure, 76–77
    salinity, 80
    sediments, 82–83
    temperature, air, 77
    temperature, water, 80–81
    tides, 78–79
    turbidity, 81–82
    waves, 77–78
    wind, 77
Atherina, 131
Atherinidae, family, 131
Aulostomidae, family, 130